W0091224

Springer Series on
ATOMIC, OPTICAL, AND PLASMA PHYSICS   46

Springer Series on
ATOMIC, OPTICAL, AND PLASMA PHYSICS

The Springer Series on Atomic, Optical, and Plasma Physics covers in a comprehensive manner theory and experiment in the entire field of atoms and molecules and their interaction with electromagnetic radiation. Books in the series provide a rich source of new ideas and techniques with wide applications in fields such as chemistry, materials science, astrophysics, surface science, plasma technology, advanced optics, aeronomy, and engineering. Laser physics is a particular connecting theme that has provided much of the continuing impetus for new developments in the field. The purpose of the series is to cover the gap between standard undergraduate textbooks and the research literature with emphasis on the fundamental ideas, methods, techniques, and results in the field.

Bernd Lohmann

Angle and Spin Resolved Auger Emission

Theory and Applications to Atoms and Molecules

With 45 Figures

 Springer

Dr. Bernd Lohmann
Universität Münster, Institut für Theoretische Physik
Wilhelm-Klemm-Str. 9, 48149 Münster, Germany
E-mail: lohmanb@uni-muenster.de

The Wombat, poem by Claudia Lehmann, © Claudia Lehmann, 1994

Springer Series on Atomic, Optical, and Plasma Physics   ISSN 1615-5653
ISBN 978-3-540-74629-4   e-ISBN 978-3-540-74630-0

Library of Congress Control Number: 2008929557
© Springer-Verlag Berlin Heidelberg 2009

Typesetting and production: VTEX
Cover concept: eStudio Calmar Steinen
Cover design: WMX Design GmbH, Heidelberg

SPIN 11673309   57/3180/VTEX
Printed on acid-free paper

9 8 7 6 5 4 3 2 1

springer.com

– To my wife and to Tanya, our little Mausebär. –
– And to all the Wombats of Australia. –

The Wombat
Cutest little thing
on Earth
I met here –
which really makes it worth
to care for !!

Claudia Lehmann,
Melbourne, 1994

Foreword

It is widely recognised by the international scientific community that in the early decades of the twentieth century remarkable advances were made in laying the foundations for understanding the fundamental properties of atoms and molecules. There was a rich interaction between experimental and theoretical developments leading to some fundamental discoveries. Scientists that are synonymous with that era include Einstein, Bohr, Planck, Schrödinger and Dirac. The photoionization of an atom, molecule or solid ejecting an electron was explained on the basis of the Einstein *Photoelectric Effect*, discovered in 1905, and the subsequent de-excitation of the species following inner shell ionisation by a radiative X-ray emission process. It was not until the pioneering work of Pierre Auger in the 1920s that the importance of an alternative competitive pathway to X-ray radiative de-excitation was recognised. This alternative pathway, now known as the *Auger Effect*, involves the ejection of a second electron via a non-radiative emission process resulting in the de-excitation of the species.

Many decades later following the refinement of quantum theory and the availability of powerful computers and tuneable synchrotron light sources, the richness to the electronic structure information imbedded in the correlated emission of the photoelectron and the Auger electron from an atom or molecule was able to be recognised and understood. The earliest Auger electron spectral studies were largely confined to the investigation of Auger energies and decay rates; however, since the 1980s many on the investigations have focused on angular and spin resolved effects. In this book Bernd Lohmann provides a detailed account of theoretical developments in the study of angular and spin resolved Auger processes, principally associated with free atoms, with one chapter devoted to diatomic molecules. Most of this work has been reported in the international literature over the past 30 years and is directly linked to the understanding of experimental data.

The principal emphasis of the present work, therefore, is the gathering together for the current status of our knowledge of the Auger effect through a rigorous presentation of theoretical approaches coupled with numerical methods to calculate Auger energies, decay rates, angular distributions and spin polarisation properties. This work will be invaluable to experienced theoreticians and experimentalists seeking to gain a deeper understanding of some properties of atoms and molecules associated with multiple electron emission.

The rare gas atoms are the principal reference points for many of the theoretical and experimental studies reported. They are used in this work to illustrate the theoretical methods principally developed or refined by Lohmann and his collaborators. What one has learned is the importance of accounting for relativistic and correlation effects when accurately modeling the behaviour of electrons in atomic systems. Extensions to open shell systems, while generally more complex, can be treated within the same framework. The elucidation of molecular Auger spectra is considerably more complex. An insight into this field is provided mainly through an account of Auger processes in diatomic molecules.

Many advances have been made in understanding the fundamental information contained in the two electron emission process since the original experiments of Pierre Auger in 1923. Bernd Lohmann has produced a comprehensive, well written text that will be viewed as a significant contribution to the field of Auger electron spectroscopy. This book will be a most useful reference for those researchers wishing to further advance the field.

Melbourne, May 2008   *Frank P. Larkins*

Preface

The Auger emission has been detected by Pierre Auger more than eighty years ago; short after the beginning of quantum mechanics. The observation of radiation free transitions perfectly complemented the case of electron emission due to photon impact. The emission of an electron after photoionization had been already understood and explained by the work of Einstein (1905, Nobelprize 1921 for the *photoelectric effect*). Wentzel (1927) was first by theoretically interpreting the Auger emission as a process caused by a two-particle interaction. Since photoionization had been already explained as a single-particle interaction effect, and due to the fact that first experiments mainly concentrated on the detection of the valence shell electrons, the investigation of radiation free transitions, i.e. the Auger effect, showed only little development during the early decades of the last century. Only starting in the early 1980s, scientific interest came back. This has been caused due to the availability of improved and more developed detection methods and, on the theoretical side, by the application of advanced methods of density matrix theory which allowed for more precise predictions, like angular distribution and spin polarization, of the observed particles, i.e. the Auger electrons. It is worth pointing out that both, theoretical as well as experimental progress, benefited from efforts which had been made in the field of nuclear physics. In particular the work by Fano (1957) and co-workers need to be mentioned who first adapted the nuclear physics methods and successfully applied them to electron emission problems in atomic physics. Adopting the experimental techniques of nuclear physics, like angle and spin resolved observation of particles, there has always been, and still is, need for a theoretical interpretation and prediction of angle and spin resolved data. This need is even more urgent due to the fact that, after starting investigations of angle and spin resolved Auger emission in the early 1980s, the scientific community is today, due to the availability of third generation synchrotron radiation sources and more refined detection methods, able to produce and obtain reliable data which can be compared with the theoretical predictions. This book is intended to focus on these still important areas of Auger emission theory. Our aim is to derive the theoretical framework and to provide a number of data and useful relations necessary for the interpretation of present and forthcoming experiments.

The main aim of this book is twofold. On the one hand, this book might be seen as a theoretical introduction into the broad field of angle and spin resolved electron emission processes, giving a comprehensive survey of nowadays Auger emission physics for the advanced pre- and post-graduate student. Strictly applying the methods of density matrix yields a general basis for the interpretation and analysis and leads to more advanced applications.

On the other hand, this book provides a comprehensive compendium for the active researcher on performed, current, and outlined foreseen research, related to the Auger decay, which might be useful for the planning and performing of even more exciting and sophisticated studies in a field of physics which concerns areas of not only atomic and molecular physics, but also emerging fields like chemo- or biophysical applications, or even quantum information. In addition, the kind of obtained data is important for the knowledge and interpretation of different types of radiation spectra observed from NASA, ESA and the Hubble Space Telescope which will provide a database on cross sections for electron impact excitation, of highly charged ions, electron-ion recombination, photoionization, charge transfers, and X-ray emission.

An introduction into the field of angle and spin resolved Auger emission will be given in the first chapter of the book and the particular focus of the subsequent chapters will be discussed. The general theory of angle and spin resolved Auger emission and the relevant equations will be derived in Chap. 2. Numerical methods and related program packages will be introduced in Chap. 3. Chapter 4 covers the broad field of examples and applications and the comparison of experimental and numerical data and their interpretation. Eventually, Chap. 5 deals with the field of molecular Auger processes, while the last chapter gives an outlook of current, planned and foreseen research related to Auger emission. A number of useful equations and relations will be given or derived in the subsequent Appendix.

Fortunately, research is still ongoing and never ends, thus, I am delighted to receive the outcomes of new researchers.

At this point I would like to say thanks to the numerous and helpful people over all these years who contributed with their efforts, expertise, and valuable discussions in realizing this book. The many discussions with my co-workers and collaborators often stimulated the beginning of a new research project or, on the other hand, led investigations into the right direction.

First of all I would like to express my sincere thanks to Prof. Dr. K. Blum of the University of Münster. I have been a member of his group since many years and we often had inspiring discussions which have been most helpful in the formulation and clarification of the general theory. Besides, he always tried to support me financially. I am also thankful to the former members of his group, Prof. Dr. K. Bartschat, and Drs. S. and K. Bonhoff, A. Busalla, A. Dellen, R. Fandreyer, U. Kleiman, J. Lehmann, M. Musigmann, R.P. Nordbeck, A. Raeker, E. Taute, and G. Wöste for many helpful discussions and the nice atmosphere.

During my years in Münster, there has been always a close collaboration with the experimental groups and I am thankful to Prof. Dr. J. Kessler and Prof. Dr. G.F. Hanne for their ongoing interest in my research. In particular, I would like to thank Prof. Dr. H. Merz. In his group the first experiments on spin polarized Auger emission have been performed, and our common discussions have been often stimulating for the further development of both, experiment and theory.

I am also thankful to Prof. Dr. W. Mehlhorn of the University of Freiburg for his ongoing interest in my work, and, for inviting me to Freiburg where I first met Prof. Dr. N.M. Kabachnik of Moscow State University with whom I started a fruitful and long collaboration in the field of Auger emission. This collaboration has been later extended by including Prof. Dr. A. Grum-Grzhmailo of Moscow State University and still yields a fruitful output.

I am particularly indebted to Prof. Dr. U. Becker and also to Priv.-Doz. Dr. U. Hergenhahn of the Fritz-Haber Institut (FHI) of the Max-Planck-Gesellschaft (MPG) with whom I have a fruitful and still ongoing collaboration. During my various stays at the FHI in Berlin the theory of resonant Auger transitions has been further developed. I am very grateful to Prof. Dr. U. Becker for many inspiring discussions and for many times providing financial support which eventually enabled for the preparation of this book. I am also thankful to his current and former group members Drs. B. Langer, J. Viefhaus and M. Braune for discussions and help with part of the figures and for providing most of the necessary software which has come into effect during the preparation of the manuscript.

I am most thankful to Prof. Dr. F.P. Larkins who invited me to Australia and Dr. T.E. Meehan with whom I worked in School of Chemistry during my stay at the University of Melbourne, Australia. Working interdisciplinary, as a physicist in theoretical chemistry, has been proofed to be most fruitful for both sides. First ideas in the field of molecular Auger emission have been invented during my time in Melbourne, and parts of the results presented in this book have been obtained. I very much appreciate the hospitality which has been extended to me during my stay, and also the nice introduction to the Australian life-style.

There is an ongoing collaboration with Apl. Prof. Dr. S. Fritzsche of the University of Kassel with whom I started the work on open shell atoms and I am thankful for the ongoing and inspiring discussions.

Further, I would like to thank Prof. Dr. A.K. Edwards of the University of Georgia for stimulating discussions in the field of molecular Auger emission.

I am also thankful to Drs. B. Nestmann and B. Schimmelpfennig of the University of Bonn for valuable discussions and support with the numerical results obtained for molecular Auger processes.

Eventually, I am indebted to Prof. Dr. H. Kleinpoppen of the University of Stirling for his support and for inviting me to Stirling. Besides our common research interests he has been always interested in my work and we had long and most helpful discussions. I often appreciated his talent of *bringing all things back to normal* irrespective of physics or non-physics problems.

I am most thankful to Prof. Dr. K. Blum and to Dr. U. Kleiman for carefully proof-reading the manuscript and for giving valuable suggestions and remarks. I am dearly thankful to my wife, Dipl. Bio. C. Lehmann, for the task of double-checking the index.

I am thankful to the Springer staff, particularly Dr. C. Ascheron and Ms. A. Duhm, for their help and assistance with the final form of the manuscript.

I am most grateful to the series editors Prof. Dr. Hans Kleinpoppen, FRSE, and Prof. Dr. Phil G. Burke, CBE, FRS, for their interest in this work and in the idea of publishing.

Besides of all the *boring* sciences, there is, from time to time, someone saying: "*and now to something completely different*". Alas, this has been said by the *Monty Pythons*. Keeping the Wombat in mind, I, eventually, like to give the overall and final thanks to my wife for her support, for most of the time taking care of our little Tanya, and for standing tall against all these quarrels during the preparation of the manuscript.

Thank you Claudia.

Münster, May 2008   *Bernd Lohmann*

Contents

1 Introduction
to Angle and Spin Resolved Auger Emission

Electrons and photons are reactive and abundant species in all phases of matter. They are generated and interact with each other by a multiplicity of mechanisms; e.g. photoionization/excitation, collisions involving excited species, negative ions or high energy particles, and ejection or emission from surfaces. The interaction of electrons and photons with atoms and molecules therefore plays an essential role in our understanding of many natural processes and is important for the development of many technologies.

While the detection of electromagnetic waves in the visible, as well as in the UV, or in the X-ray range had reached, at the beginning, a very high level of accuracy, the detection of charged particles, namely, low energy electrons, has only in the past three decades attained a reasonable level of efficiency and resolution.

Pierre Auger's famous discovery, the *Auger effect*, took place in 1923. During his Ph.D. he learned how to use a *Wilson chamber* which, at his time, was the only way to detect charged particles. Applying this tool, he had in mind to send X-rays inside the gas. In this kind of situation it was expected that the track of the emitted photoelectron, following a photoelectric effect during its deceleration inside the gas, would be observed. However, instead of observing the track of a single photoelectron, two or more tracks seem to originate from the point of impact. This is the observation from which he discovered the Auger effect (Auger 1923, 1924); see Fig. 1.1.

Pierre Auger understood that, after being ionized in its inner shells, an atom spontaneously emits one of its remnant electrons, an idea a few decades ahead of its time with respect to our present understanding of the electron-electron interaction (correlation) inside an atom.

Some time later he found (Auger 1926) that the decay of an atom ionized in its inner shells may proceed via either the emission of an X-ray or an Auger electron, and invented the concept of fluorescence yield. Auger further demonstrated that, after ionization of the innermost shells, X-rays were predominantly emitted by heavy elements and Auger electrons by light elements, or from the outermost shells of heavy elements. Thus, the Auger electrons can be seen as the exact counterparts of the X-rays, and are of the same fundamental importance in physics.

Auger's discovery has been further investigated over the years by many groups. Most of the experimental and theoretical work focused on the determination of the Auger energies and decay rates. Experimentally, this requires the measurement of an angle integrated spectrum, while theoretically, the absolute squared values of the

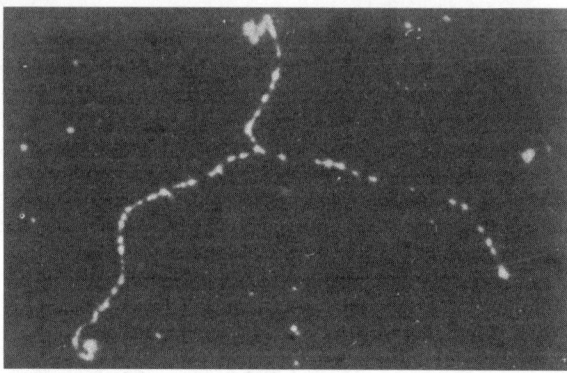

Fig. 1.1. Wilson chamber photography of the Auger effect observed in krypton; after Schpolski (1983)

transition matrix elements need to be determined, e.g. Åberg and Howat (1982); Burhop and Asaad (1972); Chattarji (1976); Siegbahn *et al.* (1967) and references therein.

A further development started in 1968 with Mehlhorn's prediction that Auger electrons emitted from an inner shell vacancy with a total angular momentum $J > 1/2$ may have a non-isotropic angular distribution (Mehlhorn 1968). The first angle resolved experiment was performed by Cleff and Mehlhorn (1971). The next step, the first angle and spin resolved Auger emission experiment has been performed by Hahn *et al.* (1985). The experimental set-up is illustrated in Fig. 1.2.

Theoretical investigations of the spin polarization of Auger electrons by Klar (1980), Kabachnik (1981) and Huang (1982), and by ourselves (Lohmann 1984; Blum *et al.* 1986), pointed out the advantages of such complicated experiments. These authors showed that the relevant observables, i.e. angular distribution and spin polarization, depend not only on the magnitudes of the transition amplitudes, but also on their relative phases. By calculating the relevant observables, and comparing them with experimental results, a wealth of information on the dynamics of Auger processes can be obtained. As pointed out by Kessler (1985) such type of investigations can be seen as a step closer towards a *complete experiment*, i.e. the determination of *all* transition amplitudes and scattering phases.

These studies were followed by a number of theoretical investigations on both, angle and spin resolved Auger emission, mainly by ourselves and by Kabachnik and co-workers, where many of these studies have been performed in a joint collaboration, e.g. Kabachnik and Sazhina (1988); Kabachnik *et al.* (1988, 1991); Lohmann (1990); Kabachnik and Lee (1989); Hergenhahn *et al.* (1991, 1993); Chen (1992, 1993); Lohmann *et al.* (1993). Further references may be found in the cited literature. A good review has been published by Mehlhorn (1990).

More recently, combined experimental and theoretical studies have been performed by a joint collaboration of the groups of Nora Berrah, Uwe Becker, and my-

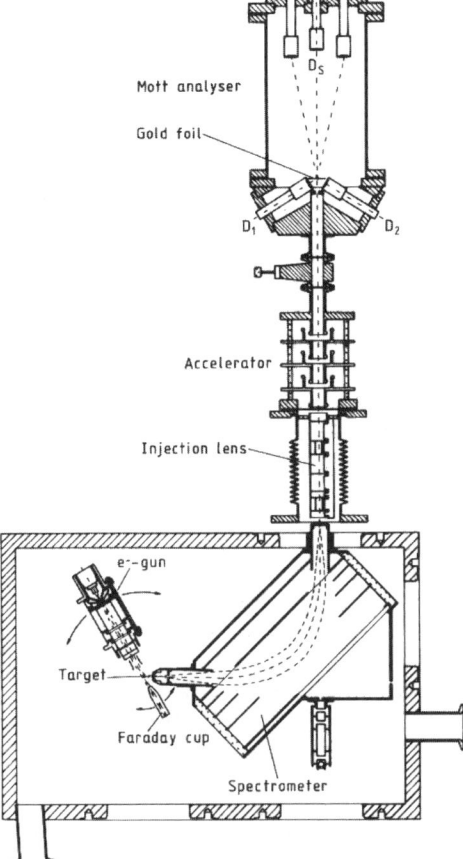

Fig. 1.2. The first experiment for measuring the spin polarization of the emitted Auger electrons after electron impact ionization on a rare gas target by means of Mott analyzer techniques; after Hahn *et al.* (1985)

self. These studies particularly focus on resonant Auger emission processes (Lohmann *et al.* 2003b, 2005).

A particular point of research has been the investigation of the two different physical mechanisms which can create a spin polarization of the Auger electron:

- In case that either the atom or the ionizing particle is polarized, e.g. circularly polarized photons or longitudinally polarized electrons, the intermediate ionic state is oriented. This orientation can be partially transferred to the electron spin. This polarization transfer can be viewed as a consequence of the conservation of angular momentum. We will therefore denote the spin polarization of the emitted Auger electron as *intrinsic*.

– From a physical point of view the other process is interesting. Here, in the intermediate ionic state an alignment is generated, which can be transformed into a spin polarization of the outgoing Auger electron via its final state interaction with the core charge distribution. In particular, we point out the importance of spin-dependent forces necessary for a non-vanishing spin polarization. In this case, the generation of spin polarization explicitly depends on the Auger decay dynamics and is therefore denoted as *dynamic* spin polarization.

The latter process has been of particular interest since a dynamic spin polarization can basically occur even for an unpolarized beam and an unpolarized target, too. This will be an important point in the discussion later on.

Important experimental progress has been made due to the availability of third-generation synchrotron radiation sources and high resolution spectrometers, e.g. Kuntze *et al.* (1993); Müller *et al.* (1995); Snell *et al.* (1996b); Langer *et al.* (1997); Hergenhahn *et al.* (1999); Schmidtke *et al.* (2000a, 2000b, 2001); Meyer *et al.* (2001); Drescher *et al.* (2003); O'Keeffe *et al.* (2003, 2004); Lohmann *et al.* (2003b, 2005); Turri *et al.* (2007). These developments stimulated further theoretical research, e.g. Lohmann and Fritzsche (1994); Lohmann and Larkins (1994); Kabachnik *et al.* (1994); Kabachnik and Schmidt (1995); Kabachnik and Ueda (1995); Hergenhahn and Becker (1995a, 1995b); Lohmann *et al.* (1996, 2002, 2003b, 2005) Kleiman *et al.* (1999a); Lohmann (1999a); Fritzsche *et al.* (2000); Schmidtke *et al.* (2000a, 2000b, 2001); Fritzsche (2001); Meyer *et al.* (2001); Kabachnik and Grum-Grzhimailo (2001); Kabachnik and Sazhina (2002); O'Keeffe *et al.* (2003, 2004); Kabachnik (2005); Kleinpoppen *et al.* (2005); Turri *et al.* (2007).

The analysis and physical interpretation of the experimental results requires necessarily developments into two directions:

1. Numerical calculations must be performed for all relevant Auger transition amplitudes. The Auger process is a complicated many-particle problem. Electron-exchange, electron-electron correlation and explicit spin-dependent effects must be taken into account. Of particular importance is the final state interaction between the Auger electron and the residual ion, that is, the scattering of the Auger electron in the field of the ion. Here, the methods of scattering theory must be used.

2. In order to compare numerical and experimental results general equations must be derived which relate the basic numerical quantities, i.e. transition amplitudes and scattering phases, to the experimental observables as angular distribution and spin polarization of the emitted Auger electron. These equations are the starting point for the numerical calculations, and, on the other side, are essential for any analysis of experimental results. In addition, such equations are often useful for the planning of experiments as well as for their subsequent analysis.

Both aspects will be discussed in this book.

In Chap. 2 we will consider the general assumptions necessary for a detailed investigation of angle and spin resolved Auger emission. Particularly, we introduce the two-step model, that is, the primary ionization and the Auger emission are treated as independent processes. The validity of the two-step model has been demonstrated in many experiments investigating either the Auger emission of free atoms and molecules or their emission from solid surfaces, e.g. Åberg and Howat (1982); Burhop and Asaad (1972); Chattarji (1976); Mehlhorn (1985); Siegbahn *et al.* (1967), and references therein, see also the review by Mehlhorn (1990). The general symmetries which apply to the Auger emission are also considered.

Then, we will derive the relevant equations for angular distribution and spin polarization of the emitted Auger electrons. After the primary ionization process the ions will be generally in excited states. The shape and spatial orientation of the excited orbitals will influence the subsequent Auger emission. The first task is therefore to characterize this anisotropic ensemble. This is most conveniently done in terms of density matrices and state multipoles. Applying the two-step model of Auger emission, we will derive and discuss the essential formulas of the primary electronic or photonic excitation and ionization process, respectively. Then, the relevant equations of Auger emission are derived. The angular distribution and spin polarization of the Auger electrons will be expressed in terms of state multipoles characterizing the intermediate ionic states, and by anisotropy parameters containing the information about the Auger decay dynamics, i.e. transition amplitudes and scattering phases. Particular emphasis will be given to the point which information can be obtained in such experiments by using either electron or photon impact for the primary ionization and excitation process, respectively. We also consider the different possibilities of polarized or unpolarized ionizing electrons and photons, respectively. Assuming an unpolarized electron or photon beam, we discuss the physically important process of generation of spin polarization out of alignment.

We will also derive non-linear interrelations between the angular distribution and spin polarization parameters of the Auger emission. This is still an active field of research. The important field of so-called resonant Auger transitions will be discussed. Resonant Auger transitions can be observed after a primary photoexcitation of the target atom, resulting in a singly ionized target after the Auger emission. The resonant Auger emission yields specific advantages due to symmetry relations of the photoexcitation process. Experimentally, it has been widely investigated by utilizing 3rd generation synchrotron beam techniques which allow for using either linearly or circularly polarized light for the primary photoexcitation process. Further, we will consider some special cases of Auger emission in more detail. Particularly, spin polarization of isotropic multiplets will be discussed, and the angle dependent intensity and spin polarization for specific intermediate ionization hole states will be considered. We will also discuss the case of Auger transitions from an unresolved intermediate fine structure. As another set of observables, we will introduce asymmetry parameters of Auger emission, and we are demonstrating their dependency on spin-dependent forces. Eventually, so-called linear dichroism of Auger emission after a primary photoionization/excitation will be considered.

In Chap. 3 the numerical methods developed to evaluate the relative intensities, angular distribution and spin polarization parameters are described. In particular the two program packages used in the calculations will be discussed. Common to all parameters is the numerical calculation of Auger transition amplitudes. The main numerical difficulty in the calculation of angle and spin resolved Auger transitions is that the relevant equations of angular distribution and spin polarization are commonly functions not only of the transition amplitudes but of the scattering phases, too. Their explicit knowledge is therefore crucial for the calculation of numerical data.

Auger transitions in heavy atoms show large fine structure splitting. Thus, a relativistic approach is required for the calculation of the matrix elements. This has been taken into account by calculating the Auger transition matrix elements within a multiconfigurational Dirac–Fock (MCDF) approach.

The numerical calculations have been performed using the two program packages ANISO and RATR. They, both, have been developed in the context of scattering theory (cf. Åberg and Howat 1982) and apply self-consistent field methods with configuration interaction (Δ-SCFCI method). Thus, intermediate coupling in the many-electron wavefunctions has been accounted for in both packages.

The ANISO package has been developed and frequently extended by ourselves, e.g. see Lohmann (1988, 1990, 1997). A number of useful approximations have been introduced and applied in the ANISO package. In particular, though applying a relativistic approach, we neglect information from the *small* component of the Dirac–Fock wavefunctions. Further, an energy dependent local exchange potential is used in the calculation of the continuum wavefunctions. This enables for a calculation of a decoupled set of differential equations and avoids the problem of solving the coupled set of integro-differential equations. This method is considerably improved compared to previous calculations, e.g. Aksela *et al.* (1984b); Chen *et al.* (1990); Lohmann (1990) which, for instance, completely neglect exchange with the continuum.

The RATR package is based on a previous version (Fritzsche 1991, 1992) developed for the calculation of Auger transition rates. It has been extended by Lohmann and Fritzsche (1994) to allow for the calculation of anisotropy parameters. As pointed out, this requires explicit knowledge of the scattering phases. While a number of useful approximations have been applied in the ANISO package, RATR fully includes the relativistic framework. It takes the small component of the Dirac–Fock wavefunction into account by solving the coupled set of integro-differential equations for the evaluation of the continuum wavefunction. Particularly, the right-hand side of the integro-differential equations contain an inhomogeneous part where orthogonality needs to be ensured by the introduction of Lagrange multipliers. Exchange interaction with the continuum and spin-orbit coupling is automatically accounted for in this approach. In addition, RATR provides a number of *switches* which allow e.g. for suppressing the Breit-interaction or the continuum exchange. This is most useful for calculations using large basis sets of several 1000 configuration state functions (CSF). It also allows for a detailed investigation of the strength of relativistic or exchange effects.

In Chap. 4 we will discuss a variety of examples and apply the relations derived in Chap. 2 in combination with the numerical tools introduced in Chap. 3. Here, we particularly focus on angular distribution and spin polarization parameters and their comparison to experimental and other theoretical data. Most data are available for the rare gases. This is, on the one hand, since rare gases are comparatively easy to handle in an experiment, on the other hand, their closed shell structure shows some numerical advantages. While we have performed investigations on, for instance, the alkalis (Lohmann and Fritsche 1994), or mercury (Lohmann 1992, 1993), which will be also considered, most of the applications and examples in this chapter will concentrate on numerical results for the rare gases and their comparison to the experimental data.

We will give a detailed comparison between theoretical and experimental data of the angular anisotropy of Auger electrons emitted from noble gas atoms which has been investigated by Kabachnik *et al.* (1991).

The spin polarization of Auger electrons emitted after photoionization with circularly polarized light is discussed for Auger transitions in Ar, Kr and Xe. Since a circularly polarized photon beam generates an orientation in the intermediate ionic state, the effect of polarization transfer is considered in more detail. For this, the intrinsic in-reaction plane components of the spin polarization vector are investigated.

For a certain type of Auger transitions the intermediate ionic state can be oriented but not aligned. As a consequence, the emitted Auger electrons can be spin polarized though remain isotropic with respect to their angular distribution. We will consider the intrinsic spin polarization of the isotropic Ar L_2MM Auger multiplet in detail (Lohmann and Larkins 1994).

The spin polarization of the Xe $M_{4,5}N_{4,5}N_{4,5}$ Auger lines has been measured in a high resolution experiment applying third generation synchrotron beam techniques (Hergenhahn 1996; Snell *et al.* 1996b). We will discuss our theoretical data in comparison to the experiment.

Correlation effects in the $N_{6,7}O_{4,5}O_{4,5}$ Auger spectrum of mercury have been investigated by Lohmann (1992, 1993). This research is of interest, as it has been the first numerical approach for calculating angular anisotropy and spin polarization data, while including f-electrons in the configuration basis set. We will focus on the correlation effects by investigating initial and final state configuration interaction, applying different numerical approaches in order to obtain data for the intensities of the Auger lines as well as for the angular distribution and spin polarization parameters.

Particularly, the Auger line intensities allow for a comparison with other experimental and numerical data.

A special field of research has been the Auger angular distribution of resonantly excited transitions. Here, in contrast to the so-called diagram transitions, we have an initial excitation of an inner shell electron into a Rydberg level. Eventually, the inner shell hole decays via resonant Auger emission. Applying some simple restrictions, a *spectator model* has been developed to calculate the matrix elements for this type of Auger transitions (Lohmann 1991). We will discuss a more sophisticated version of this spectator model which has been applied by Hergenhahn *et al.* (1993).

Considering the dynamic spin polarization, their values have been found as too small for experimental scrutiny for most of the diagram transitions. Experiments (Snell *et al.* 1996a, 1996b) have given evidence of a large dynamic spin polarization for certain lines of the resonantly excited $Xe^*(4d_{5/2}^{-1}6p)_{J=1}N_5O_{2,3}O_{2,3}$ Auger spectrum. In a more recent investigation (Lohmann 1999a), we derived simple propensity rules which allow, for the first time, predictions for a large dynamic spin polarization of Auger electrons emitted in resonantly excited Auger transitions. We will give a qualitative and quantitative explanation for the derived propensity rules. Our related calculations have been found in good agreement with the experimental data (Hergenhahn *et al.* 1999).

Recently, a detailed analysis of the resonantly excited $Ar^*(2p^{-1}4s)_{J=1}$ $L_{2,3}M_{2,3}M_{2,3}$ Auger decay has been performed in a joint theoretical and experimental collaboration (Lohmann *et al.* 2003b, 2005). The resonantly excited $Ar^*(2p^{-1}4s)_{J=1}$ $L_{2,3}M_{2,3}M_{2,3}$ Auger spectrum has been hitherto understood as a showcase for vanishing dynamic but large transferred spin polarization.[1] Surprisingly, the experimental data have given evidence for a large dynamic spin polarization of specific Auger lines even in a low resolution spectrum. We have been able to explain this effect as configuration interaction (CI) induced and caused by internal selection rules based on propensity rules. This new effect will be analyzed and discussed in detail.

Our investigations have been further extended to open shell atoms. Here, our main focus has been the investigation of the angle resolved KLL Auger transitions. For closed shell atoms, like the rare gases, KLL Auger spectra have to be isotropic since no alignment can be generated. However, as has been shown by Dill *et al.* (1975) for the case of photoelectron emission, this is totally different for open shell atoms. Here, we are entering an open field of research where a number of theoretical predictions still need experimental proof. We will discuss a detailed investigation which has been performed on the KLL Auger transitions of the alkali elements Na–Cs (Lohmann and Fritzsche 1994) where absolute Auger rates and the angular distribution parameters of Auger emission have been discussed. This work has then been extended to the angle resolved analysis of laser excited Na KLL Auger transitions (Lohmann *et al.* 1996).

Eventually, we will consider the KLL Auger transitions of atomic oxygen. First data on the angle resolved O KLL spectrum have been published in an experiment by Krause *et al.* (1996), where we published theoretical data for the Auger rates and angular distribution parameters (Lohmann and Fritzsche 1996).

While the first four chapters focus on Auger emission from free atoms, we will consider angle and spin resolved molecular Auger processes in Chap. 5. This is an active field of theoretical and experimental research. We will concentrate our discussion on diatomic and small polyatomic molecules. So far, most of the experimental and theoretical studies of molecular Auger processes have concentrated on determining the energy and probability of a given Auger transition. A general theory for the angular distribution of Auger electrons following photoabsorption

---

[1] See Sects. 2.5.4 and 4.7 for a detailed explanation of these phrases.

or photoionization of molecules has already been developed by Dill *et al.* (1980) and by Chandra and Chakraborty (1992). These authors have shown that vacancies produced in molecular photoionization behave differently than their atomic counterparts. In particular, Dill *et al.* (1980) showed that Auger electrons, emitted in the decay of a K-shell vacancy in a diatomic molecule produced by photons, can have a non-isotropic angular distribution, contrary to the atomic case. This is due to the anisotropic nature of photon-molecule interaction.

For an analysis of experimental results for freely rotating molecules, and also for numerical calculations, it is necessary to develop a general theoretical framework. We will derive the general formulas for the angular distribution and spin polarization of Auger emission after electron impact ionization, exceeding the treatments of Dill *et al.* (1980) and Chandra and Chakraborty (1992). For electron impact ionization, the number of independent parameters is no longer restricted by dipole selection rules and thus, additional parameters must be introduced. From the state multipoles of the emitted Auger electrons we will derive the general expressions for the Cartesian components of the spin polarization vector. The general equations of angular distribution and spin polarization will be discussed. They can be simplified for certain polarization states of the ionizing electron beam and we consider some special cases of interest.

An important point is the occurrence of coherence in the primary ionization process which has been first discussed by Bonhoff *et al.* (1996). This results in a further increase in the number of independent parameters necessary to correctly interpret the experimental data, even for photoionization as initial ionizing process which has been overlooked in the literature before.

We will give a brief discussion of the numerical methods for the calculation of molecular anisotropy parameters which have been developed by Schimmelpfennig *et al.* (1995) and further extended by Bonhoff *et al.* (1997). As for the atomic case, the calculation of molecular anisotropy parameters requires explicit knowledge of the scattering phase. Molecular calculations for Auger rates (e.g. Schimmelpfennig 1994) however, are commonly done in a multi-center basis expansion. Thus, a basic difficulty in the calculation of molecular anisotropy parameters is the determination of the scattering phases of the emitted partial waves given in a one-center basis. This problem has been solved employing Greens operator methods. We will present some results for the HF Auger spectrum and for the related angular distribution parameters.

The main focus of Chap. 5 will be on diatomic, or more generally, linear molecules. On the other hand, introducing comparatively simple extensions applying group theoretical methods, the derived theory can be generalized to polyatomic molecules in a similar way. A discussion of the angular distribution of Auger electrons emitted from non-linear polyatomic molecules may be found in Lehmann *et al.* (1997) where results for the H_2O molecule have been presented. A more detailed investigation of the angular distribution of Auger electrons emitted from polyatomic molecules applying group-theoretical methods has been given by Lehmann and Blum (1997). We will adopt their formalism and investigate the anisotropy of polyatomic molecular Auger decay for the case of non-degenerate and degenerate point groups. The consequences of specific molecular symmetries for the angular distribution of the emitted molecular Auger electrons are pointed out for some examples.

Eventually, we will give a conclusion and an outline of planned and possible future work.

2 Theory

The main goal of this chapter is to derive the relevant equations for angular distribution and spin polarization of the emitted Auger electrons. The derived equations are of physical importance in a two-fold manner. On the one hand, they form the starting point for the realization of numerical calculations for angle and spin resolved Auger transitions. On the other hand, they are the necessary basis for the planning and performing of experiments as well as for their subsequent analysis.

In Sect. 2.1 we will consider the general assumptions necessary for a detailed investigation of angle and spin resolved Auger emission. Particularly, we will introduce the two-step model, that is, the primary ionization and the Auger emission are treated as independent processes. The validity of the two-step model has been demonstrated in many experiments (e.g. see the review by Mehlhorn 1990). The general symmetries which apply to the Auger emission are also considered. After the primary ionization process the ions A^{+*} will generally be in excited states, and the shape and spatial orientation of the excited orbitals will influence the subsequent Auger emission. Therefore, the first task is to characterize this anisotropic ensemble. This is most conveniently done in terms of density matrices and state multipoles. In Sect. 2.2 we will give a brief outline of the formalism and derive the basic properties.[1] Applying the two-step model of Auger emission discussed in Sect. 2.1, we will derive the essential formulas of the primary ionization process in Sect. 2.3 and, in more detail, discuss the cases of electronic or photonic excitation and ionization, respectively. The relevant equations of Auger emission will then be derived in Sect. 2.4. In Sect. 2.5 angular distribution and spin polarization of the Auger electrons will be expressed in terms of state multipoles characterizing the intermediate ionic states A^{+*}, and by anisotropy parameters containing the information about the Auger decay dynamics, i.e. transition amplitudes and scattering phases, respectively. Particular emphasis will be given to the point which information can be obtained in such experiments by using either electron or photon impact for the primary ionization process. We will also consider the different possibilities of polarized or unpolarized electrons and photons, respectively. Assuming an unpolarized electron or photon beam, we will discuss the physically important process of generating spin polarization out of alignment. We will further derive interrelations between the angular distribution and spin polarization parameters of the Auger

---

[1] For a more detailed introduction into the field of density matrices and state multipoles we refer to the book by Blum (1996).

emission, resulting in a variety of linear and non-linear equations which are still an active field of research. In Sect. 2.6 we will focus on the important field of so-called resonant Auger transitions which can be observed after a primary photoexcitation of the target atom, resulting in a singly ionized target after the Auger emission. The resonant Auger emission yields specific advantages due to symmetries in the photoexcitation process, and, experimentally, has been widely investigated by utilizing 3rd generation synchrotron beam techniques which allow for using either linearly or circularly polarized light. In the last section, we will consider some special cases of Auger emission in more detail. Particularly, spin polarization of isotropic multiplets will be discussed, and angle dependent intensities and spin polarization for specific intermediate ionized hole states will be considered. We will also discuss the case of Auger transitions from an unresolved intermediate fine structure, and, we will introduce and discuss so-called asymmetry parameters of Auger emission. Eventually, we will consider linear dichroism of Auger emission.

2.1 The Auger Effect

2.1.1 General Considerations

Our present understanding of the Auger effect is that of a two-step process (e.g. Mehlhorn 1990),

$$e^- + A \longrightarrow A^+ + e_e^- + e_s^- \tag{2.1}$$

$$\longmapsto A^+ \longrightarrow A^{++} + e_{Auger}^-. \tag{2.2}$$

Here, the target atom A is ionized in a first step, for instance by electron impact. The primary emitted electron e_e^-, and the scattered electron e_s^-, are usually not observed. In the second step, the Auger emission takes place. The essential assumption is that both processes are considered as independent.

In many experiments photons are used instead for the primary ionization.

$$\gamma + A \longrightarrow A^+ + e_{Phot}^-. \tag{2.3}$$

The case of photoionization is illustrated in Fig. 2.1.

The resonantly excited intermediate ionic state A^+ is usually referred to as the initial state of the Auger decay and is assumed to be independent of the specific ionization process. Of course, the physical generation of the state A^+ depends on the primary process. The intermediate ions A^+ are generally in excited states. Shape and spatial orientation of the relevant orbitals will influence the subsequent Auger decay (2.2). However, the dynamics of the Auger emission are not influenced by the dynamics of the primary ionization process (2.1) and (2.3), respectively. The Auger emission process is illustrated in Fig. 2.2.

Applying the two-step model the following experimental conditions must be fulfilled

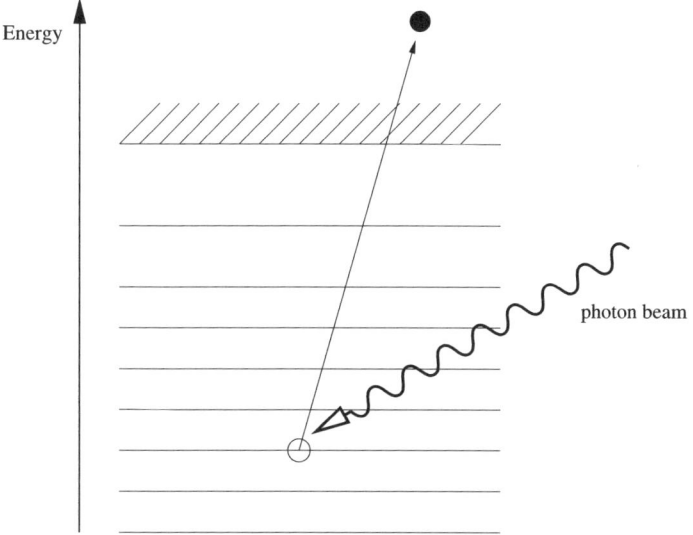

Fig. 2.1. The primary photoionization

- The width of the singly ionized intermediate excited state is small compared to the double photoionization decay width.
- Post-collision interaction (PCI) is negligible during the Auger emission.

With this, the excitation/ionization of the atom, i.e. (2.1) or (2.3) and the Auger emission (2.2) can be treated as independent processes. This has been discussed in more detail in a review by Sandner (1985). The validity of the two-step model for Auger transitions has been demonstrated in many experiments investing either the Auger emission of free atoms and molecules or the emission from solid surfaces (e.g. Åberg and Howat 1982; Burhop and Asaad 1972; Chattarji 1976; Mehlhorn 1985; Siegbahn *et al.* 1967, and references therein).

It should be noted that in the description of the secondary process (2.2) final state interaction must be taken into account, that is, the scattering of the emitted Auger electron in the field of the residual ion. This is an important point in the development of the theory and relates the description of Auger emission processes to scattering physics and most of its principles can be applied. As a matter of fact the Auger emission can be considered as a *half collision process* (Bergmann and Schaefer 1992).

2.1.2 Angle and Spin Resolved Auger Emission

As has been outlined e.g. in the book by Kessler (1985), one has to go beyond the well-known experimental techniques of determining the energy levels and cross sections to obtain more information from a scattering experiment.

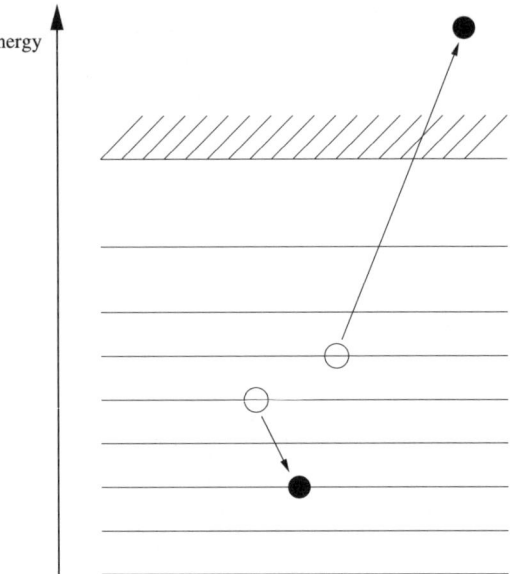

Energy

Fig. 2.2. The Auger emission process

Many experimental and theoretical results have been published on Auger spectra (e.g. see Mehlhorn 1990, and references therein). Of increasing interest have been studies on the angular distribution and spin polarization of the emitted Auger electrons. These investigations allow to obtain more detailed information on the dynamics of the Auger process and therefore can be seen as a step closer to a complete experiment, i.e. determining all transition amplitudes and scattering phases scattering phase of the combined particle + target system. We will concentrate on these topics and consider processes which are related to recent experiments.

In the following we allow the incoming electrons or photons, i.e. (2.1) or (2.3), to be arbitrarily polarized. The initial atomic ensemble is assumed to be unpolarized. Throughout this book it will be assumed that the primary scattered or the photoelectrons are not observed. The first aim is to derive general formulas which relate the experimental observables of angular distribution and spin polarization to the polarization of the incident projectiles, and to the transition matrix elements which contain the information on the dynamics. The obtained equations are important for the planning of experiments, as well as for an analysis of experimental results. In addition, they are the basis for the numerical calculations.

2.1.3 Coordinate Frame and General Symmetries

If the incoming electron or photon beam is in a pure polarization state, we can always choose an appropriate coordinate frame, in order to reduce the number of parameters necessary for a full interpretation of the experimental results. Considering the general case of an arbitrarily polarized ionizing beam the problem of choosing

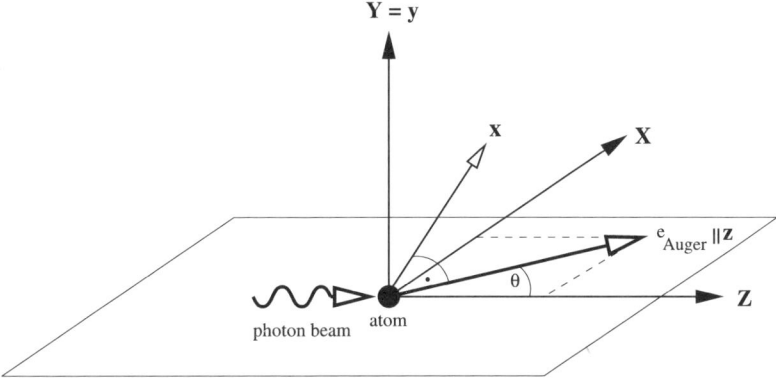

Fig. 2.3. The reaction plane

a most simple coordinate system occurs. This is because the polarization density matrix cannot be diagonalized for an arbitrarily polarized beam. We discussed this problem for electron impact ionization (Lohmann 1984), and for the case of photo-ionization with an arbitrarily polarized synchrotron beam (Kleiman *et al.* 1999a).

We will choose the coordinate system shown in Fig. 2.3. The laboratory frame is denoted by XYZ, the Z-axis is parallel to the incident beam direction. The reaction plane is spanned by the axis of the incoming projectiles and the direction of the observed Auger electrons. With respect to this axis we define a second coordinate system, the *helicity system* of the observed Auger electrons, denoted by xyz. The z-axis is parallel to the momentum of the Auger electrons and the y-axis is perpendicular to the reaction plane. For the following, we will choose the X-axis to be in the reaction plane and let the Y-axis coincide with y-axis of the helicity frame.

Let us now discuss the symmetries which can be generally obtained by using either polarized electrons or photons as ionizing particles.

Electron and Photon Polarization

Supposing a polarized electron beam the spin polarization vector **P** can be described by its Cartesian components P_i, $i = X, Y, Z$. Thus, for a longitudinally polarized electron beam the spin polarization vector is parallel to the electron beam axis, while a transversely polarized electron beam can be described by a component perpendicular to the reaction plane and an in-plane component, respectively. In the helicity frame, the components of the spin polarization vector can be described analogously.

For the case of photoionization the photon polarization can usually be expressed in terms of *Stokes parameters* η_i, $i = 1, 2, 3$. A detailed description of these parameters can be found in Born and Wolf (1970).

A circularly polarized photon beam is described by the parameter η_2

$$\eta_2 = \frac{I_+ - I_-}{I_{tot}}, \qquad (2.4)$$

where $I_\pm$ denotes the intensity of right and left handed circularly polarized light, respectively. I_{tot} denotes the total photon beam intensity.

The degree of linear polarization is given by the Stokes parameters η_3 and η_1

$$\eta_3 = \frac{I(0^o) - I(90^o)}{I_{tot}}, \tag{2.5}$$

$$\eta_1 = \frac{I(45^o) - I(135^o)}{I_{tot}}. \tag{2.6}$$

For our choice of coordinate frame $I(\beta)$ denotes the intensity transmitted by a Nicol prism oriented at an angle β with respect to the X-axis.

In the following, the spin polarization of the Auger electrons is discussed with respect to general symmetry principles. A few simple results can be immediately obtained from these principles.

Ionization with Unpolarized Electrons or Photons

The Auger process is invariant under reflection in the X–Z plane (reaction plane). Due to spin-orbit interaction, the outgoing Auger electron may however be polarized. Since the spin polarization vector transforms under reflection as an axial vector, only its y-component can be expected to be non-zero. The transformation properties of the spin polarization vector of the emitted Auger electrons are shown in Fig. 2.4.

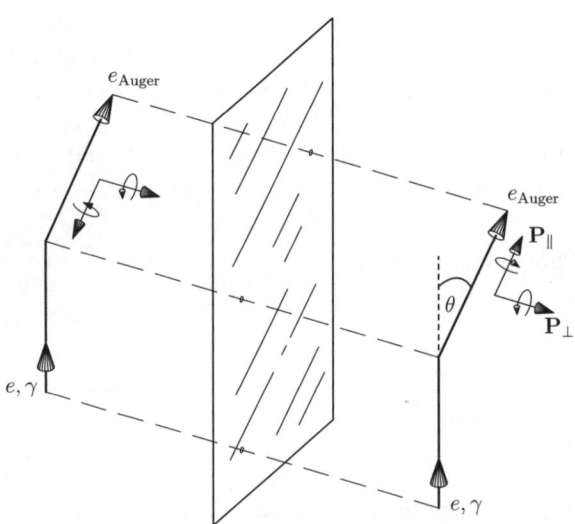

Fig. 2.4. Transformation properties of the spin polarization vector under mirror symmetry of the Auger process

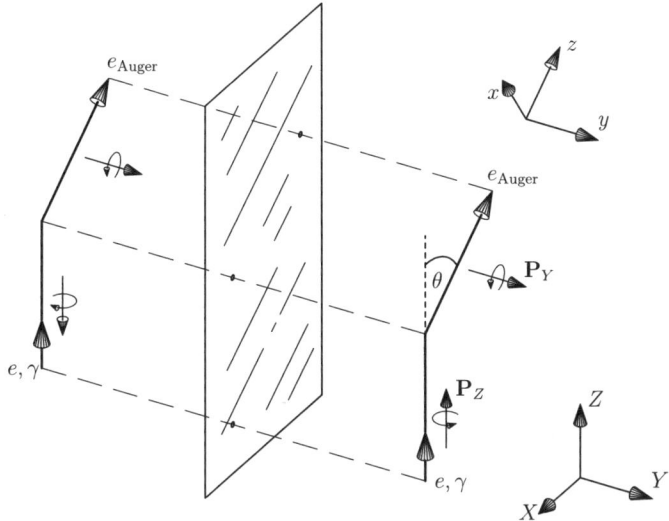

Fig. 2.5. The Auger process with longitudinally polarized electrons and circularly polarized photons, respectively. Transformation properties of the spin polarization vectors of longitudinally polarized electrons or circularly polarized photons and of the y-component of the spin polarization of the Auger electron

Ionization with Longitudinally Polarized Electrons or Circularly Polarized Photons

The reaction plane is no longer a symmetry plane due to the axial nature of the spin polarization vector. This is illustrated in Fig. 2.5. Thus, all components of the Auger electron spin polarization vector can be different from zero.

Ionization with Transversely Polarized Electrons

Here, we need to consider two cases:

$P_X \neq 0$: Again, the reaction plane is not a symmetry plane, and thus, all components of the Auger electron spin polarization vector can be different from zero.

$P_Y \neq 0$: Here, the X–Z plane shows again reflection symmetry since the spin polarization of the incoming electron beam remains unchanged. Thus, only the y-component of the spin polarization vector of the Auger electrons can be nonzero.

Ionization with Linearly Polarized Photons

Here, we make use of the law of superposition. An electric field vector **E** oscillating in an arbitrary but fixed direction perpendicular to the reaction plane can always be split into two components. One component oscillating perpendicular and one oscillating in the reaction plane (see Fig. 2.6). It should be noted that due to its rapid

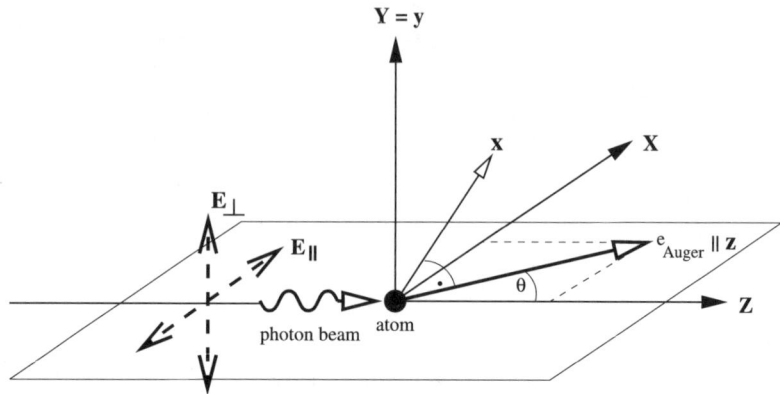

Fig. 2.6. Superposition of the electric field vector **E**

Table 2.1. Predictions for the spin polarization vector of the Auger electron

Polarization of projectiles	Auger electrons p_x	p_y	p_z
unpol. e, γ, $\mathbf{P} = 0$	0	$\neq 0$	0
trans. pol. e, $P_x \neq 0$	$\neq 0$	0	$\neq 0$
trans. pol. e, $P_y \neq 0$	0	$\neq 0$	0
long. pol. e, $P_z \neq 0$	$\neq 0$	0	$\neq 0$
circ. pol. γ, $\eta_2 \neq 0$	$\neq 0$	0	$\neq 0$
lin. pol. γ, $\eta_3 \neq 0$	0	$\neq 0$	0
lin. pol. γ, $\eta_1 \neq 0$	$\neq 0$	0	0

oscillations the direction of the **E** vector is physically unimportant and only its axis is physically defined.

Thus, for a fully linearly polarized photon beam it is sufficient to consider the two cases of an in-plane and a perpendicular to the plane linearly polarized beam. For both cases, the reaction plane is showing reflection invariance since the two components of the electric field vector are transformed onto themselves. Thus, we have the same case as for an unpolarized ionizing beam and therefore only the y-component of the spin polarization vector of the Auger electrons can be expected to be different from zero.

Our predictions for the spin polarization vector of the Auger electrons derived from general symmetry principles are presented in Table 2.1 which must be read as follows. The components of the spin polarization vector which can be non-zero if an unpolarized beam is used are always occurring, e.g. see the first line of Table 2.1. The following lines show which additional components can be expected to be different from zero if a polarized beam is used for the primary ionization process.

2.2 General Density Matrix Theory

The application of the framework of density matrices to scattering problems in the field of atomic and molecular physics has been first introduced by Fano and co-workers (e.g. see Fano 1957). Meanwhile, the general formalism has been outlined and applied to a large variety of experiments. In the following we will only point out some properties of the density matrix and state multipoles. For a detailed introduction into density matrix theory we refer to the book by Blum (1996).

2.2.1 Definition and Basic Relations

Consider a mixture of independently prepared states $|\psi_n\rangle$, where $n = 1, 2, \ldots$, with statistical weight W_n. The density operator describing the mixture is then defined as

$$\hat{\rho} = \sum_n W_n |\psi_n\rangle\langle\psi_n|, \tag{2.7}$$

where the sum extends over all states present in the mixture. $\hat{\rho}$ is also referred to as *statistical operator*.

The density matrix can be expanded into a basis of orthonormalized states, e.g. $|\phi_n\rangle$, and its matrix elements can be written as

$$\langle\phi_i|\hat{\rho}|\phi_j\rangle = \sum_n W_n a_i^{(n)} a_j^{(n)*}, \tag{2.8}$$

where $a_i^{(n)}$ are the expansion coefficients of the basis states. Equation (2.8) is called the $\{|\phi_n\rangle\}$ representation of the density matrix.

Form (2.8) it is evident that $\hat{\rho}$ is Hermitian, i.e. the matrix (2.8) satisfies the condition

$$\langle\phi_i|\hat{\rho}|\phi_j\rangle = \langle\phi_j|\hat{\rho}|\phi_i\rangle^*. \tag{2.9}$$

The probability of finding the system in the state $|\phi_m\rangle$ is given by the diagonal elements of the density matrix. Thus, it follows

$$\langle\phi_m|\hat{\rho}|\phi_m\rangle \geq 0. \tag{2.10}$$

The expectation value of any operator $\hat{O}$ is given by the trace of the product of $\hat{\rho}$ and $\hat{O}$

$$\langle\hat{O}\rangle = \frac{\mathrm{tr}(\hat{\rho}\,\hat{O})}{\mathrm{tr}\,\hat{\rho}}. \tag{2.11}$$

This relation is an important result. As we learned from quantum mechanics all information on the behaviour of a given system can be expressed in terms of expectation values of suitably chosen operators. Thus, the basic problem is to calculate the expectation values. Since the expectation value of any operator can be obtained by use of (2.11) the density matrix contains all physically significant information on the system.

Considering two interacting quantum systems, any state vector $|\psi\rangle$ representing the state of a coupled system, can always be written as a sum of direct products

$$|\psi\rangle = \sum_{i,j} a(i, j)|\Phi_i\rangle|\varphi_j\rangle. \tag{2.12}$$

In many cases, only one of the two systems is of interest where the other one remains undetected. Suppose $\hat{\mathbf{Q}}(M)$ being an operator acting only on the variables of the system $|M\rangle$, its matrix elements are given by

$$\langle M', m'|\hat{\mathbf{Q}}(M)|M, m\rangle = \langle M'|\hat{\mathbf{Q}}(M)|M\rangle\langle m'|m\rangle, \tag{2.13}$$

where the states $|m\rangle$ have been assumed as orthogonal. With this, we define the *reduced density matrix* $\hat{\rho}(M)$ describing the M system alone as

$$\langle M'|\hat{\rho}(M)|M\rangle = \sum_m \langle M', m|\hat{\rho}(M)|M, m\rangle. \tag{2.14}$$

Thus, the reduced density matrix contains all information on the M system. Essentially, the total density matrix is calculated and then projected onto the subspace of interest. This is a powerful property of the density matrix and has been considerably applied in the theoretical description of coupled ensembles.

2.2.2 Tensor Operators and State Multipoles

Consider an ensemble of particles in various angular momentum states $|JM\rangle$ characterized by a density matrix $\hat{\rho}$ with elements $\langle J'M'|\hat{\rho}|JM\rangle$. The density operator in the $\{|JM\rangle\}$ representation can then be written in the form

$$\hat{\rho} = \sum_{J'JM'M} \langle J'M'|\hat{\rho}|JM\rangle|J'M'\rangle\langle JM|. \tag{2.15}$$

Using the definition of tensor operators (e.g. see Blum 1996),

$$T(J'J)_{KQ} = \sum_{M'M} (-1)^{J'-M'}\sqrt{2K+1} \begin{pmatrix} J' & J & K \\ M' & -M & -Q \end{pmatrix} |J'M'\rangle\langle JM|, \tag{2.16}$$

we obtain the *state multipoles* or *statistical tensors* as the expectation values of irreducible tensor operators

$$\left\langle T(J'J)^+_{KQ}\right\rangle = \sum_{M'M} (-1)^{J'-M'}\sqrt{2K+1} \begin{pmatrix} J' & J & K \\ M' & -M & -Q \end{pmatrix} \langle J'M'|\hat{\rho}|JM\rangle. \tag{2.17}$$

The inverse relations can be easily obtained by multiplying both sides with a Clebsch–Gordan coefficient $\left(J'N', J - N|KQ\right)$, summing over all values of K

and Q, and using the orthogonality properties of the Clebsch–Gordan coefficients. Expressing the Clebsch–Gordan coefficient via a $3j$-symbol yields

$$\langle J'N'|T(J'J)_{KQ}|JN\rangle = (-1)^{J'-N'}\sqrt{2K+1}\begin{pmatrix} J' & J & K \\ N' & -N & Q \end{pmatrix}, \qquad (2.18)$$

and expressing the density matrix elements in terms of state multipoles gives

$$\langle J'N'|\hat{\rho}|JN\rangle = \sum_{KQ}(-1)^{J'-M'}\sqrt{2K+1}\begin{pmatrix} J' & J & K \\ N' & -N & -Q \end{pmatrix}\langle T(J'J)_{KQ}^{+}\rangle. \quad (2.19)$$

Substituting (2.18) and (2.19) into (2.15) gives the expansion of the density operator in terms of irreducible tensor operators

$$\hat{\rho} = \sum_{J'JKQ}\langle T(J'J)_{KQ}^{+}\rangle T(J'J)_{KQ}. \qquad (2.20)$$

Multiplying both sides of (2.20) by $T(J'J)_{KQ}^{+}$, taking the trace, and using the orthonormality conditions of the tensor operators yields the inverse relation

$$\langle T(J'J)_{KQ}^{+}\rangle = \mathrm{tr}\big(\hat{\rho}\,T(J'J)_{KQ}^{+}\big), \qquad (2.21)$$

which is equivalent to (2.17).

If the ensemble of interest is an incoherent mixture of J states the density matrix is diagonal in J, and (2.20) then reduces to

$$\hat{\rho} = \sum_{JKQ}\langle T(J)_{KQ}^{+}\rangle T(J)_{KQ}. \qquad (2.22)$$

This result shows that state multipoles $\langle T(J'J)_{KQ}^{+}\rangle$ with $J' \neq J$ describe the coherence between states with different angular momentum J.

If the ensemble is an incoherent superposition of states with different quantum numbers M, then the density matrix is diagonal in M and (2.17) shows that all multipoles with $Q \neq 0$ vanish. The corresponding density operator is then given by

$$\hat{\rho} = \sum_{J'JK}\langle T(J'J)_{K0}^{+}\rangle T(J'J)_{K0}. \qquad (2.23)$$

Hence, the coherence between states with different quantum number M is characterized by the non-vanishing multipoles with $Q \neq 0$.

2.2.3 Properties of State Multipoles

The state multipoles show some useful properties (again we refer to Blum 1996, for their derivation). From the Hermiticity condition we get

$$\langle T(J'J)_{KQ}^{+}\rangle^{*} = (-1)^{J'-J+Q}\langle T(J'J)_{K-Q}^{+}\rangle. \qquad (2.24)$$

This implies for sharp angular momentum $J' = J$

$$\left(T(J)_{KQ}^+\right)^* = (-1)^Q\left(T(J)_{K-Q}^+\right),\tag{2.25}$$

which relates the components Q and $-Q$ of the multipoles to each other. In particular, (2.25) ensures that the *state multipoles* $\left(T(J)_{K0}^+\right)$ *are real numbers*.

Most convenient for a theoretical description of a physical problem are the transformation properties of the state multipoles under rotation. Defining a set of state multipoles $\left(T(J'J)_{KQ}^+\right)$ with respect to a coordinate system with axes X, Y, Z with corresponding quantum numbers M', M and Q, a second set of multipoles $\left(T(J'J)_{Kq}^+\right)$ can be defined with respect to a coordinate system with axes x, y, and z. The two sets are then related to each other by

$$\left(T(J'J)_{KQ}^+\right) = \sum_q\left(T(J'J)_{Kq}^+\right)\mathcal{D}_{qQ}^{(K)}(\omega)^*,\tag{2.26}$$

where $\mathcal{D}_{qQ}^{(K)}(\omega)$ denotes the rotation matrices and ω is the spatial angle between the two coordinate frames. This result shows that the state multipoles transform as irreducible tensors of rank K and component Q. In particular, their rank K remains unchanged under rotation.

2.2.4 Wigner–Eckart Theorem

An important tool in angular momentum theory is the Wigner–Eckart theorem. It allows for decoupling those quantities which depend explicitly on the dynamics of the interaction from those which are purely geometrical. Applying the Wigner–Eckart theorem to the tensor operators yields

$$\left(J'M'|T(J'J)_{KQ}|JM\right) = (-1)^{J'-M'}\begin{pmatrix} J' & K & J \\ -M' & Q & M \end{pmatrix}\left(J'\|T_K\|J\right),\tag{2.27}$$

where $\left(J'\|T_K\|J\right)$ denotes the corresponding reduced matrix element. Comparing (2.27) with (2.18) and using the symmetry properties of the $3j$-symbols under interchanging of columns yields

$$\left(J'\|T_K\|J\right) = \sqrt{2K+1}.\tag{2.28}$$

Inserting (2.28) back into (2.27), we see that the tensor operators $T(J'J)_{KQ}$ are purely geometrical quantities.

2.3 Primary Ionization–Excitation

The general density matrix theory which has been briefly outlined in the previous section is now applied to the first part of the Auger process, i.e. the creation of an inner shell hole. Particularly, we consider the ionization and excitation process with arbitrarily polarized photons and electrons, respectively.

2.3.1 Photoionization

Let us consider a primary photoionization process, e.g. by absorption of Synchrotron radiation,

$$\gamma_{Syn} + A \longrightarrow A^{+^*} + e^-_{Phot}, \tag{2.29}$$

i.e. a photoelectron is emitted into the continuum which leaves the atom in a singly ionized excited state. For the following, we assume the incoming photon beam as arbitrarily polarized, and the initial atomic ensemble as unpolarized. Since we are not considering coincidence experiments the emitted photoelectron is not detected.

The density matrix $\hat{\rho}_{out}$ describes the combined ensemble of the intermediate excited ionic state A^{+^*} and the photoelectron e^-_{Phot}, i.e.

$$\hat{\rho}_{out} = \hat{\rho}_{out}\left(A^{+^*} + e^-_{Phot}\right). \tag{2.30}$$

For describing the subsequent Auger emission process we require information about the intermediate excited ionic state, only. Thus, we need to determine the reduced density matrix of the intermediate ion.

The total density matrix describing the uncorrelated ensemble of the initial photon γ_{Syn} and the atomic state A can be written as a direct product

$$\hat{\rho}_{in} = \hat{\rho}(A) \times \hat{\rho}(\gamma_{Syn}). \tag{2.31}$$

In particular, assuming an arbitrarily polarized synchrotron beam and choosing the quantization axis along the incoming beam direction, the density matrix of the synchrotron photon can be expressed in terms of *Stokes parameters*,

$$\hat{\rho}(\gamma_{Syn}) = \frac{I}{2}\begin{pmatrix} 1+\eta_2 & -\eta_3+i\eta_1 \\ -\eta_3-i\eta_1 & 1-\eta_2 \end{pmatrix}, \tag{2.32}$$

where the notation of Blum (1996) has been adopted. η_1 and η_3 describe the linear polarization state of the photon beam, whereas η_2 gives the degree of circular polarization. The Stokes parameters have been discussed in Sect. 2.1.3.

The two total ensembles are related to each other by the transition operator[2] T_i

$$\hat{\rho}_{out} = T_i \, \hat{\rho}_{in} \, T_i^+. \tag{2.33}$$

Now, we are expressing the density matrix $\hat{\rho}_{out}$ of the combined ensemble of the ionic and electronic states in terms of statistical tensorial sets (e.g. see Blum 1996). As stated above we only need information about the ionic ensemble, i.e. we need to determine the reduced density matrix of the ions. Thus, (2.30) may be expressed in terms of state multipoles as

$$\left\langle T(J)^+_{K'Q'}\right\rangle = \text{tr}\left(\hat{\rho}_{out} T(J)^+_{K'Q'}\right), \tag{2.34}$$

where tr denotes the trace. The statistical tensorial set $T(J)^+_{K'Q'}$ refers to the intermediate excited ionic state.

---

[2] The special form of T_i will be discussed later.

Table 2.2. The state multipoles of an arbitrarily polarized photon beam and their connection to the Stokes parameters. The photon beam axis has been chosen as quantization axis. Multipoles of rank $K > 2$ must be zero due to dipole selection rules

State multipoles	Stokes parameters	Stokes parameters	State multipoles
$\langle T_{00}^+ \rangle = \frac{I}{\sqrt{3}}$		$I = \sqrt{3}\langle T_{00}^+ \rangle$	
$\langle T_{10}^+ \rangle = \frac{I}{\sqrt{2}}\eta_2$		$I\eta_2 = \sqrt{2}\langle T_{10}^+ \rangle$	
$\langle T_{1\pm1}^+ \rangle = 0$			
$\langle T_{20}^+ \rangle = \frac{1}{\sqrt{2}}\langle T_{00}^+ \rangle = \frac{I}{\sqrt{6}}$		$I = \sqrt{6}\langle T_{20}^+ \rangle$	
$\langle T_{2\pm1}^+ \rangle = 0$			
$\langle T_{22}^+ \rangle = \frac{1}{2}(-\eta_3 + i\eta_1)$		$I\eta_3 = -2Re\langle T_{22}^+ \rangle$	
$\langle T_{2-2}^+ \rangle = \langle T_{22}^+ \rangle^* = \frac{1}{2}(-\eta_3 - i\eta_1)$		$I\eta_1 = 2Im\langle T_{22}^+ \rangle$	

Analogously, we use tensorial sets for the initial photonic, $T_{\Gamma\gamma}^+$, and atomic states, $T(J_0)_{K_0 Q_0}^+$. We write

$$\hat{\rho}(\gamma_{Syn}) = \sum_{\Gamma\gamma} \langle T_{\Gamma\gamma}^+ \rangle T_{\Gamma\gamma} \qquad (2.35)$$

for the photonic state. With this, the state multipoles $\langle T_{\Gamma\gamma}^+ \rangle$ describing the Synchrotron photon can be connected to the Stokes parameters which is shown in Table 2.2. In particular, we note that for any arbitrarily polarized photon beam only the tensors $\langle T_{\Gamma 0}^+ \rangle$, with $\Gamma \le 2$ and $\langle T_{2\pm2}^+ \rangle$ can be non-zero. However, this depends on the choice of quantization axis. As can be seen from Table 2.2 the photon alignment $\langle T_{20}^+ \rangle$ is directly proportional to the monopole. From a physical point of view, this demonstrates the fact that a photon beam is always aligned due to the transverse character of the electromagnetic field.

The initial atomic state has been assumed as unpolarized, i.e.

$$\hat{\rho}(A) = \frac{1}{2J_0 + 1}\mathbf{1} \qquad (2.36)$$

where $\mathbf{1}$ is the $(2J_0 + 1)$-dimensional unit matrix.

Inserting the above two equations into (2.34) and using (2.33) we obtain

$$\left\langle T(J)_{K'Q'}^+ \right\rangle = \sum_{\Gamma\gamma} \left\langle T_{\Gamma\gamma}^+ \right\rangle B_{phot}(K'Q', \Gamma\gamma), \qquad (2.37)$$

where the anisotropy parameter B_{phot} is defined as

$$B_{phot}(K'Q', \Gamma\gamma) = \frac{1}{2J_0 + 1} \, tr\big(T_i \, \mathcal{T}_{\Gamma\gamma} \, T_i^+ \, T(J)_{K'Q'}^+\big). \qquad (2.38)$$

Now, we insert complete basis sets. Using ket vectors, the initial atomic state is expressed as $|J_0 M_0\rangle$ where J_0 and M_0 denote the total angular momentum and magnetic quantum number, respectively. The incoming photon is characterized as $|\omega \mathbf{n} \lambda\rangle$ where ω is the photon energy, $\mathbf{n}$ the incoming beam direction, and λ the helicity of the photon state. Analogously, the intermediate ionic state is described as $|J M\rangle$ and the emitted photoelectron as $|\mathbf{p}^{(-)} m_s\rangle$ where $\mathbf{p}$ is the electronic momentum and m_s its magnetic spin quantum number. The upper index $(-)$ on $\mathbf{p}$ indicates that we are dealing with scattering solutions with incoming boundary conditions. With this, the anisotropy parameter B_{phot} can be expressed as

$$B_{phot}(K'Q', \Gamma\gamma) = \frac{1}{2J_0 + 1} \int d\mathbf{p} \sum_{\substack{M M' m_s \\ \lambda \lambda' M_0}} \langle \lambda | \mathcal{T}_{\Gamma\gamma} | \lambda' \rangle$$

$$\times \langle J M' | T(J)^+_{K'Q'} | J M \rangle$$

$$\times \langle J M \mathbf{p}^{(-)} m_s | T_i | J_0 M_0 \omega \mathbf{n} \lambda \rangle$$

$$\times \langle J M' \mathbf{p}^{(-)} m_s | T_i^+ | J_0 M_0 \omega \mathbf{n} \lambda' \rangle^* \qquad (2.39)$$

where the asterisk denotes the complex conjugate matrix element.

The matrix elements of the irreducible tensors may be expressed in terms of $3j$-symbols

$$B_{phot}(K'Q', \Gamma\gamma) = \frac{\sqrt{(2K'+1)(2\Gamma+1)}}{2J_0 + 1} \int d\mathbf{p} \sum_{\substack{M M' m_s \\ \lambda \lambda' M_0}} (-1)^{J - M + 1 - \lambda}$$

$$\times \langle J M \mathbf{p}^{(-)} m_s | T_i | J_0 M_0 \omega \mathbf{n} \lambda \rangle$$

$$\times \langle J M' \mathbf{p}^{(-)} m_s | T_i^+ | J_0 M_0 \omega \mathbf{n} \lambda' \rangle^*$$

$$\times \begin{pmatrix} 1 & 1 & \Gamma \\ \lambda & -\lambda' & -\gamma \end{pmatrix} \begin{pmatrix} J & J & K' \\ M & -M' & -Q' \end{pmatrix}. \qquad (2.40)$$

The anisotropy parameter B_{phot} is the most general for photoionization experiments as described in (2.29). It fully contains the dynamics of the ionization process. The information on the excited ionic ensemble is given by the state multipoles $\langle T(J)^+_{K'Q'}\rangle$, see (2.37). The emitted photoelectrons are not observed.

Following some textbooks on scattering theory and photoionization, e.g. Amusia (1990), we expand the transition matrix elements in terms of dipole matrix elements; see Appendix D.

$$\langle J M \mathbf{p}^{(-)} m_s | T_i | J_0 M_0 \omega \mathbf{n} \lambda \rangle = \langle J M \mathbf{p}^{(-)} m_s | d_\lambda | J_0 M_0 \rangle$$

$$= \frac{1}{|\mathbf{p}|} \sum_{\substack{\ell m j m_j \\ J_1 M_1}} (-i)^\ell \, e^{i\sigma_\ell^j} \, Y_{\ell m}(\hat{\mathbf{p}}) \, \langle (Jj) J_1 \| d \| J_0 \rangle$$

$$\times (-1)^{-\ell + 1/2 - m_j - J + j - J_1} \sqrt{(2j+1)(2J_1+1)}$$

$$\times \begin{pmatrix} \ell & 1/2 & j \\ m & m_s & -m_j \end{pmatrix} \begin{pmatrix} J & j & J_1 \\ M & m_j & -M_1 \end{pmatrix} \begin{pmatrix} J_1 & 1 & J_0 \\ -M_1 & \lambda & M_0 \end{pmatrix}. \quad (2.41)$$

Here, d_λ denotes the irreducible components of the dipole operator. The reduced matrix elements can be obtained by applying the Wigner–Eckart theorem, i.e. (2.27), and are written as $\langle (Jj) J_1 \| d \| J_0 \rangle$. The scattering phase is denoted as σ_ℓ^j and $|\mathbf{p}|$ is the absolute value of the electronic momentum. The functions $Y_{\ell m}(\hat{\mathbf{p}})$ are the spherical harmonics where the argument $\hat{\mathbf{p}}$ denotes the unit vector in the direction $\mathbf{p}$.

Inserting (2.41) twice into (2.40) and rearranging the phase factors the coefficient B_{phot} can be contracted. The calculation is analogously to the case of electronic impact; see Appendix E and we remain with

$$B_{phot}(K'Q', \Gamma\gamma) = \delta_{K',\Gamma} \, \delta_{Q',\gamma} \frac{1}{2J_0+1} \sum_{\ell j J_1 J_1'} (-1)^{j-J-J_1}$$

$$\times \langle (Jj) J_1 \| d \| J_0 \rangle \langle (Jj) J_1' \| d \| J_0 \rangle^*$$

$$\times \sqrt{(2J_1+1)(2J_1'+1)}$$

$$\times \begin{Bmatrix} 1 & 1 & K' \\ J_1' & J_1 & J_0 \end{Bmatrix} \begin{Bmatrix} J_1' & J_1 & K' \\ J & J & j \end{Bmatrix}. \quad (2.42)$$

Thus, the anisotropy parameter B_{phot} becomes independent of the components Q' and γ and depends on the tensorial rank K', only.

$$B_{phot}(K') = B_{phot}(K'Q', \Gamma\gamma) \delta_{K',\Gamma} \, \delta_{Q',\gamma}. \quad (2.43)$$

Equation (2.42) represents the contracted form of the general anisotropy parameter B_{phot}. So far, an arbitrarily polarized synchrotron beam and an unpolarized target have been assumed. Besides the fact that we are considering transitions with a resolved initial and final state fine structure, no further assumptions have been made.

Inspecting the complex conjugate of (2.42) and interchanging the summation over J_1 and J_1' it can be shown that the anisotropy parameter B_{phot} is a real number

$$B_{phot}(K') = B_{phot}(K')^*. \quad (2.44)$$

In particular, we obtain

$$B_{phot}(0) = \frac{1}{(2J_0+1)\sqrt{3(2J+1)}} \sum_{\ell j J_1} (-1)^{J_0+J_1} |\langle (Jj) J_1 \| d \| J_0 \rangle|^2. \quad (2.45)$$

The parameter $B_{phot}(0)$ can be seen as a normalization parameter for introducing relative parameters which is often an advantage for the interpretation of experimental data. It is directly proportional to the total cross section of the intermediate ionic states generated during the photoionization.

Inserting the result (2.42) into (2.37) the summation over Γ and γ vanishes and we remain with

$$\left\langle T(J)^+_{K'Q'} \right\rangle = B_{phot}(K') \left\langle T^+_{\Gamma\gamma} \right\rangle \delta_{K',\Gamma}\, \delta_{Q',\gamma}. \tag{2.46}$$

Thus, for a primary photoionization we have the general selection rules

$$K' = \Gamma \quad \text{and} \quad Q' = \gamma. \tag{2.47}$$

Therefore, the rank and component of the state multipoles $\left\langle T(J)^+_{K'Q'} \right\rangle$ describing the ionic state must be the same as for the state multipoles of the photonic system. This results in general restrictions for the quantum numbers of the state multipoles describing the intermediate ionic ensemble

$$0 \le K' \le 2 \quad \text{and} \quad Q' = 0, \pm 2. \tag{2.48}$$

Due to parity reasons, the emitted partial waves of the photoelectron must have the same parity. As a result of the dipole approximation we find that either

$$J_1 = J_0 \quad \text{or} \quad J_1 = J_0 \pm 1 \tag{2.49}$$

must be fulfilled.

Equation (2.42) can be further reduced if the initial atomic state has a vanishing total angular momentum. This is the usual case for photoionization of closed shell atoms, e.g. the rare gases or the earth alkalis. Inserting $J_0 = 0$, the symmetry relations of the $6j$-symbols yield

$$J_1 = J'_1 = 1. \tag{2.50}$$

Thus, the summation over J_1 and J'_1 can be omitted and we obtain

$$B^{J_0=0}_{phot}(K') = \sum_{\ell j} (-1)^{j-J} \begin{Bmatrix} 1 & 1 & K' \\ J & J & j \end{Bmatrix} \left| \langle (Jj)1 \| d \| 0 \rangle \right|^2, \tag{2.51}$$

and the normalization parameter as $B^{J_0=0}_{phot}(0)$

$$B^{J_0=0}_{phot}(0) = \frac{1}{\sqrt{3(2J+1)}} \sum_{\ell j} \left| \langle (Jj)1 \| d \| 0 \rangle \right|^2. \tag{2.52}$$

We keep in mind that (2.46) connects the state multipoles $\left\langle T(J)^+_{K'Q'} \right\rangle$ characterizing the ionic ensemble A^{+*}, with the polarization parameters $\left\langle T^+_{\Gamma\gamma} \right\rangle$ describing the incoming photon beam via the anisotropy parameters $B_{phot}(K')$, (2.42), which contain the dynamics of the photoionization process.

2.3.2 Photoexcitation

It is of interest to also consider the primary photoexcitation process,

$$\gamma_{Syn} + A \longrightarrow A^*, \tag{2.53}$$

i.e. the neutral initial atomic state is excited via photoabsorption. This has been particularly used in the investigation of Auger emission from excited rare gases; the so-called *resonant Auger transitions*. This type of Auger transitions will be discussed in more detail in Sect. 2.6. Applying the same formalism as in the previous section we obtain

$$\left\langle T(J)^+_{K'Q'} \right\rangle = \sum_{\Gamma\gamma} \left\langle T^+_{\Gamma\gamma} \right\rangle B_{exci}(K'Q', \Gamma\gamma), \tag{2.54}$$

and the anisotropy parameter of photoexcitation as

$$B_{exci}(K'Q', \Gamma\gamma) = \frac{\sqrt{(2K'+1)(2\Gamma+1)}}{2J_0+1} \sum_{\substack{MM'M_0 \\ \lambda\lambda'}} (-1)^{J-M+1-\lambda}$$

$$\times \langle JM|T_i|J_0M_0\omega\mathbf{n}\lambda\rangle\langle JM'|T_i^+|J_0M_0\omega\mathbf{n}\lambda'\rangle^*$$

$$\times \begin{pmatrix} 1 & 1 & \Gamma \\ \lambda & -\lambda' & -\gamma \end{pmatrix} \begin{pmatrix} J & J & K' \\ M & -M' & -Q' \end{pmatrix}. \tag{2.55}$$

Applying the dipole approximation and using standard methods of angular momentum algebra we are able to eliminate the summation over all magnetic quantum numbers. This yields

$$B_{exci}(K') = B_{exci}(K'Q', \Gamma\gamma) = \frac{1}{2J_0+1} \left| \langle J\|d\|J_0\rangle \right|^2$$

$$\times (-1)^{1+J+J_0+K} \begin{Bmatrix} 1 & 1 & K' \\ J & J & J_0 \end{Bmatrix} \delta_{K',\Gamma} \, \delta_{Q',\gamma}. \tag{2.56}$$

From this, it is evident that B_{exci} is a real quantity and that we obtain the same selection rules as for a primary photoionization, i.e. (2.47) and (2.48) apply. The restriction $K \leq 2$ is again caused by the dipole approximation. In particular, we obtain the normalization parameter

$$B_{exci}(0) = \frac{1}{(2J_0+1)\sqrt{3(2J+1)}} \left| \langle J\|d\|J_0\rangle \right|^2. \tag{2.57}$$

The state multipoles describing the intermediate excited atomic ensemble are therefore related to the state multipoles of the incoming photon beam by a similar relation as has been obtained for the case of photoionization. Analogously to (2.46) we obtain

$$\left\langle T(J)^+_{K'Q'} \right\rangle = B_{exci}(K') \left\langle T^+_{\Gamma\gamma} \right\rangle \delta_{K',\Gamma} \, \delta_{Q',\gamma}. \tag{2.58}$$

Considering an initial atomic state with total angular momentum $J_0 = 0$ the symmetry relations of the $6j$-symbols yield $J = 1$ and (2.56) can be further reduced

$$B_{exci}^{J_0=0}(K') = B_{exci}^{J_0=0}(0) = \frac{1}{3}\left|\langle 1\|d\|0\rangle\right|^2. \tag{2.59}$$

Thus, the anisotropy parameters become a constant number, independent of their rank K'. This case is of importance e.g. for the photoexcitation of the rare gases and their subsequent Auger decay.

2.3.3 Electron Impact Excitation

The electron impact excitation process can be described as

$$e^- + A \longrightarrow A^* + e_s^-, \tag{2.60}$$

where e_s^- denotes the inelastically scattered electron.

The state multipoles are connected with the density matrix of the combined excited final electron-atom system by

$$\left\langle T(J)_{K'Q'}^+\right\rangle = \mathrm{tr}\left(\hat{\rho}_{out}T(J)_{K'Q'}^+\right), \tag{2.61}$$

where the final state density matrix is connected to the initial state via the transition operator T_i by

$$\hat{\rho}_{out} = T_i\,\hat{\rho}_{in}\,T_i^+. \tag{2.62}$$

Applying first order Born approximation we get $T_i = V = V^+ = T_i^+$ which yields

$$\hat{\rho}_{out} = V\,\hat{\rho}_{in}\,V^+ = V\,\hat{\rho}_{in}\,V, \tag{2.63}$$

where V denotes the Coulomb operator, and for the state multipoles we get

$$\left\langle T(J)_{K'Q'}^+\right\rangle = \mathrm{tr}\left(V\,\hat{\rho}_{in}\,V\,T(J)_{K'Q'}^+\right). \tag{2.64}$$

The total density matrix describing the uncorrelated ensemble of initial electron e^- and atomic state A can be written as a direct product

$$\hat{\rho}_{in} = \hat{\rho}(A) \times \hat{\rho}(e^-). \tag{2.65}$$

The initial atomic ensemble is supposed as unpolarized and is therefore described by (2.36). The electron beam is assumed as arbitrarily polarized, i.e.

$$\hat{\rho}(e^-) = \sum_{kq}\left\langle t_{kq}^+\right\rangle t_{kq}. \tag{2.66}$$

The state multipoles $\langle t_{kq}^+\rangle$ describing the incoming electronic beam can be connected to the Cartesian components of the spin polarization vector which is shown in Table 2.2. In particular, we note that for an arbitrarily polarized electron beam $k \leq 1$, and thus, $q = 0, \pm 1$, only.

Table 2.3. The state multipoles of an arbitrarily polarized electron beam and their connection to the Cartesian components of the spin polarization vector. The electron beam axis has been chosen as quantization axis. The incoming beam intensity has been normalized to unity

State multipoles	Spin polarization
$\langle t_{00}^+ \rangle = \frac{1}{\sqrt{2}}$	
$\langle t_{10}^+ \rangle = \frac{1}{\sqrt{2}} p_z$	
$\langle t_{11}^+ \rangle = -\frac{1}{2}(p_x - i p_y)$	
$\langle t_{1-1}^+ \rangle = -\langle t_{11}^+ \rangle^* = \frac{1}{2}(p_x + i p_y)$	

Inserting (2.36) and the above equation into (2.64) we obtain

$$\left\langle T(J)_{K'Q'}^+ \right\rangle = \sum_{kq} \left\langle t_{kq}^+ \right\rangle B_{scat}(K'Q', kq), \tag{2.67}$$

where the anisotropy parameter B_{scat} is defined as

$$B_{scat}(K'Q', kq) = \frac{1}{2J_0 + 1} \, \mathrm{tr}\!\left(V \, t_{kq} \, V \, T(J)_{K'Q'}^+\right). \tag{2.68}$$

For describing the initial and final states we adopt the notation of the previous sections. The initial electron is characterized as $|\mathbf{p}_0^{(+)} m_{s_0}\rangle$ where $\mathbf{p}_0^{(+)}$ denotes, in the sense of scattering theory, an incoming electron with impact $\mathbf{p}_0$ and outgoing boundary conditions (cf. Starace 1982), and m_{s_0} denotes its magnetic spin quantum number. Inserting complete basis sets and expressing the tensors in terms of $3j$-symbols yields

$$B_{scat}(K'Q', kq) = \frac{\sqrt{(2k+1)(2K+1)}}{2J_0 + 1} \int \mathrm{d}p \sum_{\substack{MM'm_s \\ m_{s_0} m_{s_0}' M_0}} (-1)^{J-M+1/2-m_{s_0}}$$

$$\times \langle JM\mathbf{p}^{(-)}m_s | V | J_0 M_0 \mathbf{p}_0^{(+)} m_{s_0}\rangle$$

$$\times \langle JM'\mathbf{p}^{(-)}m_s | V | J_0 M_0 \mathbf{p}_0^{(+)} m_{s_0}'\rangle^*$$

$$\times \begin{pmatrix} 1/2 & 1/2 & k \\ m_{s_0} & -m_{s_0}' & -q \end{pmatrix} \begin{pmatrix} J & J & K' \\ M & -M' & -Q' \end{pmatrix}. \tag{2.69}$$

Equation (2.69) is the general anisotropy parameter for electron impact excitation experiments as described in (2.60) and fully contains the excitation dynamics. As in the previous sections, no assumptions have been made on the degree of polarization of the incoming electron beam. The information on the excited atomic ensemble is given by the state multipoles $\langle T(J)^+_{K'Q'}\rangle$. The scattered electrons are not observed.

Carrying out the partial wave expansion of the incoming electron $|\mathbf{p}_0^{(+)}m_{s0}\rangle$ and of the outgoing electron $|\mathbf{p}^{(-)}m_s\rangle$, respectively, and applying the same standard methods of angular momentum algebra as in the previous sections, also see Sect. E.1, the summation over the magnetic quantum numbers can be carried out and the anisotropy parameter may be expressed as

$$
B_{scat}(K'kq) = \frac{\Delta E}{4\pi|\mathbf{p}_0|^2}\frac{\sqrt{(2k+1)(2K'+1)}}{2J_0+1}\sum_{\substack{\ell_0\ell'_0 j_0 j'_0 \\ \ell j J_1 J'_1 b}} i^{\ell_0-\ell'_0}\,e^{i(\sigma^{j_0}_{\ell_0}-\sigma^{j'_0}_{\ell'_0})}
$$

$$
\times (-1)^{J+J'_1+J_1+J_0+j-j_0+\ell_0-1-K'+q}
$$

$$
\times \langle (Jj)J_1\|V\|(J_0j_0)J_1\rangle\langle (Jj)J'_1\|V\|(J_0j'_0)J'_1\rangle
$$

$$
\times \sqrt{(2\ell_0+1)(2\ell'_0+1)(2j_0+1)(2j'_0+1)(2J_1+1)(2J'_1+1)}
$$

$$
\times (2b+1)\begin{pmatrix} b & \ell_0 & \ell'_0 \\ 0 & 0 & 0 \end{pmatrix}\begin{pmatrix} b & k & K' \\ 0 & -q & q \end{pmatrix}
$$

$$
\times \begin{Bmatrix} j'_0 & j_0 & K' \\ J_1 & J'_1 & J_0 \end{Bmatrix}\begin{Bmatrix} J & J & K' \\ J_1 & J'_1 & j \end{Bmatrix}\begin{Bmatrix} \ell'_0 & b & \ell_0 \\ 1/2 & k & 1/2 \\ j'_0 & K' & j_0 \end{Bmatrix}. \tag{2.70}
$$

Here, ℓ_0 and j_0 are the orbital and total angular momenta of the partial wave of the incoming electron, and $\sigma^{j_0}_{\ell_0}$ denotes the scattering phase. The angular momentum b is artificial and has been introduced to carry out the summation over the magnetic quantum numbers. The total angular momentum of the initial and final states is denoted by J_1. Since the transition operator is proportional to a zero-rank tensor, J_1 must be conserved.

From the second $3j$-symbol we get the important selection rule

$$
q = Q', \tag{2.71}
$$

and thus, the anisotropy parameter B_{scat} can be re-defined as

$$
B_{scat}(K'kq) = B_{scat}(K'Q',kq)\delta_{qQ'}. \tag{2.72}
$$

Comparing the anisotropy parameter B_{scat}, i.e. (2.70), with those for photoionization/excitation we point out two major differences. First, the parameter B_{scat} does not solely depend on the rank K' as in the photon case but is also a function of rank k and magnetic component q of the state multipoles $\langle t^+_{kq}\rangle$ describing the polarization state of the incoming electron beam. Secondly, while the parameter B_{phot} does only depend on the reduced matrix elements, for electron impact excitation, the parameter B_{scat} depends on the scattering phase of the incoming partial waves, too. Thus, its numerical calculation becomes more tedious than in the photonic case.

Using the selection rule (2.71) and the re-defined anisotropy parameter, (2.67) can be written as

$$\left\langle T(J)^+_{K'q} \right\rangle = \sum_k \left\langle t^+_{kq} \right\rangle B_{scat}(K'kq). \tag{2.73}$$

This equation yields a physically important result. Due to the selection rule $q = Q'$ we find that the state multipoles $\left\langle T(J)^+_{K'Q'} \right\rangle$ describing the excited atomic state must have the same z-component as the state multipoles $\left\langle t^+_{kq} \right\rangle$ describing the incoming e-lectron beam. I.e. the magnetic component q of the electronic tensors is conserved during the excitation process and is directly transferred to the excited atomic ensemble.

Since electrons are spin $1/2$ particles their corresponding state multipoles $\left\langle t^+_{kq} \right\rangle$ are generally restricted by angular momentum coupling rules

$$k \le 1 \quad \text{and} \quad -k \le q \le k \quad \text{i.e.} \quad q = 0, \pm 1. \tag{2.74}$$

Thus, the atomic excited state is described by tensors with magnetic component $|q| \le 1$, only. This is directly caused by the fact that an electron beam, or more generally speaking, a beam of spin $1/2$ particles has been used for the excitation process. As has been shown in the previous sections a primary photoionization/ex-citation does result in different selection rules.

The anisotropy parameter B_{scat} shows some useful symmetries. Considering the anisotropy parameter $B_{scat}(K'k-q)$ it differs from the coefficient $B_{scat}(K'kq)$ in the phase factor and in the second $3j$-symbol, only. This yields the relation

$$B_{scat}(K'k - q) = (-1)^{K'+k+b} B_{scat}(K'kq). \tag{2.75}$$

Another symmetry can be obtained by inspecting (2.73). Inserting $-q$ and using the Hermiticity conditions of the state multipoles we get

$$B_{scat}(K'k - q) = B_{scat}(K'kq)^*. \tag{2.76}$$

This results in the fact that all parameters $B_{scat}(K'k0)$ are real quantities.

In the following, we make use of the fact that the Coulomb interaction is invari-ant under reflection of the total system, i.e.

$$V = R^+ V R \tag{2.77}$$

where R denotes a reflection operator containing the electron beam axis. Applying R to the electronic and atomic bra- and ket-states yields

$$\left\langle JM\mathbf{p}m_s | R^+ V R | J_0 M_0 \mathbf{p}_0 m_{s_0} \right\rangle = (-1)^{L+J-M+L_0+J_0-M_0+1-m_s-m_{s_0}}$$
$$\times \left\langle J - M\mathbf{p} - m_s | V | J_0 - M_0 \mathbf{p}_0 - m_{s_0} \right\rangle. \tag{2.78}$$

Equation (2.69) is symmetric in summing over the magnetic quantum numbers. Therefore, we are able to interchange their signs. After re-arranging the phase factor and using the symmetry conditions of the $3j$-symbols we get

$$B_{scat}(K'kq) = (-1)^{K+k} B_{scat}(K'k - q). \tag{2.79}$$

Combining the symmetry relations (2.76) and (2.79) we obtain

$$B_{scat}(K'kq) = \begin{cases} \text{real, if} & K+k = \text{even,} \\ \text{imaginary, if} & K+k = \text{odd.} \end{cases} \qquad (2.80)$$

Thus, the anisotropy parameters B_{scat} are either solely real or imaginary numbers depending on the rank K' and k of the state multipoles.

Further, combining relations (2.75) and (2.79) we obtain the selection rule

$$(-1)^b = 1, \qquad (2.81)$$

i.e. the angular momentum b must be even. With this, another physically important result can be obtained. Inspecting the first $3j$-symbol of (2.70) we get

$$(-1)^{\ell_0+\ell_0'} = 1. \qquad (2.82)$$

Equation (2.82) requires that only partial waves of the same parity can take part in the interference terms, i.e. the bi-linear products of matrix elements, of the excitation process. This result is somewhat surprising since it is a well known fact that a free electron beam does not have a determined parity. The symmetry conditions of the considered experiment however, restrict the number of partial waves of the incoming electron beam which can take part in the excitation process to those with identical parity. Partial waves with a different parity are excluded. Thus, though still without any fixed parity, the incoming electron beam behaves as if it has a fixed parity.

In order to obtain relative parameters we need to consider the case $K' = 0$ in more detail. From the selection rule (2.71) we immediately get $q = 0$. Therefore, we need to consider the coefficients $B_{scat}(0k0)$, only. From the Hermiticity condition (2.76) we find the parameters $B_{scat}(0k0)$ to be real quantities and applying (2.80) we eventually get $k = 0$. Thus, the parameter $B_{scat}(000)$ can be interpreted as the general normalization parameter.

Inserting $K' = k = q = 0$ into (2.70) we immediately get the selection rule $b = 0$ from the second $3j$-symbol. Thus, the sum over b can be omitted which yields for the normalization parameter

$$\begin{aligned} B_{scat}(000) = \frac{\Delta E}{4\pi |\mathbf{p}_0|^2} \frac{1}{2J_0 + 1} \sum_{\substack{\ell_0 \ell_0' j_0 j_0' \\ \ell j J_1 J_1'}} i^{\ell_0 - \ell_0'} e^{i(\sigma_{\ell_0}^{j_0} - \sigma_{\ell_0'}^{j_0'})} \\ \times (-1)^{J+J_1'+J_1+J_0+j-j_0+\ell_0-1} \\ \times \langle (Jj)J_1 \| V \| (J_0 j_0)J_1 \rangle \langle (Jj)J_1' \| V \| (J_0 j_0')J_1' \rangle \\ \times \sqrt{(2\ell_0+1)(2\ell_0'+1)(2j_0+1)(2j_0'+1)(2J_1+1)(2J_1'+1)} \\ \times \begin{pmatrix} 0 & \ell_0 & \ell_0' \\ 0 & 0 & 0 \end{pmatrix} \begin{Bmatrix} j_0' & j_0 & 0 \\ J_1 & J_1' & J_0 \end{Bmatrix} \begin{Bmatrix} J & J & 0 \\ J_1 & J_1' & j \end{Bmatrix} \begin{Bmatrix} \ell_0' & 0 & \ell_0 \\ 1/2 & 0 & 1/2 \\ j_0' & 0 & j_0 \end{Bmatrix}. \qquad (2.83) \end{aligned}$$

Applying the symmetry relations of the nj-symbols we get additional selection rules for the normalization parameter

$$\ell_0 = \ell_0', \quad j_0 = j_0', \quad \text{and} \quad J_1 = J_1'. \tag{2.84}$$

With this (2.83) can be further reduced and we obtain the general normalization parameter $B_{scat}(000)$ as

$$B_{scat}(000) = \frac{\Delta E}{4\pi |\mathbf{p}_0|^2} \frac{1}{2J_0 + 1} \frac{1}{\sqrt{2(2J + 1)}}$$
$$\times \sum_{\ell_0 j_0 \ell j J_1} \left| \langle (Jj)J_1 \| V \| (J_0 j_0)J_1 \rangle \right|^2. \tag{2.85}$$

On the other hand, (2.70) can be simplified if we consider electron impact excitation of a closed shell atom, e.g. the rare gases. Inserting $J_0 = 0$, the symmetry relations of the $6j$-symbols yield the selection rules

$$j_0' = J_1' \quad \text{and} \quad j_0 = J_1, \tag{2.86}$$

which eventually results in

$$B_{scat}^{J_0=0}(K'kq) = \Delta E \frac{\sqrt{(2k + 1)(2K' + 1)}}{4\pi |\mathbf{p}_0|^2} \sum_{\substack{\ell_0 \ell_0' j_0 j_0' \\ \ell j b}} \mathrm{i}^{\ell_0 - \ell_0'} \, \mathrm{e}^{\mathrm{i}(\sigma_{\ell_0}^{j_0} - \sigma_{\ell_0'}^{j_0'})}$$

$$\times (-1)^{J + j + j_0 + \ell_0 + q}$$
$$\times \langle (Jj)j_0 \| V \| (0j_0) j_0 \rangle \langle (Jj)j_0' \| V \| (0j_0') j_0' \rangle$$
$$\times (2b + 1)\sqrt{(2\ell_0 + 1)(2\ell_0' + 1)(2j_0 + 1)(2j_0' + 1)}$$

$$\times \begin{pmatrix} \ell_0 & \ell_0' & b \\ 0 & 0 & 0 \end{pmatrix} \begin{pmatrix} b & k & K' \\ 0 & -q & q \end{pmatrix} \begin{Bmatrix} J & J & K' \\ j_0 & j_0' & j \end{Bmatrix} \begin{Bmatrix} \ell_0 & \ell_0' & b \\ 1/2 & 1/2 & k \\ j_0 & j_0' & K' \end{Bmatrix}. \tag{2.87}$$

For this particular case the normalization parameter reduces to

$$B_{scat}^{J_0=0}(000) = \frac{\Delta E}{4\pi |\mathbf{p}_0|^2} \frac{1}{\sqrt{2(2J + 1)}} \sum_{\ell_0 j_0 \ell j} \left| \langle (Jj)j_0 \| V \| (0j_0) j_0 \rangle \right|^2. \tag{2.88}$$

2.3.4 Electron Impact Ionization

Now, let us consider a primary electron impact ionization

$$e^- + A \longrightarrow A^{+*} + e_s^- + e_e^-, \tag{2.89}$$

where e_s^- and e_e^- denote the inelastically scattered and the emitted electron, respectively.

Applying the same standard methods as in the previous section and adopting the notation, we obtain

$$\langle T(J)_{K'Q'}^+ \rangle = \sum_{kq} \langle t_{kq}^+ \rangle B_e(K'Q', kq), \tag{2.90}$$

with the anisotropy parameter B_e defined as

$$B_e(K'Q', kq) = \frac{1}{2J_0 + 1} \; \mathrm{tr}\!\left(T_i\, t_{kq}\, T_i^+\, T(J)_{K'Q'}^+\right). \tag{2.91}$$

The information on the excited ionic ensemble is given by the state multipoles $\langle T(J)_{KQ}^+ \rangle$. The scattered electrons are not observed.

Again, the atomic ensemble has been assumed as unpolarized in its initial state, and the incoming electron beam as arbitrarily polarized. The scattered and the emitted electrons are not observed. Thus, inserting complete basis sets and expressing the tensors in terms of $3j$-symbols the anisotropy parameter B_e can be written as

$$B_e(K'Q', kq) = \frac{\sqrt{(2k+1)(2K'+1)}}{2J_0 + 1}$$

$$\times \int d\mathbf{p}_1 \int d\mathbf{p}_2 \sum_{\substack{MM'm_{s1}m_{s2} \\ m_{s0}m'_{s0}M_0}} (-1)^{J-M+1/2-m_{s0}}$$

$$\times \langle JM\mathbf{p}_1^{(-)}m_{s1}\mathbf{p}_2^{(-)}m_{s2} | T_i | J_0 M_0 \mathbf{p}_0^{(+)}m_{s0}\rangle$$

$$\times \langle JM'\mathbf{p}_1^{(-)}m_{s1}\mathbf{p}_2^{(-)}m_{s2} | T_i^+ | J_0 M_0 \mathbf{p}_0^{(+)}m'_{s0}\rangle^*$$

$$\times \begin{pmatrix} 1/2 & 1/2 & k \\ m_{s0} & -m'_{s0} & -q \end{pmatrix} \begin{pmatrix} J & J & K' \\ M & -M' & -Q' \end{pmatrix}. \tag{2.92}$$

Equation (2.92) is the general anisotropy parameter for electron impact ionization experiments as described in (2.89) and contains the dynamics of the ionization process. No assumptions have been made on the degree of polarization of the incoming electron beam.

Carrying out the partial wave expansions for the incoming $|\mathbf{p}_0^{(+)}m'_{s0}\rangle$ electron and for the scattered $|\mathbf{p}_1^{(-)}m_{s1}\rangle$ and emitted $|\mathbf{p}_2^{(-)}m_{s2}\rangle$ electrons, and applying some tedious but straightforward methods of angular momentum algebra, see Appendix E, the anisotropy parameter B_e becomes independent of the magnetic component Q'

$$B_e(K'kq) = B_e(K'Q', kq)\,\delta_{Q',q}\,, \tag{2.93}$$

and, using the results of Appendix E, can be eventually expressed as

$$
\begin{aligned}
B_e(K'kq) = {} & \frac{\Delta E_{12}}{4\pi |\mathbf{p}_0|^2} \frac{\sqrt{(2k+1)(2K'+1)}}{2J_0+1} \sum_{\substack{\ell_0 j_0 \ell_0' j_0' \\ \ell_1 j_1 \ell_2 j_2}} \sum_{\substack{J_1 J_f \\ J_1' J_f' b}} i^{\ell_0 - \ell_0'} e^{i(\sigma_{\ell_0}^{j_0} - \sigma_{\ell_0'}^{j_0'})} \\
& \times (-1)^{J_1 - J_1' - J + j_1 + j_2 + J_f + J_f' + J_0 + \ell_0 - j_0' + 1 - q} \\
& \times \overline{\langle ([Jj_1]J_1 j_2) J_f \| V \| (J_0 j_0) J_f \rangle \langle ([Jj_1]J_1' j_2) J_f' \| V \| (J_0 j_0') J_f' \rangle} \\
& \times \sqrt{(2\ell_0+1)(2j_0+1)(2J_1+1)(2\ell_0'+1)(2j_0'+1)(2J_1'+1)} \\
& \times \sqrt{(2J_f+1)(2J_f'+1)}\,(2b+1) \begin{pmatrix} b & k & K' \\ 0 & -q & q \end{pmatrix} \begin{pmatrix} b & \ell_0 & \ell_0' \\ 0 & 0 & 0 \end{pmatrix} \\
& \times \begin{Bmatrix} J_1' & J_1 & K' \\ J & J & j_1 \end{Bmatrix} \begin{Bmatrix} J_f & J_f' & K' \\ J_1' & J_1 & j_2 \end{Bmatrix} \begin{Bmatrix} j_0' & j_0 & K' \\ J_f & J_f' & J_0 \end{Bmatrix} \begin{Bmatrix} K' & b & k \\ j_0 & \ell_0 & 1/2 \\ j_0' & \ell_0' & 1/2 \end{Bmatrix}. \quad (2.94)
\end{aligned}
$$

Thus, (2.90) can be reduced in perfect analogy to the electron impact excitation process which yields

$$
\left\langle T(J)_{K'Q'}^+ \right\rangle = \sum_k \left\langle t_{kq}^+ \right\rangle B_e(K'kq), \quad (2.95)
$$

and the same selection rules, i.e. (2.74)–(2.76) as well as (2.79)–(2.82) do apply.

As in the case of electron impact excitation we introduce a normalization parameter. Following the same arguments and inserting $K' = k = q = 0$ we end up with a similar result for the normalization parameter of electron impact ionization

$$
\begin{aligned}
B_e(000) = {} & \frac{\Delta E_{12}}{4\pi |\mathbf{p}_0|^2} \frac{1}{2J_0+1} \sum_{\substack{\ell_0 j_0 \ell_0' j_0' \\ \ell_1 j_1 \ell_2 j_2}} \sum_{J_1 J_f J_1' J_f'} i^{\ell_0 - \ell_0'} e^{i(\sigma_{\ell_0}^{j_0} - \sigma_{\ell_0'}^{j_0'})} \\
& \times (-1)^{J_1 - J_1' - J + j_1 + j_2 + J_f + J_f' + J_0 + \ell_0 - j_0' + 1} \\
& \times \overline{\langle ([Jj_1]J_1 j_2) J_f \| V \| (J_0 j_0) J_f \rangle \langle ([Jj_1]J_1' j_2) J_f' \| V \| (J_0 j_0') J_f' \rangle} \\
& \times \sqrt{(2\ell_0+1)(2j_0+1)(2J_1+1)(2\ell_0'+1)(2j_0'+1)(2J_1'+1)} \\
& \times \sqrt{(2J_f+1)(2J_f'+1)} \begin{pmatrix} 0 & \ell_0 & \ell_0' \\ 0 & 0 & 0 \end{pmatrix} \begin{Bmatrix} 0 & 0 & 0 \\ j_0 & \ell_0 & 1/2 \\ j_0' & \ell_0' & 1/2 \end{Bmatrix} \\
& \times \begin{Bmatrix} J_1' & J_1 & 0 \\ J & J & j_1 \end{Bmatrix} \begin{Bmatrix} J_f & J_f' & 0 \\ J_1' & J_1 & j_2 \end{Bmatrix} \begin{Bmatrix} j_0' & j_0 & 0 \\ J_f & J_f' & J_0 \end{Bmatrix}. \quad (2.96)
\end{aligned}
$$

Again, this yields additional selection rules

$$
\ell_0 = \ell_0', \quad j_0 = j_0', \quad J_f = J_f', \quad \text{and} \quad J_1 = J_1', \quad (2.97)
$$

which allow for a further reduction of (2.96) and we eventually obtain

$$B_e(000) = \frac{\Delta E_{12}}{4\pi|\mathbf{p}_0|^2} \frac{-1}{(2J_0 + 1)\sqrt{2(2J + 1)}}$$
$$\times \sum_{\substack{\ell_0 j_0 J_1 J_f \\ \ell_1 j_1 \ell_2 j_2}} \left|\left\langle([Jj_1]J_1 j_2)J_f\|V\|(J_0 j_0)J_f\right\rangle\right|^2, \qquad (2.98)$$

as the general normalization parameter.

Considering the special case of the rare gases, i.e. $J_0 = 0$, we get the same selection rules as in the previous case of electron impact excitation

$$j_0' = J_f' \quad \text{and} \quad j_0 = J_f, \qquad (2.99)$$

and the coefficient B_e can be further contracted which yields

$$B_e^{J_0=0}(K'kq) = \frac{\Delta E_{12}}{4\pi|\mathbf{p}_0|^2} \frac{\sqrt{(2k+1)(2K'+1)}}{2J_0+1} \sum_{\substack{\ell_0 j_0 \ell_0' j_0' \\ \ell_1 j_1 \ell_2 j_2}} \sum_{\substack{J_1 J_f \\ J_1' J_f' b}} i^{\ell_0 - \ell_0'} e^{i(\sigma_{\ell_0}^{j_0} - \sigma_{\ell_0'}^{j_0'})}$$

$$\times (-1)^{J_1 - J_1' - J + j_1 + j_2 + \ell_0 - j_0' + 1 - K' - q}$$

$$\times \left\langle([Jj_1]J_1 j_2)j_0\|V\|(0j_0)j_0\right\rangle\left\langle([Jj_1]J_1' j_2)j_0'\|V\|(0j_0')j_0'\right\rangle$$

$$\times \sqrt{(2\ell_0 + 1)(2j_0 + 1)(2\ell_0' + 1)(2j_0' + 1)}$$

$$\times \sqrt{(2J_1 + 1)(2J_1' + 1)}\,(2b+1)\begin{pmatrix} b & k & K' \\ 0 & -q & q \end{pmatrix}\begin{pmatrix} b & \ell_0 & \ell_0' \\ 0 & 0 & 0 \end{pmatrix}$$

$$\times \begin{Bmatrix} J_1' & J_1 & K' \\ J & J & j_1 \end{Bmatrix}\begin{Bmatrix} j_0 & j_0' & K' \\ J_1' & J_1 & j_2 \end{Bmatrix}\begin{Bmatrix} K' & b & k \\ j_0 & \ell_0 & 1/2 \\ j_0' & \ell_0' & 1/2 \end{Bmatrix}. \qquad (2.100)$$

Analogously, the normalization parameter reduces to

$$B_e^{J_0=0}(000) = \frac{\Delta E_{12}}{4\pi|\mathbf{p}_0|^2} \frac{-1}{\sqrt{2(2J + 1)}}$$
$$\times \sum_{\substack{\ell_0 j_0 J_1 J_f \\ \ell_1 j_1 \ell_2 j_2}} \left|\left\langle([Jj_1]J_1 j_2)J_f\|V\|(0j_0)J_f\right\rangle\right|^2, \qquad (2.101)$$

for this case.

2.4 The Auger Emission

2.4.1 General Formalism

In this section we will consider the Auger decay.

$$A^{+*} \longrightarrow A^{++} + e_{Auger}^-. \qquad (2.102)$$

In principle, the Auger emission may take place after a primary ionization or excitation process and the subsequent results can be applied in either case. For the case of photoexcitation, this is known as so-called *resonant Auger transitions* which will be discussed in detail in Sect. 2.6. For the following, we are supposing the Auger decay as depicted in (2.102) in order to simplify the discussion.

After some certain lifetime the intermediate anisotropic ensemble of ions A^{+*}, discussed in the previous sections, decays via emission of Auger electrons. Using ket-vectors the Auger transition can be expressed as

$$|JM\rangle \longrightarrow |J_f M_f \mathbf{p}^{(-)} m_s\rangle, \tag{2.103}$$

where $|JM\rangle$ denotes the intermediate ionic state A^{+*}. The final ionic state is described by $|J_f M_f\rangle$ and $|\mathbf{p}^{(-)} m_s\rangle$ denotes the emitted Auger electron with momentum $\mathbf{p}$ and spin m_s. The upper index $(-)$ on $\mathbf{p}$ indicates that we are dealing with scattering solutions with incoming boundary conditions.

Choosing the helicity system as coordinate frame, the density matrix $\hat{\rho}$ of the ion A^{+*} can be expanded into a series of state multipoles $\langle T(J)^+_{KQ}\rangle$

$$\hat{\rho} = \sum_{KQ} \langle T(J)^+_{KQ}\rangle T(J)_{KQ}. \tag{2.104}$$

Applying first order perturbation theory the density matrix of the intermediate ionic state can be connected with the density matrix $\hat{\rho}_{out}$ of the final state,

$$\hat{\rho}_{out} = V \hat{\rho} V^+, \tag{2.105}$$

(e.g. Blum 1996) where V denotes the Coulomb operator. Since we are only interested in the ensemble of the emitted Auger electrons, we have to evaluate the reduced density matrix of the electronic system (see Sect. 2.2).

In the following we apply the normalization

$$I = \text{tr}(\hat{\rho}_{out}), \tag{2.106}$$

where tr denotes the trace of the density matrix and I is the intensity. With this, the tensors $\langle t^+_{kq}\rangle$, describing the emitted Auger electron and $\langle T(J)^+_{KQ}\rangle$ of the intermediate singly ionized ionic state are related to each other by

$$I\langle t^+_{kq}\rangle = \text{tr}(V\hat{\rho} V^+ t^+_{kq}) = \sum_{KQ} \langle T(J)^+_{KQ}\rangle \text{tr}(V T(J)_{KQ} V^+ t^+_{kq}). \tag{2.107}$$

Inserting complete basis sets for the initial and final states we obtain for the trace

$$\text{tr}(V T(J)_{KQ} V^+ t^+_{kq}) = \sum_{\substack{M_f M M' \\ m_s m'_s}} \langle J_f M_f \mathbf{p}^{(-)} m_s |V| JM\rangle$$
$$\times \langle JM|T(J)_{KQ}|JM'\rangle$$
$$\times \langle JM'|V^+|J_f M_f \mathbf{p}^{(-)} m'_s\rangle \langle m'_s|t^+_{kq}|m_s\rangle. \tag{2.108}$$

The doubly ionized final state A^{++} is not detected. Therefore, we have $M_f = M'_f$. The tensor operators $T(J)_{KQ}$ and t_{kq} can be expressed in terms of $3j$-symbols which yields

$$
\begin{aligned}
\mathrm{tr}\left(V T(J)_{KQ} V^+ t_{kq}^+\right) = {} & \sqrt{(2K+1)(2k+1)} \\
& \times \sum_{\substack{M_f M M' \\ m_s m'_s}} (-1)^{J-M+1/2-m_s} \\
& \times \langle J_f M_f \mathbf{p}^{(-)} m_s | V | J M \rangle \\
& \times \langle J_f M_f \mathbf{p}^{(-)} m'_s | V | J M' \rangle^* \\
& \times \begin{pmatrix} J & J & K \\ M & -M' & -Q \end{pmatrix} \begin{pmatrix} 1/2 & 1/2 & k \\ m_s & -m'_s & -q \end{pmatrix},
\end{aligned}
\tag{2.109}
$$

where the asterisk denotes the complex conjugate.

Since we have chosen the helicity system, the angular momentum of the Auger electron cannot have a magnetic component in its direction of propagation, i.e. parallel to its momentum. Thus, the Auger transition matrix elements vanish if the selection rules

$$
M_f + m_s = M \quad \text{and} \quad M_f + m'_s = M' \tag{2.110}
$$

are not fulfilled simultaneously. Applying the symmetry relations of the $3j$-symbols we obtain the important selection rule

$$
Q = q. \tag{2.111}
$$

I.e. the magnetic component of the state multipoles $\langle T(J)_{KQ}^+ \rangle$ describing the intermediate ionic state must be the same as for the state multipoles $\langle t_{kq}^+ \rangle$ of the emitted Auger electrons. Considering the fact that electrons are spin $1/2$ particles, the rank of the state multipoles $\langle t_{kq}^+ \rangle$ must be restricted to

$$
k \le 1. \tag{2.112}
$$

Thus, its magnetic component is restricted and with (2.111) we get the important limitation

$$
|Q| = |q| \le 1. \tag{2.113}
$$

With this and using Lohmann (1990) we define the generalized anisotropy parameter $A(KkQ)$ as

$$
A(KkQ) = \mathrm{tr}\left(V T(J)_{KQ} V^+ t_{kq}^+\right) \delta_{Q,q}. \tag{2.114}
$$

It follows that the summation over Q in (2.107) vanishes and using the anisotropy parameters we eventually get

$$
I\langle t_{kQ}^+ \rangle = \sum_K \langle T(J)_{KQ}^+ \rangle A(KkQ). \tag{2.115}
$$

The anisotropy parameters $A(KkQ)$ show some useful symmetry relations. With the symmetries of the $3j$-symbols we find

$$A(Kk-Q) = A(KkQ)^*. \tag{2.116}$$

Especially, all parameters $A(Kk0)$ must be real numbers.

Another relation can be obtained from reflection invariance. Since the Coulomb operator transforms as an irreducible tensor of zero rank it remains unchanged under reflection in the reaction plane

$$V = R_{xz}^+ V R_{xz}, \tag{2.117}$$

where R_{xz} denotes the reflection operator. This yields (Lohmann 1990)

$$A(KkQ) = (-1)^{K+k} A(Kk-Q). \tag{2.118}$$

With the above symmetries we find that, depending on the rank of the state multipoles, the anisotropy parameters are either solely real or imaginary numbers,

$$A(KkQ) = \begin{cases} \text{real, if} & K+k = \text{even}, \\ \text{imaginary, if} & K+k = \text{odd}. \end{cases} \tag{2.119}$$

Equation (2.115) refers to the helicity system. Transforming to the laboratory frame with the beam axis as quantization axis and denoting the state multipoles in the laboratory frame as $\langle T(J)_{KQ'}^+ \rangle$, they are related to the state multipoles of the helicity system by

$$\left\langle T(J)_{KQ}^+ \right\rangle = \sum_{Q'} \left\langle T(J)_{KQ'}^+ \right\rangle d_{Q'Q}^{(K)}(\theta), \tag{2.120}$$

where $d_{Q'Q}^{(K)}(\theta)$ denote the reduced rotation matrices and θ is the angle between the beam axis and the direction of Auger emission, see Fig. 2.3. Note, that generally, the two irreducible tensorial sets of state multipoles are related to each other via the complex rotation matrices $\mathcal{D}_{Q'Q}^{(K)}(\omega)$. For our chosen symmetry, i.e. the definition of the reaction plane, the two sets of state multipoles can be simply transformed by using the reduced rotation matrices $d_{Q'Q}^{(K)}(\theta)$. They are real numbers and solely depend on the polar angle θ. The reduced rotation matrices show some useful symmetries which are explicitly discussed in Appendix B.

Inserting (2.120) into (2.115) we eventually get

$$I(\theta)\left\langle t_{kQ}^+ \right\rangle = \sum_{KQ'} A(KkQ) \left\langle T(J)_{KQ'}^+ \right\rangle d_{Q'Q}^{(K)}(\theta). \tag{2.121}$$

Carrying out the summation over all magnetic quantum numbers the generalized anisotropy parameter $A(KkQ)$ can be expressed as (Lohmann 1990)

$$A(KkQ) = \frac{\sqrt{(2K+1)(2k+1)}}{4\pi|\mathbf{p}|^2} \sum_{\ell\ell'jj'b} i^{\ell+\ell'} e^{i(\sigma_{\ell'}^{j'}-\sigma_{\ell}^{j})} (-1)^{J_f+J+j'}$$

$$\times \langle J_f\|V\|(jJ)J_f\rangle\langle J_f\|V\|(j'J)J_f\rangle$$

$$\times (2b+1)\sqrt{(2\ell+1)(2\ell'+1)(2j+1)(2j'+1)}$$

$$\times \begin{pmatrix} K & b & k \\ -Q & 0 & Q \end{pmatrix} \begin{pmatrix} b & \ell' & \ell \\ 0 & 0 & 0 \end{pmatrix} \begin{Bmatrix} j' & j & K \\ J & J & J_f \end{Bmatrix} \begin{Bmatrix} K & b & k \\ j' & \ell' & 1/2 \\ j & \ell & 1/2 \end{Bmatrix}. \quad (2.122)$$

Here, J and J_f denote the initial and final state total angular momenta, whereas j and ℓ are total and orbital angular momenta of the partial waves, and σ_ℓ^j is the scattering phase, while $\mathbf{p}$ is the momentum of the emitted Auger electrons. K denotes the rank of the tensor components describing the initial ionic state and $k = 0, 1$ refers to the spin tensors of the Auger electron. The z-component Q is restricted by both tensors and thus only $|Q| = 0, 1$ is possible. The angular momentum b was introduced to eliminate all magnetic quantum numbers. As has been shown (Lohmann 1990), the parameter b is restricted to even values

$$(-1)^b = 1. \quad (2.123)$$

In order to express the physical quantities in terms of relative parameters we consider the normalization parameter $A(000)$. The first $3j$-symbol of (2.122) yields the restriction $b = 0$. Therefore, the summation over b vanishes and we get

$$A(000) = \frac{1}{4\pi|\mathbf{p}|^2} \sum_{\ell\ell'jj'} i^{\ell+\ell'} e^{i(\sigma_{\ell'}^{j'}-\sigma_{\ell}^{j})} (-1)^{J_f+J+j'}$$

$$\times \langle J_f\|V\|(jJ)J_f\rangle\langle J_f\|V\|(j'J)J_f\rangle$$

$$\times \sqrt{(2\ell+1)(2\ell'+1)(2j+1)(2j'+1)}$$

$$\times \begin{pmatrix} 0 & \ell' & \ell \\ 0 & 0 & 0 \end{pmatrix} \begin{Bmatrix} j' & j & 0 \\ J & J & J_f \end{Bmatrix} \begin{Bmatrix} 0 & 0 & 0 \\ j' & \ell' & 1/2 \\ j & \ell & 1/2 \end{Bmatrix}. \quad (2.124)$$

From the nj-symbols we have the additional selection rules

$$\ell = \ell' \quad \text{and} \quad j = j', \quad (2.125)$$

and reducing (2.124) eventually yields

$$A(000) = \frac{1}{4\pi|\mathbf{p}|^2} \frac{1}{\sqrt{2(2J+1)}} \sum_{\ell j} |\langle J_f\|V\|(jJ)J_f\rangle|^2. \quad (2.126)$$

Within the framework of the two-step model, (2.121) is the most general expression to describe the process of a primary ionization/excitation, and its subsequent Auger decay. On the left-hand side of (2.121), the state multipoles $\langle t_{kQ}^+ \rangle$ yield the angular distribution and the degree of spin polarization of the emitted Auger electrons. The right-hand side is a summation over three factors. The generalized anisotropy parameters $A(KkQ)$ contain the information on the dynamics of the Auger emission. Note, that its magnetic component Q must be the same as for the state multipoles $\langle t_{kQ}^+ \rangle$, describing the emitted Auger electrons. The dynamics of the primary ionization/excitation is fully described by the state multipoles $\langle T(J)_{KQ'}^+ \rangle$. They have been extensively discussed in Sect. 2.3 for a variety of excitation and ionization processes. Both, the state multipoles $\langle t_{kQ}^+ \rangle$ and the anisotropy parameters $A(KkQ)$ refer to the helicity system, i.e. the quantization axis is the direction of Auger emission. The state multipoles $\langle T(J)_{KQ'}^+ \rangle$ describing the primary ionization/excitation process refer to the primary beam axis as quantization axis. The angular dependency of both frames is expressed in terms of reduced rotation matrices $d_{Q'Q}^{(K)}(\theta)$. One should note, that the tensorial rank K of all three parameters is the same for each term of the sum over K. This is a direct consequence of the transformation properties of the state multipoles. Depending on the explicit primary ionization/excitation process, the summation over K and Q' in (2.121) can be restricted. This will be discussed in more detail in the next sections.

2.5 Angular Distribution and Spin Polarization

2.5.1 General Equations

Considering the Auger emission in the helicity system, the Cartesian components of the spin polarization vector $\mathbf{P}$ of the Auger electrons can be expressed in terms of state multipoles $\langle t_{kQ}^+ \rangle$ as

$$p_x = -\left(\langle t_{11}^+ \rangle - \langle t_{1-1}^+ \rangle\right) = -2Re\langle t_{11}^+ \rangle, \qquad (2.127)$$

$$p_y = -i\left(\langle t_{11}^+ \rangle + \langle t_{1-1}^+ \rangle\right) = 2Im\langle t_{11}^+ \rangle, \qquad (2.128)$$

$$p_z = \sqrt{2}\langle t_{10}^+ \rangle, \qquad (2.129)$$

and the zero rank tensor is a normalization factor

$$\langle t_{00}^+ \rangle = \frac{1}{\sqrt{2}}. \qquad (2.130)$$

Inserting (2.121) into these equations, applying the symmetries of the reduced rotation matrices and of the anisotropy parameters $A(KkQ)$, and using the Hermiticity condition of the state multipoles, we obtain for the angular distribution of the emitted Auger electrons

$$I(\theta) = \sqrt{2} \sum_{K \, even} A(K00) \left(\left\langle T(J)^+_{K0} \right\rangle d^{(K)}_{00}(\theta) \right.$$

$$\left. + \sum_{Q'>0} 2 \, Re \left\langle T(J)^+_{KQ'} \right\rangle d^{(K)}_{Q'0}(\theta) \right). \quad (2.131)$$

Integrating both sides over the solid angle $\hat{\omega} = (\theta, \phi)$ we find the Auger rate and the total intensity, respectively to be related to the coefficient $A(000)$ which is proportional to the squared modulus of the matrix elements

$$I_0 = \int d\hat{\omega} \, I(\theta) = 4\pi \sqrt{2} \, A(000) \left\langle T(J)^+_{00} \right\rangle$$

$$= \frac{1}{|\mathbf{p}|^2} \frac{1}{2J+1} \sum_{\ell j} |\langle J_f \| V \| (Jj)J_f \rangle|^2. \quad (2.132)$$

Analogously we get for the Cartesian components of the spin polarization vector

$$I(\theta) \, p_x = 2 \sum_{K \, even} Im A(K11) \left(\sum_{\substack{Q'>0 \\ even}} Im \left\langle T(J)^+_{KQ'} \right\rangle \left[d^{(K)}_{Q'1}(\theta) - d^{(K)}_{-Q'1}(\theta) \right] \right.$$

$$\left. + \sum_{\substack{Q'>0 \\ odd}} Im \left\langle T(J)^+_{KQ'} \right\rangle \left[d^{(K)}_{Q'1}(\theta) + d^{(K)}_{-Q'1}(\theta) \right] \right)$$

$$- 2 \sum_{K \, odd} Re A(K11) \left(\left\langle T(J)^+_{K0} \right\rangle d^{(K)}_{01}(\theta) \right.$$

$$+ \sum_{\substack{Q'>0 \\ even}} Re \left\langle T(J)^+_{KQ'} \right\rangle \left[d^{(K)}_{Q'1}(\theta) + d^{(K)}_{-Q'1}(\theta) \right]$$

$$\left. + \sum_{\substack{Q'>0 \\ odd}} Re \left\langle T(J)^+_{KQ'} \right\rangle \left[d^{(K)}_{Q'1}(\theta) - d^{(K)}_{-Q'1}(\theta) \right] \right), \quad (2.133)$$

$$I(\theta)\, p_y = 2 \sum_{K\,even} Im\, A(K11) \Big(\big\langle T(J)^+_{K0} \big\rangle d^{(K)}_{01}(\theta)$$

$$+ \sum_{\substack{Q'>0 \\ even}} Re\, \big\langle T(J)^+_{KQ'} \big\rangle \Big[d^{(K)}_{Q'1}(\theta) + d^{(K)}_{-Q'1}(\theta) \Big]$$

$$+ \sum_{\substack{Q'>0 \\ odd}} Re\, \big\langle T(J)^+_{KQ'} \big\rangle \Big[d^{(K)}_{Q'1}(\theta) - d^{(K)}_{-Q'1}(\theta) \Big] \Big)$$

$$+ 2 \sum_{K\,odd} Re\, A(K11) \Big(\sum_{\substack{Q'>0 \\ even}} Im\, \big\langle T(J)^+_{KQ'} \big\rangle \Big[d^{(K)}_{Q'1}(\theta) - d^{(K)}_{-Q'1}(\theta) \Big]$$

$$+ \sum_{\substack{Q'>0 \\ odd}} Im\, \big\langle T(J)^+_{KQ'} \big\rangle \Big[d^{(K)}_{Q'1}(\theta) + d^{(K)}_{-Q'1}(\theta) \Big] \Big), \qquad (2.134)$$

and

$$I(\theta)\, p_z = \sqrt{2} \sum_{K\,odd} A(K10) \Big(\big\langle T(J)^+_{K0} \big\rangle d^{(K)}_{00}(\theta)$$

$$+ \sum_{Q'>0} 2\, Re\, \big\langle T(J)^+_{KQ'} \big\rangle d^{(K)}_{Q'0}(\theta) \Big). \qquad (2.135)$$

Equations (2.131)–(2.135) are the most general expressions for angular distribution and spin polarization of the Auger electrons. So far, no assumptions have been made on the primary ionization/excitation process. Though the above equations are somewhat lengthy, one should note, that they only contain real quantities; i.e. all anisotropy parameters and state multipoles have been split in their real and imaginary parts, respectively. Further, applying the symmetry properties of the state multipoles, all summations over the magnetic quantum number Q' are reduced to positive values, only. This allows for an easy discussion of the different types of primary ionization and excitation processes, respectively.

Only certain state multipoles $\big\langle T(J)^+_{KQ'} \big\rangle$ can be generated, depending on the specific type of the primary ionization/excitation process. Analogously, the number of anisotropy parameters $A(KkQ)$ and thus, the number of measurements determining these quantities is restricted. This is discussed in more detail in the following sections.

2.5.2 Relative Parameters

It is often useful to express physical quantities in terms of relative parameters. Adopting the usual notations (e.g. see Kabachnik and Sazhina 1986; Kabachnik *et al.* 1991; Lohmann 1996b), we identify the parameters of alignment and orientation, containing the information on the primary ionization and excitation dynamics, respectively, as

$$\mathcal{A}_{KQ} = \frac{\left\langle T(J)_{KQ}^{+} \right\rangle}{\left\langle T(J)_{00}^{+} \right\rangle} \quad \text{for} \quad K = \text{even}, \tag{2.136}$$

and

$$\mathcal{O}_{KQ} = \frac{\left\langle T(J)_{KQ}^{+} \right\rangle}{\left\langle T(J)_{00}^{+} \right\rangle} \quad \text{for} \quad K = \text{odd}, \tag{2.137}$$

respectively. We will not give explicit expressions for the relative parameters, here. They can be easily derived from the general equations describing the primary i-onization or excitation process as discussed in Sect. 2.3. We will discuss a special case later in Sect. 2.6. Numerical investigations on deep inner shell alignment and orientation have been performed by Berezhko *et al.* (1978b) for the case of photo-ionization, almost 30 years ago. A large variety of alignment and orientation data for closed shell atoms and cations have been published, only recently (Kleiman and Lohmann 2003; Kleiman and Becker 2005), while Lohmann and Kleiman (2006) developed a theoretical model for calculating alignment and orientation of photoionized open shell atoms and provided first numerical open shell data and predictions for inner shell alignment and orientation, respectively.

Now, considering the Auger decay dynamics, besides the total intensity I_0 we need to introduce the parameter

$$\alpha_K = \frac{A(K00)}{A(000)} \tag{2.138}$$

which is the so-called angular distribution anisotropy parameter of Auger decay. From the symmetry relations of (2.119) we find the anisotropy parameters restricted to even values of K.

The coefficients $A(KkQ)$ have been discussed in Sect. 2.4. A general expression for the coefficient α_K in terms of matrix elements may be found in Kabachnik and Sazhina (1984) or Lohmann (1990) and can be written as

$$\alpha_K = \left(\sum_{\ell j} |\langle (J_f j)J \| V \| J \rangle|^2 \right)^{-1}$$
$$\times \sqrt{(2K+1)(2J+1)} \sum_{\ell \ell' j j'} (-1)^{J_f + J + 1/2 + j + j' + \ell' } \mathrm{i}^{\ell + \ell'}$$
$$\times \cos\left(\sigma_{\ell'}^{j'} - \sigma_{\ell}^{j}\right) C(K)_{jj'} \langle (J_f j)J \| V \| J \rangle \langle (J_f j')J \| V \| J \rangle, \tag{2.139}$$

where we introduced the coefficients

$$C(K)_{jj'} = -\sqrt{(2j+1)(2j'+1)} \begin{pmatrix} j' & j & K \\ 1/2 & -1/2 & 0 \end{pmatrix} \begin{Bmatrix} j' & j & K \\ J & J & J_f \end{Bmatrix}. \tag{2.140}$$

Table 2.4. The anisotropy parameters α_2 and α_4 for Auger transitions from an initial state with total angular momentum J to a final state with $J_f = 0$

Initial	Anisotropy parameters	
state J	α_2	α_4
3/2	-1	0
5/2	$-\sqrt{8/7} \sim -1.069$	$\sqrt{6/7} \sim 0.926$
7/2	$-\sqrt{25/21} \sim -1.091$	$\sqrt{81/77} \sim 1.026$

The C-coefficients show a useful symmetry which can be used to reduce the numerical effort

$$C(K)_{jj'} = C(K)_{j'j}. \tag{2.141}$$

Here, J and J_f denote the total angular momenta of the intermediate singly ionized state A^+ and the final ionic state A^{++}, respectively. The total and orbital angular momenta of the Auger electron are given by j, j' and ℓ, ℓ', respectively where the prime refers to the possible interference of different partial waves.

We point out that for Auger transitions to a final state with total angular momentum $J_f = 0$ the anisotropy coefficients become independent of the transition matrix elements, which eventually yields the simple expression

$$\alpha_K(J_f = 0) = (-1)^{K+J+3/2} \sqrt{(2K+1)(2J+1)} \begin{pmatrix} J & J & K \\ 1/2 & -1/2 & 0 \end{pmatrix}. \tag{2.142}$$

Thus, as has been first shown by Cleff and Mehlhorn (1974a), the anisotropy parameter α_K is purely analytic for this type of Auger transitions. Some explicit values are given in Table 2.4.

Analogously, introducing the spin polarization parameters of Auger emission we define the parameter

$$\delta_K = \frac{A(K10)}{A(000)}. \tag{2.143}$$

Here, the symmetries (2.116) and (2.119) restrict K to odd values. Inserting the expressions for $A(000)$, and $A(K10)$ we remain with

$$\delta_K = \left(\sum_{\ell j} |\langle (J_f j)J \| V \| J \rangle|^2 \right)^{-1}$$
$$\times (-1)^K \sqrt{(2K+1)(2J+1)} \sum_{\ell \ell' jj'} (-1)^{J_f+J+1/2+j+j'+\ell'+\ell'} i^{\ell+\ell'}$$
$$\times \cos\left(\sigma_{\ell'}^{j'} - \sigma_\ell^j\right) C(K)_{jj'} \langle (J_f j)J \| V \| J \rangle \langle (J_f j')J \| V \| J \rangle. \tag{2.144}$$

The above equation shows an interesting structure. Comparing the expression obtained for the spin polarization parameter δ_K with the anisotropy parameter α_K

connected with the angular distribution of the Auger electrons, we obtain not a physically relevant but rather a useful mathematical relation between the two parameters

$$\delta_K = (-1)^K \alpha_K.$$ (2.145)

At a first glance, this might lead to the assumption that the physical information which can be obtained from the measurement of the angular distribution and the z-component of the spin polarization vector is the same, apart from a phase factor. However, one should remember that due to symmetry requirements the rank of the parameters α_K of angular distribution must be even whereas the spin polarization parameter δ_K is restricted to odd values of K. Thus, the measurement of both coefficients leads to different physical information, but gives a simple mathematical relation in order to reduce the computational effort. As will be seen in Sect. 2.5.7, this relation is also useful in order to derive linear and non-linear interrelations between the angular distribution and spin polarizations, see (2.190) and (2.191) later on.

Again, we find that the spin polarization parameter δ_K becomes independent of the transition matrix elements for Auger transitions to a final state with total angular momentum $J_f = 0$ which leads to the simple expression

$$\delta_K (J_f = 0) = (-1)^{J+3/2} \sqrt{(2K+1)(2J+1)} \begin{pmatrix} J & J & K \\ 1/2 & -1/2 & 0 \end{pmatrix}.$$ (2.146)

This equation is identical to the one for α_K but with a different phase factor. However, one should remember that K is restricted to odd numbers in the above equation. Especially, for the case that $K = 1$ we obtain

$$\delta_1 (J_f = 0) = \sqrt{\frac{3}{2J(2J+2)}}.$$ (2.147)

Thus, $\delta_1 (J_f = 0)$ is generally a positive number. Applying the monotonicity theorems of analysis one can show that δ_1 is monotonic decreasing as a function of increasing initial total angular momentum J. In addition to the upper limit of δ_1 which is equal to 1.0 for the case that $J = 1/2$, a lower limit can be given for the case that $J \rightarrow \infty$. Applying the limiting value theorems on (2.147) we get

$$1 \geq \delta_1 (J_f = 0) > 0 \quad \text{if} \quad J \geq 1/2, \quad J \quad \text{half integer},$$ (2.148)

as general limits for the parameter δ_1. Some special values for δ_1 and δ_3 are given in Table 2.5.

Eventually, we introduce the spin polarization parameter

$$\xi_K = \frac{-\sqrt{2}}{\sqrt{K(K+1)}} \frac{A(K11)}{A(000)},$$ (2.149)

where the anisotropy coefficient $A(K11)$ is solely real for $K = $ odd and fully imaginary for $K = $ even. Both parameters ξ_K are again defined with respect to the helicity

Table 2.5. Numerical values for the spin polarization parameters δ_1 and δ_3 for Auger transitions with $J_f = 0$

Initial	Spin polarization parameters	
state J	δ_1	δ_3
1/2	1	0
3/2	$\sqrt{1/5} \sim 0.447$	$-\sqrt{9/5} \sim -1.342$
5/2	$\sqrt{3/35} \sim 0.293$	$-\sqrt{8/15} \sim -0.730$
7/2	$\sqrt{1/21} \sim 0.218$	$-\sqrt{3/11} \sim -0.522$

system of the emitted Auger electrons. We obtain the general equations for the coefficients ξ_K as

$$
\xi_K^{odd} = -\left(\sum_{\ell j} |\langle (J_f j)J \| V \| J \rangle|^2 \right)^{-1}
$$

$$
\times (-1)^K \frac{\sqrt{(2K+1)(2J+1)}}{K(K+1)} \sum_{\ell \ell' jj'} (-1)^{J_f+J+1/2+j+j'+\ell'} i^{\ell+\ell'}
$$

$$
\times \cos(\sigma_{\ell'}^{j'} - \sigma_{\ell}^{j}) C(K)_{jj'} \langle (J_f j)J \| V \| J \rangle \langle (J_f j')J \| V \| J \rangle
$$

$$
\times \left[(\ell' - j')(2j'+1) + (\ell - j)(2j+1) \right], \tag{2.150}
$$

for K odd, and

$$
\xi_K^{even} = -\left(\sum_{\ell j} |\langle (J_f j)J \| V \| J \rangle|^2 \right)^{-1}
$$

$$
\times (2K+1) \sqrt{\frac{2(2J+1)}{K(K+1)}} \sum_{\ell \ell' jj'} (-1)^{J_f+J+j'} i^{\ell+\ell'}
$$

$$
\times \sin(\sigma_{\ell'}^{j'} - \sigma_{\ell}^{j}) H(K)_{jj'} \langle (J_f j)J \| V \| J \rangle \langle (J_f j')J \| V \| J \rangle, \tag{2.151}
$$

for K even, where the coefficients

$$
H(K)_{jj'} = \sqrt{(2j+1)(2j'+1)}
$$

$$
\times \begin{pmatrix} K & K & 1 \\ -1 & 0 & 1 \end{pmatrix} \begin{pmatrix} j' & j & K \\ 1/2 & 1/2 & -1 \end{pmatrix} \begin{Bmatrix} j' & j & K \\ J & J & J_f \end{Bmatrix}, \tag{2.152}
$$

have been introduced which satisfy the symmetry condition

$$
H(K)_{jj'} = (-1)^{j+j'} H(K)_{j'j} . \tag{2.153}
$$

Note, that according to (2.149), the definition of the relative anisotropy parameters ξ_K is unique. However, the physical meaning, as well as the structure of (2.150) and (2.151), is totally different. In order to clarify this point let us suppose ionization or excitation with a longitudinally polarized electron or a circularly polarized photon beam. In both cases, from (2.150) (K odd), we receive information on the anisotropy of the x-component of the spin polarization vector (see (2.166) and (2.179) later on), i.e. spin polarization in the reaction plane, where for the other case (2.151) (K even), the connection is with the y-component of the spin polarization vector, see (2.167) and (2.180), which is perpendicular to the reaction plane. Most important, from a physical point of view, is the different behaviour with respect to the scattering phases. The parameters α_K, δ_K and the coefficients ξ_K^{odd} depend on the *cosine* of the phase difference. Since the cosine remains unaffected against interchanging of the scattering phases, these coefficients allow for the determination of the modulus of the phase differences, only. The spin polarization parameter ξ_K^{even}, however, depends on the *sine* of the phase difference and therefore allows to determine the sign of the phase differences, too. Thus, the spin polarization parameter ξ_K^{even} can yield even more refined information on the decay dynamics than can be obtained from the angular anisotropy parameter and from the other spin polarization parameters, respectively. Further, like α_K and δ_K, ξ_K^{odd} depends on the C-coefficients, which reduces the computational effort, whereas the parameter ξ_K^{even} is a function of the introduced H-coefficients.

Considering the case of an Auger transition to a final state with $J_f = 0$ the parameter ξ_K^{odd} becomes independent of the matrix elements, too. Inserting J_f in (2.150) some short calculation yields

$$\xi_K^{odd}(J_f = 0) = (-1)^{J+1/2} \frac{2(\ell - J)(2J + 1)}{K(K + 1)}$$
$$\times \sqrt{(2K + 1)(2J + 1)} \begin{pmatrix} J & J & K \\ 1/2 & -1/2 & 0 \end{pmatrix}, \quad (2.154)$$

where

$$\ell - J = \begin{cases} 1/2 & \text{for } \ell > J, \\ -1/2 & \text{for } \ell < J. \end{cases} \quad (2.155)$$

Note that due to selection rules $j = j' = J$ and $\ell = \ell'$ are fulfilled in any case. However, $\ell \neq L$ is still possible, where L denotes the orbital angular momentum of the intermediate ionic state. Thus, ξ_K^{odd} alternates depending on $\ell - J$, but upper and lower limits can be given for its absolute value.

Especially, for the case that $K = 1$ we obtain

$$\xi_1(J_f = 0) = -(\ell - J) \frac{\sqrt{3}(2J + 1)}{\sqrt{2J(2J + 2)}}. \quad (2.156)$$

Using the monotonicity theorems we find

$$1 \geq |\xi_1(J_f = 0)| > \frac{\sqrt{3}}{2} \simeq 0.8660 \quad \text{if} \quad J \geq 1/2, \quad J \quad \text{half integer}, \quad (2.157)$$

as general limits for the anisotropy parameter ξ_1. Some special values for ξ_1 and ξ_3 are given in Table 2.6.

Table 2.6. Numerical values for the spin polarization parameters ξ_1 and ξ_3 for the case $J_f = 0$. The $\pm$ depends on the even and odd parity of the emitted partial wave, respectively. The parity is determined by the angular momentum ℓ of the emitted partial wave

Initial state		Spin polarization parameters	
J	ℓ	ξ_1	ξ_3
1/2	0, 1	± 1	0
3/2	1, 2	$\pm 2/\sqrt{5} \sim \pm 0.894$	$\mp 1/\sqrt{5} \sim \mp 0.447$
5/2	2, 3	$\pm\sqrt{27/35} \sim \pm 0.878$	$\pm\sqrt{5/24} \sim \pm 0.456$
7/2	3, 4	$\pm\sqrt{16/21} \sim \pm 0.873$	$\mp 2/\sqrt{33} \sim \mp 0.348$

Table 2.7. Connection of our anisotropy coefficients with those defined by Kabachnik and Lee (1989)

This book	Kabachnik and Lee	Kabachnik and Lee	This book
θ	$2\pi - \vartheta$	ϑ	$2\pi - \theta$
α_2	α_2	α_2	α_2
ξ_2	$-\frac{2}{3}\xi_2$	ξ_2	$-\frac{3}{2}\xi_2$
ξ_1	$\beta_1 - \frac{1}{2}\gamma_1$	β_1	$\frac{1}{3}(\delta_1 - 2\xi_1)$
$\delta_1 = -\alpha_1$	$\beta_1 + \gamma_1$	γ_1	$\frac{2}{3}(\delta_1 + \xi_1)$

Eventually, we remark that ξ_K^{even} vanishes for transitions to $J_f = 0$, because of the symmetry of the H-coefficients.

$$\xi_K^{even} = 0. \tag{2.158}$$

The notation of the spin polarization parameters is not unique. Where our definition has some advantages with respect to the mathematical treatment it is also useful to define a set of parameters describing the Auger emission with respect to the incoming beam axis as quantization axis. While this leaves the angular distribution parameter α_K and, besides a factor, the dynamic spin polarization parameter ξ_K^{even} undisturbed, we obtain different expressions for the two spin polarization parameters δ_K and ξ_K^{odd} associated with the in-reaction plane components of the spin polarization vector.

The connection of these two different sets of parameters is shown in Table 2.7. Here, the spin polarization parameters β_1 and γ_1 refer to the incoming beam axis as quantization axis, whereas the set of parameters introduced before refers to the helicity system of the emitted Auger electrons. Explicit expressions for β_1 and γ_1 have been derived by Kabachnik and Lee (1989). Adopting the notation of Lohmann *et al.* (1993), and denoting the matrix element as $\langle J_f(\ell\frac{1}{2})j : J \|V\| J\rangle$ we get

$$
\beta_1 = -\left(\sum_{\ell j}|\langle J_f(\ell\tfrac{1}{2})j : J\|V\|J\rangle|^2\right)^{-1}
$$
$$
\times \sqrt{2(2J+1)} \sum_{\ell j j'}(-1)^{J+J_f+1/2+\ell+j+j'}
$$
$$
\times \sqrt{(2j+1)(2j'+1)}\begin{Bmatrix} J & J & 1 \\ j & j' & J_f \end{Bmatrix}\begin{Bmatrix} \tfrac{1}{2} & \tfrac{1}{2} & 1 \\ j & j' & \ell \end{Bmatrix}
$$
$$
\times \langle J_f(\ell\tfrac{1}{2})j : J\|V\|J\rangle\langle J_f(\ell\tfrac{1}{2})j' : J\|V\|J\rangle^*, \qquad (2.159)
$$

and

$$
\gamma_1 = \left(\sum_{\ell j}|\langle J_f(\ell\tfrac{1}{2})j : J\|V\|J\rangle|^2\right)^{-1}
$$
$$
\times 2\sqrt{15(2J+1)} \sum_{\ell\ell' j j'}(-1)^{J+J_f-\ell+j'}\sqrt{(2\ell+1)(2\ell'+1)}
$$
$$
\times \sqrt{(2j+1)(2j'+1)}\begin{pmatrix} \ell & \ell' & 2 \\ 0 & 0 & 0 \end{pmatrix}\begin{Bmatrix} J & J & 1 \\ j & j' & J_f \end{Bmatrix}\begin{Bmatrix} \ell & \ell' & 2 \\ \tfrac{1}{2} & \tfrac{1}{2} & 1 \\ j & j' & 1 \end{Bmatrix}
$$
$$
\times \langle J_f(\ell\tfrac{1}{2})j : J\|V\|J\rangle\langle J_f(\ell\tfrac{1}{2})j' : J\|V\|J\rangle^*
$$
$$
= \left(\sum_{\ell j}|\langle J_f(\ell\tfrac{1}{2})j : J\|V\|J\rangle|^2\right)^{-1}
$$
$$
\times (-1)^{J+J_f-1/2}\sqrt{2J+1} \sum_{\ell\ell' j j'}\sqrt{(2j+1)(2j'+1)}\begin{pmatrix} j & j' & 1 \\ \tfrac{1}{2} & \tfrac{-1}{2} & 0 \end{pmatrix}
$$
$$
\times \begin{Bmatrix} J & J & 1 \\ j & j' & J_f \end{Bmatrix}\frac{(\ell-j)(2j+1)+(\ell'-j')(2j'+1)+2}{\sqrt{3}}
$$
$$
\times \langle J_f(\ell\tfrac{1}{2})j : J\|V\|J\rangle\langle J_f(\ell\tfrac{1}{2})j' : J\|V\|J\rangle^*. \qquad (2.160)
$$

As for δ_1 and ξ_1, the parameters β_1 and γ_1 become independent of the Auger matrix elements for transitions to a final state with $J_f = 0$. Their numbers are given in Table 2.8.

2.5.3 The Case of Electron Impact

For a certain Auger transition with a fixed total angular momentum J of the intermediate ionic state A^+ the summation over K is generally restricted to $2J + 1$ possible values

$$
0 \le K \le 2J. \qquad (2.161)
$$

Table 2.8. Numerical values for the spin polarization parameters β_1 and γ_1 defined by Kabachnik and Lee (1989) for the case $J_f = 0$

J	ℓ	β_1	γ_1
1/2	0	1	0
1/2	1	$-1/3$	$4/3$
3/2	1	$\sqrt{5}/3$	$-2/3\sqrt{5}$
3/2	2	$-1/\sqrt{5}$	$2/\sqrt{5}$
5/2	2	$\sqrt{7/15}$	$-4/\sqrt{105}$
5/2	3	$-\sqrt{5/21}$	$8/\sqrt{105}$
7/2	3	$\sqrt{3/7}$	$-2/\sqrt{21}$
7/2	4	$-\sqrt{7/27}$	$2\sqrt{5}/3\sqrt{7}$

Considering a primary ionization/excitation with electrons, the discussion of Sects. 2.3.3 and 2.3.4 additionally yields the important selection rule

$$|Q'| \leq 1. \tag{2.162}$$

Inserting this result into (2.131)–(2.135) we obtain for the angular distribution of the Auger electrons

$$I(\theta) = \sqrt{2} \sum_{K\,even} A(K00) \left(\left\langle T(J)_{K0}^+ \right\rangle d_{00}^{(K)}(\theta) + \right.$$

$$\left. \times\, 2\,Re \left\langle T(J)_{K1}^+ \right\rangle d_{10}^{(K)}(\theta) \right). \tag{2.163}$$

As pointed out, it is useful to express physical quantities in terms of relative parameters. Applying the notation introduced in Sect. 2.5.2 (2.163) may be eventually expressed as

$$I(\theta)^e = \frac{I_0}{4\pi} \left(1 + \sum_{\substack{K\geq 2 \\ even}} \alpha_K \left[\mathcal{A}_{K0}\, d_{00}^{(K)}(\theta) + 2\,Re\,\mathcal{A}_{K1}\, d_{10}^{(K)}(\theta) \right] \right). \tag{2.164}$$

As has been shown, the value of Q' is directly proportional to the magnetic quantum number q of the state multipoles $\langle t_{kq}^+ \rangle$ describing the degree of spin polarization of the incoming ionizing electron beam. Thus, for an ionization with unpolarized electrons the alignment parameter $\mathcal{A}_{K0}$ survives, only. The second alignment parameter $Re\,\mathcal{A}_{K1}$ can only be non-zero if a polarized electron beam is used, which is transversely polarized perpendicular to the reaction plane (see Fig. 2.3).

Abbreviating the parameter N^e as

$$N^e = 1 + \sum_{\substack{K\geq 2 \\ even}} \alpha_K \left[\mathcal{A}_{K0}\, d_{00}^{(K)}(\theta) + 2\,Re\,\mathcal{A}_{K1}\, d_{10}^{(K)}(\theta) \right], \tag{2.165}$$

and again using relative quantities, the transverse in-reaction plane component p_x of the spin polarization vector, i.e. (2.133), reduces to

$$
p_x(\theta)^e = \frac{1}{N^e} \left(- \sum_{K\,even} \sqrt{K(K+1)}\, \xi_K \, Im\, \mathcal{A}_{K1} \left[d_{11}^{(K)}(\theta) + d_{-11}^{(K)}(\theta) \right] \right.
$$
$$
+ \sum_{K\,odd} \sqrt{K(K+1)}\, \xi_K \left\{ \mathcal{O}_{K0}\, d_{01}^{(K)}(\theta) \right.
$$
$$
\left. \left. + Re\, \mathcal{O}_{K1} \left[d_{11}^{(K)}(\theta) - d_{-11}^{(K)}(\theta) \right] \right\} \right). \tag{2.166}
$$

Here, the orientation parameter $\mathcal{O}_{K0}$ can only occur if the primary ionizing electron beam shows a non-vanishing longitudinally spin polarization (see Table 2.3 and the discussion in Sects. 2.3.3 and 2.3.4). The other terms can only be non-zero if an electron beam is used which is transversely polarized in the reaction plane. For an unpolarized beam or an electron beam, transversely and perpendicularly polarized to the reaction plane, the transverse spin polarization p_x^e of the Auger electrons vanishes. Equation (2.166) depends on the real and imaginary parts of the anisotropy parameter ξ_K. For even K values ξ_K can be only determined if a transversely polarized electron beam with an intrinsic spin polarization $p_x \neq 0$ is used. For odd values of K a longitudinally polarized electron beam, i.e. with an intrinsic spin polarization $p_z \neq 0$, must be used.

From (2.134) we get the transverse spin polarization p_y^e, perpendicular to the reaction plane, as

$$
p_y(\theta)^e = \frac{-1}{N^e} \left(\sum_{K\,even} \sqrt{K(K+1)}\, \xi_K \left\{ \mathcal{A}_{K0}\, d_{01}^{(K)}(\theta) \right. \right.
$$
$$
\left. + Re\, \mathcal{A}_{K1} \left[d_{11}^{(K)}(\theta) - d_{-11}^{(K)}(\theta) \right] \right\}
$$
$$
\left. + \sum_{K\,odd} \sqrt{K(K+1)}\, \xi_K \, Im\, \mathcal{O}_{K1} \left[d_{11}^{(K)}(\theta) + d_{-11}^{(K)}(\theta) \right] \right), \tag{2.167}
$$

In this case, the alignment parameters $\mathcal{A}_{K0}$ can be different from zero even if an unpolarized electron beam is used. The other terms can only occur if an intrinsic, transversely, perpendicularly to the reaction plane polarized electron beam is used. Thus, for even values of K, the spin polarization parameters ξ_K can be determined even if an unpolarized electron beam is used. The determination of ξ_K for odd values of K requires however, an ionization with an electron beam which is transversely and perpendicularly polarized to the reaction plane.

The longitudinally spin polarization of the Auger electrons which is given by (2.135) reduces to

$$
p_z(\theta)^e = \frac{1}{N^e} \sum_{K\,odd} \delta_K \left(\mathcal{O}_{K0}\, d_{00}^{(K)}(\theta) + 2\, Re\, \mathcal{O}_{K1}\, d_{10}^{(K)}(\theta) \right). \tag{2.168}
$$

Here, the summation is restricted to odd values of K. The first term only occurs if the incoming electron beam is longitudinally polarized. The second term can only be non-zero if the ionizing electron beam shows a transverse spin polarization in the reaction plane. Thus, a longitudinally spin polarization of the emitted Auger e- lectrons can only occur if the ionizing electron beam has an intrinsic non-vanishing in-reaction plane polarization, which can be either longitudinally or transversely.

2.5.4 Generation of Spin Polarization out of Alignment

Let us now concentrate on the spin polarization of the emitted Auger electrons. While it is not surprising to obtain a possible non-vanishing spin polarization for the Auger electrons if polarized particles are used for the primary ionization – the physics of this process is what is commonly known as polarization transfer – the process of producing spin polarization out of alignment is of physical importance. This is known as the generation of a dynamic spin polarization.

Considering a primary ionization with unpolarized electrons or photons, only multipole tensors $\mathcal{A}_{KQ}$ with $Q = 0$ contribute to the angular distribution and spin polarization. This expresses the fact that the charge distribution of the intermediate ionic states A^+ is axially symmetric with respect to the incoming beam axis. In particular, $\mathcal{A}_{20}$ is the usual alignment parameter. Then, the problem arises how spin polarization – or, more generally, orientation – can be created out of alignment (and the corresponding higher multipoles with rank $K > 2$). Necessary conditions for the Auger emission process have been given by Klar (1980) and Kabachnik and Sazhina (1986). Here, we want to make a few additional comments considering the problem from a slightly different point of view.

As has been discussed by Lombardi (1975), in general, one is first faced with the problem of finding out how the sense of rotation, defining the polarization, is defined. In our present case of interest one axis and one direction are given by the experimental conditions; the axis of alignment of the ions, i.e. the beam axis, and the direction of observation $\mathbf{p}$ of the emitted Auger electrons (with angle θ between $\mathbf{p}$ and the beam axis). For $\theta \neq 0$ both span the x–z plane of our coordinate system. By a rotation around the y axis, either to the left or to the right, the alignment axis can be brought into the direction of $\mathbf{p}$. Because these two rotations are not equivalent (if $\theta \neq 0$ and $\theta \neq \pi/2$) a sense of rotation can be defined by the experimental conditions. This is not the case, however, if $\mathbf{p}$ is parallel or perpendicular to the axis of alignment and the spin polarization therefore vanishes at these angles.

It should be noted that the existence of a preferred plane is not sufficient for obtaining a non-vanishing spin polarization. The essential point is that a sense of orientation must be given by the geometry of the experiment.

Therefore, the next question is whether the interacting system is able to *see* this sense of rotation. This clearly depends on the details of the dynamics. Klar (1980) and Kabachnik (1981) have discussed the importance of the various spin-dependent interactions in this respect. Here, we only want to select one particular point and discuss it in some more detail. The transition amplitude of Auger emission can be written as $\langle J_f M_f \mathbf{p}^{(-)} m_s | V | J M \rangle$, where the doubly ionized target A^{++} and the Auger electron are generated via the Coulomb interaction. The wavefunction of the final state $|J_f M_f \mathbf{p}^{(-)} m_s\rangle$ can be asymptotically expressed as an outgoing Coulomb wave in direction $\mathbf{p}$ and an incoming spherical wave (see e.g. Åberg and Howat 1982). I.e., in the context of scattering theory, we are dealing with a scattering solution with incoming boundary conditions. A decisive role is played by the final state interaction (FSI) between the outgoing Auger electron and the remaining ion. In order to see this clearly let us first neglect the FSI effects. In this case the state of the emitted electron is represented by a plane wave $|\mathbf{p} m_s\rangle$ and we have

$$\left| J_f M_f \mathbf{p}^{(-)} m_s \right\rangle \simeq \left| J_f M_f \mathbf{p} m_s \right\rangle. \tag{2.169}$$

We then investigate the behaviour of the transition amplitudes under time reversal. The Coulomb operator V is invariant under this transformation, i.e., we have the condition

$$V = \hat{\Theta}^+ V \hat{\Theta}, \tag{2.170}$$

where $\hat{\Theta}$ denotes the time-reversal operator and where the Hermiticity of V has been taken into account.

Inserting (2.170) into the matrix elements, applying the operators to the bra- and ket-vectors, and using the anti-unitarity of the time-reversal operator we obtain

$$\left\langle J_f M_f \mathbf{p} m_s \left| V \right| J M \right\rangle = (-1)^{J_f - M_f + 1/2 - m_s + J - M}$$
$$\times \left\langle J_f - M_f - \mathbf{p} - m_s \left| V \right| J - M \right\rangle^*, \tag{2.171}$$

where the phase conventions of Taylor (1972) have been used.

We then rotate the system around the y-axis about an angle π. Applying the corresponding operator $R(\pi)_y$ to the states we obtain

$$R(\pi)_y \left| J_f - M_f - \mathbf{p} - m_s \right\rangle = (-1)^{J_f - M_f + 1/2 - m_s} \left| J_f M_f \mathbf{p} m_s \right\rangle \tag{2.172}$$

and a similar relation for the initial state.

$$R(\pi)_y \left| J - M \right\rangle = (-1)^{J - M} \left| J M \right\rangle. \tag{2.173}$$

Since the Coulomb operator V is invariant under this rotation ($R_y V R_y = V$) we finally get from (2.171)–(2.173)

$$\left\langle J_f M_f \mathbf{p} m_s \left| V \right| J M \right\rangle = \left\langle J_f M_f \mathbf{p} m_s \left| V \right| J M \right\rangle^*. \tag{2.174}$$

Hence, in the plane-wave approximation the transition amplitudes are *real* numbers. From (2.109) and (2.114) we get that the coefficients $A(K11)$ and thus, the spin polarization parameters ξ_K^{even} are real, too and from (2.167) it follows that the dynamic component p_y of the spin polarization vector vanishes.

It is therefore the interaction, experienced by the Auger electron when it leaves the target, which produces the dynamic spin polarization. Without these effects p_y

would be zero regardless of the magnitude of the fine structure splitting of the intermediate ionic levels and the amount of the relevant alignment parameters $\mathcal{A}_{KQ}$. A determination of p_y can therefore be expected to be a direct probe of FSI. Eventually we note, that the above discussion also applies to the case of a primary photoionization with an unpolarized photon beam.

2.5.5 The Case of Photoionization–Excitation

Considering a primary photoionization/excitation the results of Sect. 2.3.1 and 2.3.2 apply. Equations (2.47) and (2.48) yield

$$K \leq 2 \quad \text{and} \quad Q' = 0, \pm 2. \tag{2.175}$$

Here, the summation over the rank K of the state multipoles is generally restricted. This is a direct consequence of the dipole approximation and independent of the actual value of the total angular momentum J which generally restricts the tensorial rank to values of $K \leq 2J$.

Applying this to (2.131)–(2.135) we get for the angular distribution of the emitted Auger electrons

$$I(\theta)^\gamma = \frac{I_0}{4\pi} \left(1 + \alpha_2 \left[\mathcal{A}_{20} \, d_{00}^{(2)}(\theta) + 2 \, Re \, \mathcal{A}_{22} \, d_{20}^{(2)}(\theta) \right] \right). \tag{2.176}$$

Here, only alignment parameters with rank $K = 2$ occur. Thus, besides the total intensity, the anisotropy parameter α_2 needs to be determined, only where the first is proportional to the total intensity which has been derived in (2.132). The first term of (2.176), which is proportional to the alignment parameter $\mathcal{A}_{20}$, occurs if an unpolarized photon beam is used for the primary photoionization/excitation. As can be derived from the discussion of Sect. 2.3.1 (see (2.46) and Table 2.2), the last term does only occur if a linearly polarized photon beam with Stokes parameter $\eta_3 \neq 0$ is used.

Inserting explicit expressions for the reduced rotation matrices we get

$$I(\theta)^\gamma = \frac{I_0}{4\pi} \left(1 + \alpha_2 \left[\mathcal{A}_{20} \, P_2(\cos\theta) + Re \, \mathcal{A}_{22} \sqrt{\frac{3}{2}} \sin^2\theta \right] \right) \tag{2.177}$$

for the angular distribution, where $P_2(\cos\theta)$ denotes the second Legendre polynomial. Analogously to the case of electron impact we introduce the abbreviation N^γ as

$$N^\gamma = 1 + \alpha_2 \left[\mathcal{A}_{20} \, P_2(\cos\theta) + Re \, \mathcal{A}_{22} \sqrt{\frac{3}{2}} \sin^2\theta \right]. \tag{2.178}$$

With this, for a primary photoionization, the transverse in-reaction plane component p_x^γ of the spin polarization vector of the Auger electrons reduces to

$$p_x(\theta)^\gamma = \frac{1}{N^\gamma} \left(\sqrt{2}\, \xi_1 \, \mathcal{O}_{10} \, d_{01}^{(1)}(\theta) - \sqrt{6}\, \xi_2 \, Im \, \mathcal{A}_{22} \left[d_{21}^{(2)}(\theta) - d_{-21}^{(2)}(\theta) \right] \right)$$

$$= \frac{1}{N^\gamma} \left(\xi_1 \, \mathcal{O}_{10} + \sqrt{6}\, \xi_2 \, Im \, \mathcal{A}_{22} \right) \sin\theta. \tag{2.179}$$

The orientation $\mathcal{O}_{10}$ can only be generated if a circularly polarized photon beam, i.e. $\eta_2 \neq 0$, is used. For the second term there must be an intrinsic, linear polarization of the photon beam. Here, in contrast to the angular distribution, the Stokes parameter η_1 must be different from zero (see Sect. 2.3.1). The dynamics of the Auger decay is expressed via the spin polarization parameters ξ_1 and ξ_2.

The transverse spin polarization p_y^γ, perpendicular to the reaction plane can be reduced from (2.134) to

$$
p_y(\theta)^\gamma = \frac{-\sqrt{6}}{N^\gamma} \xi_2 \left(A_{20} d_{01}^{(2)}(\theta) + Re\, A_{22} \left[d_{21}^{(2)}(\theta) + d_{-21}^{(2)}(\theta) \right] \right)
$$

$$
= \frac{-3}{2N^\gamma} \xi_2 \left(A_{20} - \sqrt{\frac{2}{3}} Re\, A_{22} \right) \sin(2\theta). \tag{2.180}
$$

Here, only parameters of rank $K = 2$ do occur. The Auger decay dynamics is contained in the parameter ξ_2. The first alignment parameter A_{20} generally occurs if an unpolarized photon beam is used. Again, for the second term there must be an intrinsic, linear polarization, i.e. $\eta_3 \neq 0$.

The general expression (2.135) for the longitudinally spin polarization of the Auger electrons eventually reduces to

$$
p_z(\theta)^\gamma = \frac{1}{N^\gamma} \delta_1 \mathcal{O}_{10} d_{00}^{(1)}(\theta) = \frac{1}{N^\gamma} \delta_1 \mathcal{O}_{10} \cos\theta. \tag{2.181}
$$

Equation (2.181) can only be different from zero if the primary photon beam shows an intrinsic, circular polarization, i.e. $\eta_2 \neq 0$. The spin polarization parameter δ_1 is thus directly proportional to the longitudinally spin polarization of the Auger electrons.

Eventually, we note an interesting relation between the alignment parameters A_{20} and A_{22} (Kleiman et al. 1999a),

$$
A_{22} = A_{20} \sqrt{\frac{3}{2}} (-\eta_3 + i\eta_1) . \tag{2.182}
$$

This result is of physical importance since it demonstrates the fact that, besides the intrinsic alignment A_{20}, which is generally transferred from the unpolarized photon beam to the intermediate ionic state, any additional alignment, generated from linearly polarized photons, can always be directly related to the usual alignment A_{20}.

A physical interpretation of the parameter A_{20} has been given earlier by Hertel and Stoll (1977) who have shown that a non-zero alignment A_{20} can be related to an anisotropic deformation of the ionic charge cloud which is axially symmetric to the photon beam axis. Using a linearly polarized photon beam, and thus a non-vanishing parameter A_{22}, yields a further deformation of the ionic charge cloud, i.e. the *cigar like form* ($A_{20} > 0$) or the *disc like form* ($A_{20} < 0$) of the charge cloud are further compressed with respect to one of their principal axes.

2.5.6 Photoionization–Excitation for Arbitrarily Oriented Coordinate Frames

In the previous sections we have chosen the quantization axis in the helicity frame, and defined the reaction plane by the direction of Auger emission and the incoming beam axis; see Fig. 2.3. This yields some advantages concerning the simplification of the theoretical equations as has been outlined in Sect. 2.1.3. It particularly allows for relating the helicity and the laboratory or collision frame via the one polar angle θ, only (see (2.120); we also refer to the discussion in Sects. 2.4.1 and 2.5.1).

However, experiments using synchrotron radiation for the primary photoionization or excitation process, the synchrotron beam facilities themselves define the laboratory frame whose quantization axis can be made to coincide with the collision frame. Though, in such a case, the transformation of the state multipoles into the helicity system can be only achieved by the complex rotation matrices $\mathcal{D}^{(K)}_{Q'Q}(\omega)$ via

$$\left\langle T(J)^+_{KQ}\right\rangle = \sum_{Q'}\left\langle T(J)^+_{KQ'}\right\rangle \mathcal{D}^{(K)}_{Q'Q}(0,\theta,\phi), \qquad (2.183)$$

where the solid angle ω is given by the polar angle θ and the azimuth ϕ. The third Euler angle can be generally chosen as zero. Thus, the results of Sect. 2.5.5 may be generalized by inserting (2.183) into (2.115) and performing the same procedure as in Sect. 2.5.1. Applying the results of Sects. 2.3.1 and 2.3.2, and taking the dipole selection rules (2.175) for the state multipoles into account, we obtain the Auger electron angular distribution as

$$I(\theta,\phi)^\gamma = \frac{I_0}{4\pi}\left\{1 + \alpha_2\left(\mathcal{A}_{20}P_2(\cos\theta) + \sqrt{\frac{3}{2}}\Big[Re\mathcal{A}_{22}\cos 2\phi\right.\right.$$
$$\left.\left. - Im\mathcal{A}_{22}\sin 2\phi\Big]\sin^2\theta\right)\right\}. \qquad (2.184)$$

In contrast to (2.177) the angular distribution becomes a function of both, the real and the imaginary part of the alignment parameter $\mathcal{A}_{22}$, which may be non-zero only if a linearly polarized photon beam with η_1 and/or $\eta_3 \neq 0$ is used. Even the angular dependency on the azimuthal angle ϕ occurs for the parameter $\mathcal{A}_{22}$, only.

The general equations for the angular distribution and the Cartesian components of the spin polarization vector have been derived by Kleiman et al. (1999a) in the helicity frame of the emitted Auger electrons with the axis of Auger emission as quantization axis. For such geometry the spin polarization vector can be related to an arbitrarily chosen coordinate frame, e.g. the laboratory frame, via the two angles θ and ϕ defined with respect to the quantization axis. Introducing the abbreviation N^Γ as

$$N^\Gamma = 1 + \alpha_2\left(\mathcal{A}_{20}P_2(\cos\theta) + \sqrt{\frac{3}{2}}\Big[Re\mathcal{A}_{22}\cos 2\phi\right.$$
$$\left. - Im\mathcal{A}_{22}\sin 2\phi\Big]\sin^2\theta\right), \qquad (2.185)$$

we obtain the components p_x and p_y of the spin polarization vector as

$$p_x(\theta, \phi)^\gamma = \frac{1}{N^\Gamma}\left(\xi_1 \mathcal{O}_{10} + \sqrt{6}\,\xi_2\Big[Re\mathcal{A}_{22}\sin 2\phi\right.$$
$$\left.+ Im\mathcal{A}_{22}\cos 2\phi\Big]\right)\sin\theta, \qquad (2.186)$$

and

$$p_y(\theta, \phi)^\gamma = \frac{-3}{2N^\Gamma}\,\xi_2\left(\mathcal{A}_{20} - \sqrt{\frac{2}{3}}\Big[Re\mathcal{A}_{22}\cos 2\phi\right.$$
$$\left.- Im\mathcal{A}_{22}\sin 2\phi\Big]\right)\sin(2\theta). \qquad (2.187)$$

As for the angular distribution, both expressions show the additional dependency on the azimuthal angle ϕ in the real and imaginary parts of the alignment parameter $\mathcal{A}_{22}$, only.

For the longitudinal component p_z of the spin polarization vector the azimuthal dependency enters solely in the denominator.

$$p_z(\theta, \phi)^\gamma = \frac{1}{N^\Gamma}\,\delta_1\mathcal{O}_{10}\cos\theta. \qquad (2.188)$$

The numerator remains independent of the azimuth ϕ. This can be explained by the fact that, irrespective of the specific choice of the X- and Y-axes of the laboratory frame, the z-axis of the helicity frame is always contained in the reaction plane.

The results of Sect. 2.5.5 are included in the above equations by setting $\phi = 0$ in (2.184) which gives (2.177), and by inserting $\phi = 0$ into (2.186)–(2.188) we end up with (2.179)–(2.181), respectively.

We will refer to the obtained results in Sect. 2.6 and consider an example related to a particular experiment in Sect. 4.7.

2.5.7 Interrelations Between the Angular Distribution and Spin Polarization Parameters of Auger Emission

We will now see that the angular distribution and spin polarization parameters of the Auger emission are not independent. This has been derived by Kabachnik (2005). We will adopt his formalism in deriving the general relation, which usually yields a non-linear connection between the parameters. In order to do so, we abbreviate the reduced matrix elements. Let

$$V_{\ell j} \equiv i^\ell e^{-i\sigma_\ell^j}\langle(J_f j)J\|V\|J\rangle, \qquad (2.189a)$$

and the complex conjugate gives

$$V_{\ell j}^* \equiv (-1)^\ell\, i^\ell e^{i\sigma_\ell^j}\langle(J_f j)J\|V\|J\rangle. \qquad (2.189b)$$

Inserting (2.189) twice into the angular distribution parameter α_K and the spin polarization parameter δ_K, using the C-coefficients (2.140) and remembering the derived relation (2.145) between them, we re-write (2.139) and (2.144) combining both by introducing the parameter $\tilde{\alpha}_K$

$$\tilde{\alpha}_K = \begin{cases} \alpha_K & \text{if } K \text{ even,} \\ -\delta_K & \text{if } K \text{ odd,} \end{cases} \tag{2.190}$$

where

$$\tilde{\alpha}_K = \left(\sum_{\ell j} |V_{\ell j}|^2 \right)^{-1}$$
$$\times \sqrt{(2K+1)(2J+1)} \sum_{\ell \ell' jj'} (-1)^{J_f + J - 1/2 + j + j'} V_{\ell j} V_{\ell' j'}^*$$
$$\times \sqrt{(2j+1)(2j'+1)} \begin{pmatrix} j' & j & K \\ 1/2 & -1/2 & 0 \end{pmatrix} \begin{Bmatrix} j' & j & K \\ J & J & J_f \end{Bmatrix}. \tag{2.191}$$

Somewhat more effort is necessary in combining the two spin polarization parameters ξ_K^{odd} and ξ_K^{even}. Again, inserting (2.189) twice into (2.151), using the H-coefficients (2.152) while inserting the analytic expression of the first $3j$-symbol, the spin polarization parameter ξ_K^{even} can be re-expressed as

$$\xi_K^{even} = \left(\sum_{\ell j} |V_{\ell j}|^2 \right)^{-1}$$
$$\times \sqrt{\frac{(2K+1)(2J+1)}{K(K+1)}} \sum_{\ell \ell' jj'} (-1)^{K + J_f + J + j' - \ell'} V_{\ell j} V_{\ell' j'}^*$$
$$\times \sqrt{(2j+1)(2j'+1)} \begin{pmatrix} j' & j & K \\ 1/2 & 1/2 & -1 \end{pmatrix} \begin{Bmatrix} j' & j & K \\ J & J & J_f \end{Bmatrix}. \tag{2.192}$$

Repeatedly inserting (2.189) twice into (2.150) and using (2.140), we obtain the spin polarization parameter ξ_K^{odd} as

$$\xi_K^{odd} = \left(\sum_{\ell j} |V_{\ell j}|^2 \right)^{-1}$$
$$\times \frac{\sqrt{(2K+1)(2J+1)}}{2K(K+1)} \sum_{\ell \ell' jj'} (-1)^{K + J_f + J + 3/2 + j + j'} V_{\ell j} V_{\ell' j'}^*$$
$$\times \sqrt{(2j+1)(2j'+1)} \begin{pmatrix} j' & j & K \\ 1/2 & -1/2 & 0 \end{pmatrix} \begin{Bmatrix} j' & j & K \\ J & J & J_f \end{Bmatrix}$$
$$\times \left[(-1)^{\ell' + j' + 1/2} (2j'+1) + (-1)^{\ell + j + 1/2} (2j+1) \right]. \tag{2.193}$$

Here, we made use of the fact that

$$(\ell - j) = \frac{1}{2} (-1)^{\ell + j + 1/2}. \tag{2.194}$$

Inspecting the term in the squared brackets of (2.193), it can be re-written as

$$\left[(-1)^{\ell'+j'+1/2}(2j'+1)+(-1)^{\ell+j+1/2}(2j+1)\right] = (-1)^{-K-j-1/2}$$
$$\times \left[(-1)^{\ell+K+2j+1}(2j+1)+(-1)^{K+\ell'+j+j'+1}(2j'+1)\right]. \qquad (2.195)$$

Realizing that $2j+1$ is always even, (2.195) yields

$$[\ldots] = (-1)^{\ell-j-1/2}\left[(2j+1)+(-1)^{\ell'-\ell-K+1}(-1)^{K+j+j'}(2j'+1)\right]. \qquad (2.196)$$

In (2.193) K is restricted to odd values, thus $K+1$ must be even. As the Coulomb interaction conserves parity, ℓ and ℓ' must be both even or odd integers. Thus, the second phase factor in (2.196) is always one. Inserting this into (2.193) yields

$$\xi_K^{odd} = \left(\sum_{\ell j}|V_{\ell j}|^2\right)^{-1}$$

$$\times \frac{\sqrt{(2K+1)(2J+1)}}{2K(K+1)}\sum_{\ell\ell' jj'}(-1)^{K+J_f+J+1+\ell+j'}V_{\ell j}V_{\ell'j'}^*$$

$$\times \sqrt{(2j+1)(2j'+1)}\begin{pmatrix} j' & j & K \\ 1/2 & -1/2 & 0 \end{pmatrix}\begin{Bmatrix} j' & j & K \\ J & J & J_f \end{Bmatrix}$$

$$\times \left[(2j+1)+(-1)^{K+j+j'}(2j'+1)\right]. \qquad (2.197)$$

Eventually, expanding the phase factor of (2.197) by $(-1)^{\ell-\ell'}$, and noting that $(-1)^{K+j+j'} = (-1)^{j+j'-K}$, (A.13) can be applied and we get

$$\xi_K^{odd} = \left(\sum_{\ell j}|V_{\ell j}|^2\right)^{-1}$$

$$\times \sqrt{\frac{(2K+1)(2J+1)}{K(K+1)}}\sum_{\ell\ell' jj'}(-1)^{K+J_f+J+j'-\ell'}V_{\ell j}V_{\ell'j'}^*$$

$$\times \sqrt{(2j+1)(2j'+1)}\begin{pmatrix} j' & j & K \\ 1/2 & 1/2 & -1 \end{pmatrix}\begin{Bmatrix} j' & j & K \\ J & J & J_f \end{Bmatrix}. \qquad (2.198)$$

As has been discussed in Sect. 2.5.2 the coefficients ξ_K^{odd} are real parameters while the coefficients ξ_K^{even} are solely imaginary. Introducing the parameter $\tilde{\xi}_K$ as

$$\tilde{\xi}_K = \begin{cases} i(-1)^P\sqrt{K(K+1)}\,\xi_K^{even}, & K \neq 0, \\ (-1)^{P+1}\sqrt{K(K+1)}\,\xi_K^{odd}, \end{cases} \qquad (2.199)$$

where

$$\tilde{\xi}_K = \left(\sum_{\ell j} |V_{\ell j}|^2 \right)^{-1}$$

$$\times \sqrt{(2K+1)(2J+1)} \sum_{\ell \ell' j j'} (-1)^{J_f + J + j'} V_{\ell j} V_{\ell' j'}^*$$

$$\times \sqrt{(2j+1)(2j'+1)} \begin{pmatrix} j' & j & K \\ 1/2 & 1/2 & -1 \end{pmatrix} \begin{Bmatrix} j' & j & K \\ J & J & J_f \end{Bmatrix}. \qquad (2.200)$$

Here, we have used that $(-1)^{\ell'} = (-1)^P$, where $P = \pi_i - \pi_f$ is the parity of the particular Auger transition which determines the parity of the electronic orbital angular momenta.

Having introduced the generalized parameters $\tilde{\alpha}_K$ and $\tilde{\xi}_K$, let us now consider the sum

$$S_1(M) = \sum_K (-1)^K \sqrt{2K+1} \begin{pmatrix} J & J & K \\ M & -M & 0 \end{pmatrix} N \tilde{\alpha}_K, \qquad (2.201)$$

where we abbreviated $N = \sum_{\ell j} |V_{\ell j}|^2$. Inserting (2.191) yields

$$S_1(M) = \sum_K (-1)^K (2K+1) \begin{pmatrix} J & J & K \\ M & -M & 0 \end{pmatrix}$$

$$\times \sqrt{2J+1} \sum_{\ell \ell' j j'} (-1)^{J_f + J - 1/2 + j + j'} V_{\ell j} V_{\ell' j'}^*$$

$$\times \sqrt{(2j+1)(2j'+1)} \begin{pmatrix} j' & j & K \\ 1/2 & -1/2 & 0 \end{pmatrix} \begin{Bmatrix} j' & j & K \\ J & J & J_f \end{Bmatrix}. \qquad (2.202)$$

Changing the order of summation and using (A.26b) the sum over K can be performed and we obtain

$$S_1(M) = (-1)^{J-M} \sqrt{2J+1} \sum_{\ell \ell' j j'} \sqrt{(2j+1)(2j'+1)} \, V_{\ell j} V_{\ell' j'}^*$$

$$\times \begin{pmatrix} j' & J_f & J \\ 1/2 & M-1/2 & -M \end{pmatrix} \begin{pmatrix} j & J_f & J \\ 1/2 & M-1/2 & -M \end{pmatrix}. \qquad (2.203)$$

Here, we made use of $M_f + \frac{1}{2} = M$. Besides the phase factor $(-1)^{j+j'-1}$, which can be attached to a slightly different abbreviation of the reduced matrix elements, the sum $S_1(M)$ is identical to the one defined by Kabachnik (2005).

In the same manner we consider a second sum

$$S_2(M) = \sum_K (-1)^{K+1} \sqrt{2K+1} \begin{pmatrix} J & J & K \\ M & 1-M & -1 \end{pmatrix} N \tilde{\xi}_K. \qquad (2.204)$$

Inserting (2.200) and using the permutation relations (A.11) of the $3j$-symbols yields

$$S_2(M) = \sum_K (-1)^{K+1}(2K+1) \begin{pmatrix} J & J & K \\ -M & M-1 & 1 \end{pmatrix}$$

$$\times \sqrt{2J+1} \sum_{\ell\ell'jj'} (-1)^{J_f+3J+j+2j'} V_{\ell j} V^*_{\ell'j'}$$

$$\times \sqrt{(2j+1)(2j'+1)} \begin{pmatrix} j & j' & K \\ 1/2 & 1/2 & -1 \end{pmatrix} \begin{Bmatrix} j' & j & K \\ J & J & J_f \end{Bmatrix}. \quad (2.205)$$

Changing the summation order, again applying (A.26b) for carrying out the sum over K, and using $M_f + \frac{1}{2} = M$ yields

$$S_2(M) = (-1)^{J_f-M+1/2}\sqrt{2J+1} \sum_{\ell\ell'jj'} \sqrt{(2j+1)(2j'+1)}\, V_{\ell j} V^*_{\ell'j'}$$

$$\times \begin{pmatrix} j' & J_f & J \\ 1/2 & M-1/2 & -M \end{pmatrix} \begin{pmatrix} j & J_f & J \\ 1/2 & 1/2-M & M-1 \end{pmatrix}. \quad (2.206)$$

The sum $S_2(M)$ is identical to the one by Kabachnik (2005), besides of a phase factor stemming from the reduced matrix elements; see (2.203).

Considering the squared modulus of the second sum $|S_2(M)|^2$, it can be expressed as a sum over four angular and four total angular momenta, $\ell, \ell', \bar{\ell}, \bar{\ell}'$ and $j, j', \bar{j}, \bar{j}'$, respectively. Inspecting the complex conjugate of the first sum and changing its argument from $M \longleftrightarrow 1 - M$ yields

$$S_1^*(1-M) = (-1)^{J+M-1}\sqrt{2J+1} \sum_{\ell\ell'jj'} \sqrt{(2j+1)(2j'+1)}\, V^*_{\ell j} V_{\ell'j'}$$

$$\times \begin{pmatrix} j' & J_f & J \\ 1/2 & 1/2-M & M-1 \end{pmatrix} \begin{pmatrix} j & J_f & J \\ 1/2 & 1/2-M & M-1 \end{pmatrix}. \quad (2.207)$$

Multiplying (2.203) and (2.207), the product $S_1(M)S_1^*(1-M)$ yields a sum over the four angular and four total angular momenta, too. Now, interchanging the notations $j \leftrightarrow \bar{j}$ and $\ell \leftrightarrow \bar{\ell}$ in (2.203), and in (2.207) $j \leftrightarrow \bar{j}'$, $\ell \leftrightarrow \bar{\ell}'$ and $j' \leftrightarrow j$, $\ell' \leftrightarrow \ell$ accordingly, we obtain an important identity between the two summations

$$|S_2(M)|^2 = (-1)^{2J-1}S_1(M)S_1^*(1-M). \quad (2.208)$$

Equation (2.208) states the fact, that the two parameters $\tilde{\alpha}_K$ and $\tilde{\xi}_K$ are not independent. Note, that in the sum $S_2(M)$ all terms with K odd are real, while terms with

K even are purely imaginary. Having this in mind, we may re-substitute (2.201) and (2.204) into (2.208) and eventually obtain the following basic equation

$$
\left(\sum_{K=0}^{K_{max}} (-1)^K \sqrt{2K+1} \begin{pmatrix} J & J & K \\ M & -M & 0 \end{pmatrix} \tilde{\alpha}_K \right)
$$

$$
\times \left(\sum_{K'=0}^{K'_{max}} (-1)^{K'} \sqrt{2K'+1} \begin{pmatrix} J & J & K' \\ 1-M & M-1 & 0 \end{pmatrix} \tilde{\alpha}_{K'} \right) (-1)^{2J-1}
$$

$$
= \left| \sum_{K=1,odd}^{K_{max}} \sqrt{2K+1} \begin{pmatrix} J & J & K \\ M & 1-M & -1 \end{pmatrix} \tilde{\xi}_K \right|^2
$$

$$
+ \left| \sum_{K=2,even}^{K_{max}} \sqrt{2K+1} \begin{pmatrix} J & J & K \\ M & 1-M & -1 \end{pmatrix} \tilde{\xi}_K \right|^2. \tag{2.209}
$$

Note, that (2.209) is quadratic with respect to the angular distribution and spin polarization parameters. It is valid for any *arbitrary* electronic Auger decay $J, \pi \longrightarrow J_f, \pi_f$. The magnetic projection M may be chosen arbitrarily, too. Therefore, (2.209) provides, in fact, a compact representation of several equations which differ by its values of M.

In (2.209) J may be either integer or half-integer, depending on the intermediate excited hole state of the specific atom. In case of a half-integer total angular momentum J of the initial state of the Auger emission an *additional* linear equation can be proved. Consider the two sums $S_1(M)$ and $S_2(M)$, i.e. (2.203) and (2.206), for $M = 1/2$. Their direct comparison yields

$$
S_1(1/2) = (-1)^{J_f - J + 1/2} S_2(1/2). \tag{2.210}
$$

Realizing that in $S_2(\frac{1}{2})$ all terms with K even turn to zero, which can be seen by using (A.13) in (2.204), we obtain by inserting (2.201) and (2.204) in (2.210)

$$
\sum_{K=0}^{K_{max}} (-1)^K \sqrt{2K+1} \begin{pmatrix} J & J & K \\ 1/2 & -1/2 & 0 \end{pmatrix} \tilde{\alpha}_K
$$

$$
= (-1)^{J_f - J + 1/2} \sum_{K=1,odd}^{K_{max}} \sqrt{2K+1} \begin{pmatrix} J & J & K \\ 1/2 & 1/2 & -1 \end{pmatrix} \tilde{\xi}_K. \tag{2.211}
$$

Eventually, applying (2.190) and (2.199) in (2.209) and (2.211) in order to return to the original set of angular distribution and spin polarization parameters we end up with the following two, quadratic and linear, general equations

$$\left(\sum_{\substack{K=0 \\ even}} \sqrt{2K+1} \begin{pmatrix} J & J & K \\ M & -M & 0 \end{pmatrix} \alpha_K \right.$$

$$\left. + \sum_{\substack{K=1 \\ odd}} \sqrt{2K+1} \begin{pmatrix} J & J & K \\ M & -M & 0 \end{pmatrix} \delta_K \right)$$

$$\times \left(\sum_{\substack{K'=0 \\ even}} \sqrt{2K'+1} \begin{pmatrix} J & J & K' \\ 1-M & M-1 & 0 \end{pmatrix} \alpha_{K'} \right.$$

$$\left. + \sum_{\substack{K'=1 \\ odd}} \sqrt{2K'+1} \begin{pmatrix} J & J & K' \\ 1-M & M-1 & 0 \end{pmatrix} \delta_{K'} \right)(-1)^{2J-1}$$

$$= \left| \sum_{\substack{K=1 \\ odd}} \sqrt{K(K+1)(2K+1)} \begin{pmatrix} J & J & K \\ M & 1-M & -1 \end{pmatrix} \xi_K^{odd} \right|^2$$

$$+ \left| \sum_{\substack{K=2 \\ even}} \sqrt{K(K+1)(2K+1)} \begin{pmatrix} J & J & K \\ M & 1-M & -1 \end{pmatrix} \xi_K^{even} \right|^2, \quad (2.212)$$

and

$$\sum_{\substack{K=0 \\ even}} \sqrt{2K+1} \begin{pmatrix} J & J & K \\ 1/2 & -1/2 & 0 \end{pmatrix} \alpha_K + \sum_{\substack{K=1 \\ odd}} \sqrt{2K+1} \begin{pmatrix} J & J & K \\ 1/2 & -1/2 & 0 \end{pmatrix} \delta_K$$

$$= (-1)^{J_f-J+1/2} \sum_{\substack{K=1 \\ odd}} \sqrt{K(K+1)(2K+1)}$$

$$\times \begin{pmatrix} J & J & K \\ 1/2 & 1/2 & -1 \end{pmatrix} (-1)^{P+1} \xi_K^{odd}. \quad (2.213)$$

Equations (2.212) and (2.213) have been derived using the properties of the angular momentum coupling coefficients, only. Thus, they may be considered as *kinematical* relations. No dynamics of the Auger decay is involved besides from the conservation of angular momentum and parity. Therefore, both relations are valid for any arbitrary transition in a quantum system between two states with fixed angular momenta and parities, under emission of an electron, or more generally, any spin 1/2 particle. The derived relations are exact and may be used for checking the consistency of numerical calculations or experimental results.

The quadratic equation (2.212) is independent of the total final state angular momentum J_f and transition parity P. Therefore, it is valid for any J_f and P. The linear equation (2.213) depends on $J_f + P$ in the phase factor, only. There may be several independent quadratic equations corresponding to different values of M. The possible number of quadratic equations is given by J and $J - 1/2$ for integer and half-integer values of J, respectively.

For instance, considering $J = 5/2$ there should be two quadratic equations for $M = 5/2$ and $M = 3/2$, while $M = 1/2$ gives the linear equation (2.213) squared. Negative values of M do not give any new equations.

A number of relations have been derived earlier for particular initial angular momenta J (Schmidtke *et al.* 2000a, 2000b, 2001; Kabachnik and Grum-Grzhimailo 2001; Kabachnik and Sazhina 2002). They can be identified as special cases of the general equations (2.212) and (2.213). For example, if $J = 1/2$ we obtain from (2.213) the relation

$$1 + \delta_1 = (-1)^{J_f + P} 2\, \xi_1, \qquad (2.214)$$

which coincides with that derived by Kabachnik and Grum-Grzhimailo (2001). In this case, the quadratic equation (2.212) does not give an additional relation.

The case $J = 1$ is of particular interest since it includes not only the resonant Auger decay in closed-shell atoms, but also photoionization. Here, we need to remember that from the point of view of a formal angular momentum theory, the photoionization of a closed-shell atom is equivalent to the excitation and decay of the dipole $J = 1$ state in the continuum. Remember that for integer values of J (2.213) cannot be applied. In this case we obtain only one quadratic equation valid for any J_f which is obtained from (2.212) for $J = 1$ as

$$\frac{1}{\sqrt{2}} \left(1 - \sqrt{2}\alpha_2\right) \left(\sqrt{2} + \alpha_2 + \sqrt{3}\delta_1\right) = 3\left(\xi_1\right)^2 + \left(3\xi_2\right)^2. \qquad (2.215)$$

It coincides with the relation given by Kabachnik and Grum-Grzhimailo (2001). Note, that our definition of the parameter ξ_2, e.g. see (2.192), automatically yields the correct result, while Kabachnik and Grum-Grzhimailo (2001) need to re-define their ξ_2 parameter. It is worth noting that (2.215) is equivalent to a relation obtained by Cherepkov for the case of photoionization (Schmidtke *et al.* 2000a).

For the case $J = 3/2$, the linear equation follows directly from (2.213) and can be written as

$$\sqrt{5}\left(1 - \alpha_2\right) + \delta_1 - 3\delta_3 = (-1)^{J_f + P} 4\left(3\xi_3 - \xi_1\right), \qquad (2.216)$$

which coincides with the linear expression given by Kabachnik and Sazhina (2002) who also investigated the $J = 3/2$ case.[3]

From (2.212) we obtain the quadratic equation for $J = 3/2$ and $M = 3/2$ which yields

$$\frac{1}{4}\left(\sqrt{5}[1 + \alpha_2] + 3\delta_1 + \delta_3\right)\left(\sqrt{5}[1 - \alpha_2] - \delta_1 + 3\delta_3\right)$$
$$= 3\left(\xi_1 + 2\xi_3\right)^2 + 15\left(\xi_2\right)^2. \qquad (2.217)$$

Comparing both expressions, we find the left hand sides of (2.216) and the second factor of (2.217) having the same coefficients, besides the different sign for the δ_K parameters.

---

[3] The difference in sign on the r.h.s. of (2.216) stems from a slightly different definition of the ξ_K parameters; see (2.199).

On the other hand, Kabachnik and Sazhina (2002) derived two quadratic expressions for the $J = 3/2$ case. One, depending on α_2, δ_1, δ_3, ξ_1, and ξ_2, but not on ξ_3

$$\left(\delta_1 - 2(-1)^{J_f+P}\xi_1 + \frac{3}{\sqrt{5}}\right)\left(\sqrt{5}[1-\alpha_2] - \delta_1 + 3\delta_3\right)$$
$$= 2\left(1 - \alpha_2 - \sqrt{5}(-1)^{J_f+P}\xi_1\right)^2 + 2\left(2\xi_2\right)^2, \qquad (2.218)$$

while the other depends on α_2, δ_1, and ξ_K, but not on δ_3

$$\left(\delta_1 - 2(-1)^{J_f+P}\xi_1 + \frac{3}{\sqrt{5}}\right)\left(\sqrt{5}[1-\alpha_2] - 2(-1)^{J_f+P}[\xi_1 - 3\xi_3]\right)$$
$$= \left(1 - \alpha_2 - \sqrt{5}(-1)^{J_f+P}\xi_1\right)^2 + \left(2\xi_2\right)^2, \qquad (2.219)$$

whereas (2.217) contains all possible parameters, including both, δ_3 and ξ_3. Equations (2.218) and (2.219) can be transformed into each other by using the linear expression (2.216).

It is worth noting, that the second factors on the left hand side of (2.217) and (2.218) are identical giving rise to the relation

$$\frac{1}{2}\left(\sqrt{5}[1+\alpha_2] + 3\delta_1 + \delta_3\right)\left\{\left(1 - \alpha_2 - \sqrt{5}(-1)^{J_f+P}\xi_1\right)^2 + 4\left(\xi_2\right)^2\right\}$$
$$= \left\{3\left(\xi_1 + 2\xi_3\right)^2 + 15\left(\xi_2\right)^2\right\}\left(\delta_1 - 2(-1)^{J_f+P}\xi_1 + \frac{3}{\sqrt{5}}\right). \qquad (2.220)$$

Note, that (2.220) is always fulfilled, as the other two factors on the left hand side of (2.217) and (2.218) both contain a constant factor.

Though it has been stated by Kabachnik (2005), that either of the two forms (2.218) and (2.219) of the quadratic equation can be derived from (2.217) by substituting the expressions for ξ_3, or δ_3, obtained from the linear equation (2.216), we have not been able to reproduce this result. – Kabachnik and Sazhina (2002) used mathematical software packages for obtaining their results which have not been available to us. Considering the fact that (2.214) for $J = 1/2$ and (2.215) for $J = 1$ as well as the linear expression (2.216) for $J = 3/2$ coincide with the equations given in the literature, leaves us with the possibility that either methods of computational mathematics must be used in order to obtain the results, or that the expressions given by Kabachnik and Sazhina (2002), obtained within a different approach, yield even more restrictions for this specific case of Auger decay. – Beyond realizing that the mathematical knowledge on non-linear multi-variable functions and their interdependence is sparse until today, it shows that Auger emission theory is still an active and interesting area of research. –

As last example, let us consider the case $J = 5/2$ which should give rise to one linear and two quadratic equations, respectively. This case is of interest for future research involving anisotropy and spin polarization parameters of higher rank which might be prominent in coincidence experiments involving Auger transitions or in Auger transitions proceeding from d- or f-holes; see Sect. 4.4 or 4.5. Inserting $J = 5/2$ and $M = 5/2$ into (2.212) we get the first quadratic expression for this case as

$$\left(\sqrt{7} + \frac{1}{\sqrt{2}}\left[5\alpha_2 + \sqrt{3}\alpha_4 \right] + \frac{1}{\sqrt{6}}\left[3\sqrt{10}\delta_1 + \sqrt{35}\delta_3 + 6\delta_5 \right] \right)$$
$$\left(\sqrt{7} - \frac{1}{\sqrt{2}}\left[\alpha_2 - 3\sqrt{3}\alpha_4 \right] - \frac{1}{\sqrt{30}}\left[9\sqrt{2}\delta_1 - 7\sqrt{7}\delta_3 - 5\sqrt{5}\delta_5 \right] \right)$$
$$= 6\left(\sqrt{2}\xi_1 + 2\sqrt{7}\xi_3 + \sqrt{5}\xi_5 \right)^2 + 30\left(\sqrt{3}\xi_2 + 2\xi_4 \right)^2. \quad (2.221)$$

The other quadratic equation is obtained by using $J = 5/2$ and $M = 3/2$ in (2.212) which leaves us with

$$\left(\sqrt{7} - \frac{1}{\sqrt{2}}\left[\alpha_2 - 3\sqrt{3}\alpha_4 \right] + \frac{1}{\sqrt{30}}\left[9\sqrt{2}\delta_1 - 7\sqrt{7}\delta_3 - 5\sqrt{5}\delta_5 \right] \right)$$
$$\left(\sqrt{7} - \sqrt{2}\left[2\alpha_2 - \sqrt{3}\alpha_4 \right] - \frac{1}{\sqrt{15}}\left[3\delta_1 - 2\sqrt{14}\delta_3 - 5\sqrt{10}\delta_5 \right] \right)$$
$$= \frac{2}{5}\left(2\sqrt{2}\xi_1 - \sqrt{7}\xi_3 - 5\sqrt{5}\xi_5 \right)^2 + 2\left(\sqrt{3}\xi_2 - 5\xi_4 \right)^2. \quad (2.222)$$

Both expressions are independent of each other which might be seen from the fact that the second factor of (2.221) and the first factor of (2.222) on the left hand side consist of identical coefficients, besides from the different sign for the δ_K parameters.

Eventually, we obtain the linear expression from (2.213) by inserting $J = 5/2$ as

$$\sqrt{7} - \sqrt{2}\left(2\alpha_2 - \sqrt{3}\alpha_4 \right) + \frac{1}{\sqrt{15}}\left(3\delta_1 - 2\sqrt{14}\delta_3 + 5\sqrt{10}\delta_5 \right)$$
$$= (-1)^{J_f+P}\sqrt{\frac{6}{5}}\left(3\sqrt{2}\xi_1 - 4\sqrt{7}\xi_3 + 10\sqrt{5}\xi_5 \right). \quad (2.223)$$

Again we see, besides the different sign for the parameters δ_K, that the left hand side of (2.223) and the second factor of (2.222) consist of identical coefficients.

Our discussion has shown that general equations valid for arbitrary J and J_f can be proved. These establish relations between the parameters defining the angular distribution and spin polarization of Auger or autoionization electrons. Such type of equations are important for resolving the problem of the number of independent dynamical parameters which we are able to determine from the relevant measured quantities of the emitted electrons for a particular transition.

2.6 Resonant Auger Transitions

Let us continue considering the case of photoexcitation, which is of particular interest for the investigation of the so-called *resonant* Auger transitions. In contrast to

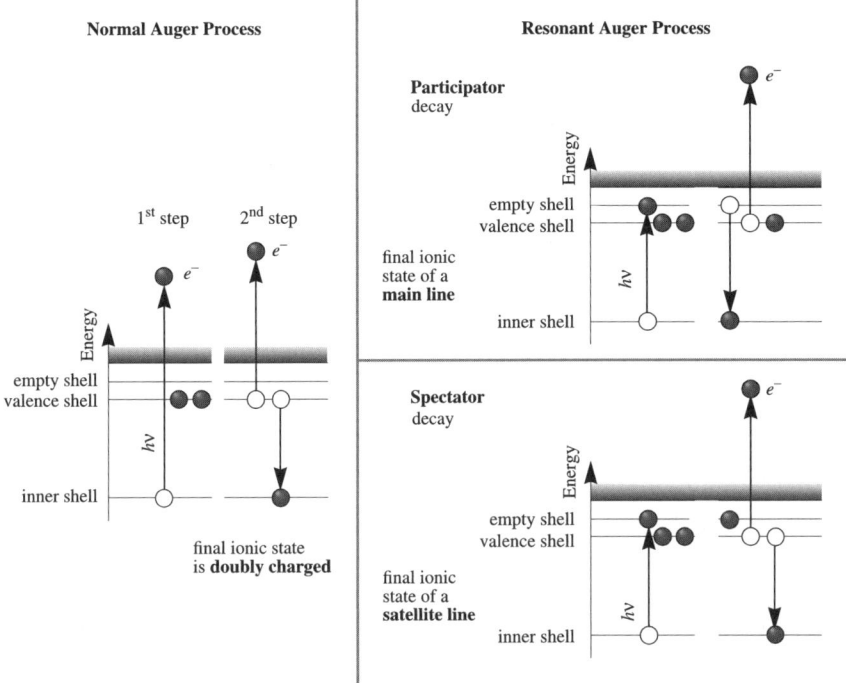

Fig. 2.7. Normal vs. resonant Auger emission; after Lohmann *et al.* (2003b)

the normal Auger decay, here, in a first step an inner shell electron is excited, for instance via synchrotron radiation, to a Rydberg state.

$$\gamma + A \longrightarrow A^* \tag{2.224a}$$
$$\longrightarrow A^{+*} + e_{\text{Auger}}. \tag{2.224b}$$

After some certain lifetime, the excited intermediate state decays via Auger emission, i.e. the inner shell hole is filled with an outer shell electron while the Auger electron is emitted and we remain with a singly ionized excited final state. In most transitions of this type the excited Rydberg electron remains at its place which is known as *spectator* decay. In the opposite case, the Rydberg electron takes part in the Auger emission, the so-called *participator* decay. Adopting the language of normal Auger transitions, the Auger lines stemming from the spectator decay therefore must be denoted as satellite lines, while the main lines originate from the participator Auger decay. The resonant Auger emission is illustrated in Fig. 2.7 versus the normal Auger transition.

2.6.1 General Equations for Arbitrarily Oriented Coordinate Frames

From the discussion in Sect. 2.3.2 we learnt that the anisotropy parameter of photo-excitation B_{exci} is independent of its tensorial rank. This leads to simple connections

Table 2.9. Connection between the alignment and orientation tensors of the photoexcited atom and the Stokes parameters

Atomic tensors	Stokes parameters
$\mathcal{O}_{10} = \sqrt{\frac{3}{2}}\eta_2$	
$\mathcal{A}_{20} = \frac{1}{\sqrt{2}}$	
$Re\,\mathcal{A}_{22} = -\frac{\sqrt{3}}{2}\eta_3$	
$Im\,\mathcal{A}_{22} = \frac{\sqrt{3}}{2}\eta_1$	

between the orientation and alignment tensors, and the Stokes parameters describing the polarization state of the synchrotron beam, respectively. For the case of photoionization this has been investigated by Kleiman *et al.* (1999a). The results for photoexcitation are given in Table 2.9.

Now, supposing the laboratory (collision) frame and the helicity frame as arbitrarily oriented to each other, as has been outlined for the general case of photoionization/excitation in Sect. 2.5.6, and inserting the relations of Table 2.9 into (2.184)–(2.188) we obtain for the angular distribution of a resonant Auger transition

$$I(\theta,\phi)^\gamma = \frac{I_0}{4\pi}\left\{1 + \frac{\alpha_2}{\sqrt{2}}\left(P_2(\cos\theta) - \frac{3}{2}[\eta_3\cos 2\phi + \eta_1\sin 2\phi]\sin^2\theta\right)\right\}. \quad (2.225)$$

Inserting the relevant expressions for the abbreviation N^Γ, see (2.185), the Cartesian components of the spin polarization vector can be expressed in terms of the Stokes parameters as

$$p_x(\theta,\phi)^\gamma = \frac{\left(\sqrt{3}\xi_1\eta_2 - 3\xi_2\left[\eta_3\sin 2\phi - \eta_1\cos 2\phi\right]\right)\sin\theta}{\sqrt{2} + \alpha_2\left(P_2(\cos\theta) - \frac{3}{2}[\eta_3\cos 2\phi + \eta_1\sin 2\phi]\sin^2\theta\right)}, \quad (2.226)$$

$$p_y(\theta,\phi)^\gamma = \frac{-\frac{3}{2}\xi_2\left(1 + [\eta_3\cos 2\phi + \eta_1\sin 2\phi]\right)\sin 2\theta}{\sqrt{2} + \alpha_2\left(P_2(\cos\theta) - \frac{3}{2}[\eta_3\cos 2\phi + \eta_1\sin 2\phi]\sin^2\theta\right)}, \quad (2.227)$$

and

$$p_z(\theta,\phi)^\gamma = \frac{\sqrt{3}\delta_1\eta_2\cos\theta}{\sqrt{2} + \alpha_2\left(P_2(\cos\theta) - \frac{3}{2}[\eta_3\cos 2\phi + \eta_1\sin 2\phi]\sin^2\theta\right)}. \quad (2.228)$$

The above equations are the general results for the angular distribution and spin polarization of an Auger emission after photoexcitation with an arbitrarily polarized photon beam while relating the helicity frame, i.e. the direction of Auger emission, and the collision frame, i.e. the direction of the exciting photon beam, to each other.

2.6.2 Simplification of Angular Distribution and Spin Polarization for $\phi = 0$ Geometry

In case that the experimental set-up allows for observing the emitted Auger electrons in the X–Z plane of the laboratory frame, i.e. the Y-axis coincides with the y-axis of the helicity frame, the azimuthal angle ϕ equals zero. From a theoretical point of view, such a geometry can always be achieved without limiting generality, see Sect. 2.1.3 and Fig. 2.3, and will be used in most parts of the book.

Inserting $\phi = 0$ into (2.225)–(2.228), or alternatively inserting the results of Table 2.9 into (2.177)–(2.181) we get for the angular distribution

$$I(\theta)^{\gamma} = \frac{I_0}{4\pi} \left(1 + \frac{\alpha_2}{\sqrt{2}} \left[P_2(\cos\theta) - \eta_3 \frac{3}{2} \sin^2\theta \right] \right). \tag{2.229}$$

Inserting for N^{γ}, see (2.178), the Cartesian components of the spin polarization vector can be contracted to

$$p_x(\theta)^{\gamma} = \frac{\sqrt{3}\left(\xi_1\,\eta_2 + \sqrt{3}\,\xi_2\,\eta_1\right)\sin\theta}{\sqrt{2} + \alpha_2\left(P_2(\cos\theta) - \eta_3\frac{3}{2}\sin^2\theta\right)}, \tag{2.230}$$

$$p_y(\theta)^{\gamma} = \frac{-\frac{3}{2}\xi_2\,(1+\eta_3)\sin 2\theta}{\sqrt{2} + \alpha_2\left(P_2(\cos\theta) - \eta_3\frac{3}{2}\sin^2\theta\right)}, \tag{2.231}$$

and

$$p_z(\theta)^{\gamma} = \frac{\sqrt{3}\,\delta_1\,\eta_2\cos\theta}{\sqrt{2} + \alpha_2\left(P_2(\cos\theta) - \eta_3\frac{3}{2}\sin^2\theta\right)}. \tag{2.232}$$

Either (2.225)–(2.228) or (2.229)–(2.232) must be seen as general results for the angular distribution and spin polarization of a resonant Auger emission photoexcited by arbitrarily polarized light. While the latter set of equations allows for a compact and easy theoretical treatment the first set might be used instead in order to obey the laboratory limitations given by the particular experiment with respect to the detection angles of the Auger electrons. In order to elucidate the general equations let us consider some special cases of photon polarization.

2.6.3 Special Cases: Unpolarized Photon Beam

If a completely unpolarized photon beam is used, i.e. $\eta_i = 0$, $i = 1, 2, 3$, we get

$$I(\theta)^{\gamma} = I(\theta, \phi)^{\gamma} = \frac{I_0}{4\pi}\left(1 + \frac{\alpha_2}{\sqrt{2}}\,P_2(\cos\theta)\right), \tag{2.233}$$

for the angular distribution, and for the Cartesian components of the spin polarization vector we remain with

$$p_x(\theta)^{\gamma} = p_x(\theta, \phi)^{\gamma} = 0, \tag{2.234}$$

$$p_y(\theta)^\gamma = p_y(\theta, \phi)^\gamma = \frac{-\frac{3}{2}\xi_2 \sin 2\theta}{\sqrt{2} + \alpha_2 P_2(\cos\theta)}, \tag{2.235}$$

and

$$p_z(\theta)^\gamma = p_z(\theta, \phi)^\gamma = 0. \tag{2.236}$$

For this case the component p_y of the spin polarization vector survives, only. The angular distribution and the spin polarization vector component p_y are independent of the azimuthal angle ϕ, irrespective of the specific orientation of the laboratory to the helicity frame. This is, because the combined photonic and target system, where the latter has been assumed as unpolarized, is axially symmetric with respect to the incoming photon beam axis.

2.6.4 Special Cases: Circularly Polarized Photon Beam

Now, assuming a fully circularly polarized photon beam, $\eta_2 = \pm 1$, we obtain the angular distribution as

$$I(\theta)^\gamma = I(\theta, \phi)^\gamma = \frac{I_0}{4\pi}\left(1 + \frac{\alpha_2}{\sqrt{2}} P_2(\cos\theta)\right). \tag{2.237}$$

The Cartesian components of the spin polarization vector can be expressed as

$$p_x(\theta)^\gamma = p_x(\theta, \phi)^\gamma = \frac{\pm\sqrt{3}\,\xi_1 \sin\theta}{\sqrt{2} + \alpha_2 P_2(\cos\theta)}, \tag{2.238}$$

$$p_y(\theta)^\gamma = p_y(\theta, \phi)^\gamma = \frac{-\frac{3}{2}\xi_2 \sin 2\theta}{\sqrt{2} + \alpha_2 P_2(\cos\theta)}, \tag{2.239}$$

and

$$p_z(\theta)^\gamma = p_z(\theta, \phi)^\gamma = \frac{\pm\sqrt{3}\,\delta_1 \cos\theta}{\sqrt{2} + \alpha_2 P_2(\cos\theta)}, \tag{2.240}$$

where the $\pm$ sign refers to right and left-handed circularly polarized light, respectively. Thus, for the case of a fully circularly polarized photon beam the angular distribution, as well as the spin polarization vector, become independent of the azimuthal angle ϕ due to the axial symmetry with respect to the photon beam axis as has been discussed for the case of a completely unpolarized photon beam. Note, that (2.237) and (2.239) are identical to the expressions derived for the case of an unpolarized photon beam, i.e. (2.235) and (2.233), respectively. This is, because both are independent of the Stokes parameter η_2 describing the degree of circular polarization.

2.6.5 Special Cases: Linearly Polarized Photon Beam

As last case, let us consider a 100% linearly polarized synchrotron beam. The linear polarization is described by the two Stokes parameters η_3 and η_1. In contrast to the previous two cases, a linear polarization of the photon beam defines an additional axis perpendicular to the photon beam axis which destroys the axial symmetry. Thus, we generally expect the spin polarization vector and the angular distribution to be functions of the polar and azimuthal angles θ and ϕ, respectively. Therefore, we have to discuss the two geometries, separately. We will focus on specific cases of interest.

Linearly Polarized Beam, $\eta_1 \neq 0$

Considering a fully linearly polarized beam with $\eta_1 = \pm 1$, i.e. its electric field vector is oriented at an angle $\chi = 45^o$ (135°) with respect to the reaction plane. For arbitrarily oriented helicity and laboratory coordinate frames we get the angular distribution from (2.225) as

$$I(\theta, \phi)^\gamma = \frac{I_0}{4\pi} \left\{ 1 + \frac{\alpha_2}{\sqrt{2}} \left(P_2(\cos\theta) \mp \frac{3}{2} \sin 2\phi \sin^2\theta \right) \right\}. \qquad (2.241)$$

For the geometry with coinciding Y- and y-axes (see Fig. 2.3) we obtain a more simpler expression

$$I(\theta)^\gamma = \frac{I_0}{4\pi} \left(1 + \frac{\alpha_2}{\sqrt{2}} P_2(\cos\theta) \right). \qquad (2.242)$$

This is due to the fact that for the latter case the reaction plane coincides with the X–Z plane of the laboratory frame, i.e. $\phi = 0$. Following the discussion in Sect 2.1.3, the electric field vector can always be expressed as a linear combination of two components. One perpendicular to and one oscillating in the reaction plane which therefore becomes a symmetry plane; see Fig. 2.6. For $\eta_1 = \pm 1$ both projections of the oscillating electric field vector onto the axes lying perpendicular to and in the reaction plane must be of equal length. Therefore, both, (2.229) and (2.242) are independent of ϕ. For the general case (2.241) the X–Z plane of the laboratory frame and the reaction plane no longer coincide which allows for different lengths of the two projections, which eventually destroys the symmetry. Note, that (2.242) is identical to the angular distribution obtained for the circularly and unpolarized cases, respectively.

For the x- and y-components of the spin polarization vector we get for the general case

$$p_x(\theta, \phi)^\gamma = \frac{\pm 3\xi_2 \cos 2\phi \sin\theta}{\sqrt{2} + \alpha_2 \left(P_2(\cos\theta) \mp \frac{3}{2} \sin 2\phi \sin^2\theta \right)}, \qquad (2.243)$$

and

$$p_y(\theta, \phi)^\gamma = \frac{-\frac{3}{2}\xi_2(1 \pm \sin 2\phi)\sin 2\theta}{\sqrt{2} + \alpha_2 \left(P_2(\cos\theta) \mp \frac{3}{2}\sin 2\phi \sin^2\theta\right)}, \qquad (2.244)$$

which is reduced for $\phi = 0$ to

$$p_x(\theta)^\gamma = \frac{\pm 3\,\xi_2\,\sin\theta}{\sqrt{2} + \alpha_2\,P_2(\cos\theta)}, \qquad (2.245)$$

and

$$p_y(\theta)^\gamma = \frac{-\frac{3}{2}\xi_2\,\sin 2\theta}{\sqrt{2} + \alpha_2\,P_2(\cos\theta)}. \qquad (2.246)$$

Note, that (2.246) coincides with the results for the unpolarized (2.235) and circularly polarized case (2.239), respectively.

As the numerator of neither (2.228) nor (2.232) depends on η_1 the z-component of the spin polarization vector vanishes

$$p_z(\theta)^\gamma = p_z(\theta, \phi)^\gamma = 0, \qquad (2.247)$$

because it can be non-zero for a circularly polarized photon beam, $\eta_2 \neq 0$, only.

Linearly Polarized Beam, $\eta_3 \neq 0$

Supposing a linearly polarized photon beam for the other case, i.e. the Stokes parameter $\eta_3 = \pm 1$, we obtain the angular distribution for the general geometry as

$$I(\theta, \phi)^\gamma = \frac{I_0}{4\pi}\left\{1 + \frac{\alpha_2}{\sqrt{2}}\left(P_2(\cos\theta) \mp \frac{3}{2}\cos 2\phi \sin^2\theta\right)\right\}, \qquad (2.248)$$

while we get for the x- and y-component of the spin polarization vector

$$p_x(\theta, \phi)^\gamma = \frac{\mp 3\xi_2 \sin 2\phi \sin\theta}{\sqrt{2} + \alpha_2\left(P_2(\cos\theta) \mp \frac{3}{2}\cos 2\phi \sin^2\theta\right)}, \qquad (2.249)$$

and

$$p_y(\theta, \phi)^\gamma = \frac{-\frac{3}{2}\xi_2(1 \pm \cos 2\phi)\sin 2\theta}{\sqrt{2} + \alpha_2\left(P_2(\cos\theta) \mp \frac{3}{2}\cos 2\phi \sin^2\theta\right)}. \qquad (2.250)$$

Irrespective of the specific choice of the coordinate frame the z-component of the spin polarization vector vanishes as its numerator depends on η_2, only.

$$p_z(\theta, \phi)^\gamma = 0. \qquad (2.251)$$

For the Stokes parameter η_3 it is of interest to consider two cases separately.

Beam Polarization $\eta_3 = 1$

If $\eta_3 = 1$, an interesting symmetry relation can be derived between (2.249) and (2.250). Considering $p_x(\theta, \phi)^\gamma$ for the polar angle $\theta = 90^o$ we get

$$p_x\left(\theta = 90^o, \phi\right)^\gamma = \frac{-6\xi_2 \sin 2\phi}{2\sqrt{2} - \alpha_2(1 + 3\cos 2\phi)}. \tag{2.252}$$

On the other hand, considering $p_y(\theta, \phi)^\gamma$ for the azimuthal angle $\phi = 0^o$ we obtain

$$p_y\left(\theta, \phi = 0^o\right)^\gamma = \frac{-3\xi_2 \sin 2\theta}{\sqrt{2} + \alpha_2\left(P_2(\cos\theta) - \frac{3}{2}\sin^2\theta\right)}, \tag{2.253}$$

which can be re-written as

$$p_y\left(\theta, \phi = 0^o\right)^\gamma = \frac{-6\xi_2 \sin 2\theta}{2\sqrt{2} - \alpha_2(1 - 3\cos 2\theta)}. \tag{2.254}$$

Comparing (2.252) and (2.254) and interchanging θ and ϕ in either of the two equations, we find both expressions to be equal, provided the cosine vanishes. This is only possible if the following relation is fulfilled

$$p_x\left(\theta = 90^o, \phi = 45^o + n\,90^o\right)^\gamma = p_y\left(\theta = 45^o + n\,90^o, \phi = 0^o\right)^\gamma, \tag{2.255}$$

where $n = 0, 1, 2, \ldots, n \in \mathbb{N}$.

Now, considering $\eta_3 = 1$ for our usual geometry (Fig. 2.3), the electric field vector oscillates in the reaction plane, which coincides with the X–Z plane of the laboratory frame. Therefore, (2.248)–(2.250) can be further reduced due to the higher symmetry which yields for the angular distribution

$$I(\theta)^\gamma = \frac{I_0}{4\pi}\left(1 + \frac{\alpha_2}{\sqrt{2}}\,[2P_2(\cos\theta) - 1]\right), \tag{2.256}$$

and for the Cartesian components of the spin polarization vector we find the x-component vanishing

$$p_x(\theta)^\gamma = 0, \tag{2.257}$$

while the y-component yields

$$p_y(\theta)^\gamma = \frac{-3\,\xi_2\,\sin 2\theta}{\sqrt{2} + \alpha_2\,(2P_2(\cos\theta) - 1)}, \tag{2.258}$$

and the z-component equals zero

$$p_z(\theta)^\gamma = 0, \tag{2.259}$$

which is clear from (2.251).

Beam Polarization $\eta_3 = -1$

For the opposite case $\eta_3 = -1$, a similar relation to (2.255) can be derived. Inserting the polar angle $\theta = 90^o$ into (2.249) yields

$$p_x \left(\theta = 90^o, \phi\right)^\gamma = \frac{6\xi_2 \sin 2\phi}{2\sqrt{2} + \alpha_2(3 \cos 2\phi - 1)}, \tag{2.260}$$

and inserting $\phi = 90^o$ into (2.250) for the azimuthal angle we get

$$p_y \left(\theta, \phi = 90^o\right)^\gamma = \frac{-6\xi_2 \sin 2\theta}{2\sqrt{2} + \alpha_2(3 \cos 2\theta - 1)}. \tag{2.261}$$

Hence, comparing (2.260) and (2.261), and interchanging ϕ and θ, both expressions show the same angular dependence but of different sign which yields the relation.

$$p_x \left(\theta = 90^o, \phi\right)^\gamma = -p_y \left(\theta, \phi = 90^o\right)^\gamma. \tag{2.262}$$

For the case of $\phi = 0^o$ geometry, we obtain an interesting and at first sight surprising result (Kleiman *et al.* 1999a, 1999b). Here, the electric field vector oscillates perpendicular to the reaction plane. As can be directly seen from the general equations (2.229)–(2.232) the angular distribution becomes isotropic within this plane

$$I(\theta)^\gamma = \frac{I_0}{4\pi} \left(1 + \frac{\alpha_2}{\sqrt{2}}\right), \tag{2.263}$$

and all components of the spin polarization vector vanish

$$p_x(\theta)^\gamma = p_y(\theta)^\gamma = p_z(\theta)^\gamma = 0. \tag{2.264}$$

Thus, the emitted Auger electrons are unpolarized and exhibit an isotropic angular distribution. This can be seen as a direct outcome of the dipole approximation. Here, any information about the origin of the incident photons vanish. Therefore, we only have to consider the axis of the electric field vector so that the system is invariant under reflection within all three planes, the x–y, the y–z and the x–z plane. Hence, all components of the spin polarization vector vanish. This is because the spin polarization vector transforms as an axial vector. Further, the system shows an axial symmetry along the y-axis. Thus, the angular distribution in the x–z plane is independent of the polar angle θ.

The effect of isotropically emitted unpolarized Auger electrons, independent of the intermediate alignment state or the dynamic of the Auger decay, is of experimental interest, too, since it can be utilized for the generation of monoenergetic unpolarized electron beams. Usually, unpolarized electron beam are produced via the emission of unpolarized photoelectrons. A process which has been utilized since years (Kleinpoppen 1997; Schmidt 1997). However, photoemission shows the disadvantage that an incomplete monochromatized photon beam reveals fluctuations in the kinetic energy of the emitted photoelectrons. A problem not present for the Auger emission, as the Auger energy is independent of the generation of the primary hole but solely depends on the specific Auger transition. However, energy conservation does not allow for accessing any electron energy experimentally required.

2.7 Special Cases of Auger Transitions

We will now discuss some special cases of interest. In particular, the different behaviour of angular distribution and spin polarization of the Auger electrons depending on the total angular momentum J of the singly ionized intermediate state, generated after the primary excitation process is considered. Here, the most simple case is that of a non-zero spin polarization for isotropic Auger multiplets. For a primary electron impact ionization the number of independent parameters is only limited by the total angular momentum J of the excited intermediate state. Thus, for a large angular momentum parameters of higher order occur. These are not accessible in experiments using synchrotron radiation for the primary ionization since the dipole approximation yields general restrictions (see Sect. 2.5.5). Three examples are discussed. As another point of physical importance, we investigate the angular distribution of Auger electrons emitted from an unresolved intermediate fine structure state. For the case of an electronic excitation or ionization, we introduce asymmetry parameters which can reveal additional information on the Auger emission not accessible in photoionization or excitation experiments. As last example, we consider linear dichroism of Auger electrons emitted from unpolarized targets.

2.7.1 Spin Polarization of Isotropic Auger Multiplets

The simple most non-trivial case is that of an excited intermediate atomic or ionic state with a total angular momentum $J = 1/2$. This type of Auger transition occurs, for instance, in the case of the rare gases where an inner shell $np_{1/2}$ hole is generated in the intermediate ionic state. For the case of a primary photoionization this has been extensively discussed for the Ar L_2MM Auger transitions by Lohmann and Larkins (1994); also see the discussion in Sect. 4.3. For such an intermediate state neither alignment nor state multipoles of rank higher than one can be generated. I.e.

$$\mathcal{A}_{KQ} = 0 \quad \text{and} \quad \mathcal{O}_{KQ} = 0 \quad \text{for} \quad K > 1. \tag{2.265}$$

Thus, the angular distribution becomes isotropic for both, a primary electronic or photonic excitation and ionization, respectively.

$$I(\theta)^e = \frac{I_0^e}{4\pi} \quad \text{and} \quad I(\theta)^\gamma = \frac{I_0^\gamma}{4\pi}. \tag{2.266}$$

The x-component of the spin polarization vector yields

$$p_x(\theta)^e = \xi_1 \left(\mathcal{O}_{10} \sin\theta + \sqrt{2}\, Re\, \mathcal{O}_{11} \cos\theta \right) \tag{2.267}$$

for an electron impact ionization or excitation and

$$p_x(\theta)^\gamma = \xi_1 \mathcal{O}_{10} \sin\theta, \tag{2.268}$$

for a primary photoionization or excitation. In both cases, the Auger decay dynamics is solely given by the spin polarization parameter ξ_1. For the case of photoionization the intrinsically transferred orientation is sufficiently described by one parameter $\mathcal{O}_{10}$ whereas for an electronic ionization the additional parameter $Re\,\mathcal{O}_{11}$ occurs.

Analogously, we obtain the y-component for the electronic case as

$$p_y(\theta)^e = \sqrt{2}\,\xi_1\,Im\,\mathcal{O}_{11}. \tag{2.269}$$

I.e. the y-component becomes independent of the angle of Auger emission and is directly proportional to the intrinsic spin polarization parameter ξ_1 and the orientation parameter $Im\,\mathcal{O}_{11}$.

For the photoionization case the y-component of the spin polarization vector vanishes

$$p_y(\theta)^\gamma = 0. \tag{2.270}$$

Eventually, we obtain for the z-component of the spin polarization vector

$$p_z(\theta)^e = \delta_1 \left(\mathcal{O}_{10}\cos\theta - \sqrt{2}\,Re\,\mathcal{O}_{11}\sin\theta\right), \tag{2.271}$$

and

$$p_z(\theta)^\gamma = \delta_1\,\mathcal{O}_{10}\cos\theta. \tag{2.272}$$

This yields the counterparts to (2.267) and (2.268). The dependency on the Auger decay dynamics is given by the parameter δ_1 whereas, for the electronic case, the intrinsic orientation is given again by the two parameters $\mathcal{O}_{10}$ and $Re\,\mathcal{O}_{11}$, and for a primary photoionization solely by the orientation parameter $\mathcal{O}_{10}$.

2.7.2 Intermediate Ionic Hole State with $J = 1$

Though there are only few investigations for the case of electron impact, for a primary photoexcitation this case is most important for the application of polarized synchrotron beam techniques. It has been often applied and investigated both, experimentally and theoretically, by several groups (e.g. Becker 1990a; Hergenhahn *et al.* 1991; Lohmann *et al.* 1993; Tulkki *et al.* 1993). Of particular interest are again the rare gases due to their $^1S_0$ ground state configuration. Here, only $J = 1$ states can be generated via photoexcitation which is a direct result of the dipole approximation. However, the dipole approximation is a general restriction in the photonic case and has been already included in our previous discussion. Thus, the general equations of angular distribution and spin polarization remain unchanged for this case.

For the electronic case, however, we get an additional restriction

$$K \leq 2, \tag{2.273}$$

i.e. only the usual components of alignment $\mathcal{A}_{2Q}$ and orientation $\mathcal{O}_{1Q}$ can be expected to be non-vanishing if an electronic excitation to an intermediate state with $J = 1$ is assumed. Thus, we obtain for the angular distribution

$$I(\theta)^e_{J=1} = \frac{I_0}{4\pi}\left(1 + \alpha_2\left[\mathcal{A}_{20}\,P_2(\cos\theta) - \sqrt{\frac{3}{2}}\,Re\,\mathcal{A}_{21}\,\sin(2\theta)\right]\right). \tag{2.274}$$

The components of the spin polarization vector can be reduced to

$$p_x(\theta)^e_{J=1} = \frac{1}{N^e_{J=1}} \left(\xi_1 \, \mathcal{O}_{10} \, \sin\theta + \sqrt{2} \left[\xi_1 \, Re \, \mathcal{O}_{11} - \sqrt{3} \, \xi_2 \, Im \, \mathcal{A}_{21} \right] \cos\theta \right),$$

(2.275)

where the parameter $N^e_{J=1}$ has been abbreviated in analogy to (2.165). For the y-component we get

$$p_y(\theta)^e_{J=1} = \frac{-1}{N^e_{J=1}} \left(\sqrt{2} \, \xi_1 \, Im \, \mathcal{O}_{11} + \frac{3}{2} \, \xi_2 \left[\mathcal{A}_{20} \, \sin(2\theta) \right. \right.$$
$$\left. \left. + \sqrt{\frac{8}{3}} \, Re \, \mathcal{A}_{21} \, \cos(2\theta) \right] \right),$$

(2.276)

and for the z-component we obtain

$$p_z(\theta)^e_{J=1} = \frac{1}{N^e_{J=1}} \, \delta_1 \left(\mathcal{O}_{10} \, \cos\theta - \sqrt{2} \, Re \, \mathcal{O}_{11} \, \sin\theta \right).$$

(2.277)

2.7.3 Intermediate Ionic Hole State with $J = 3/2$

While in this case the equations of angular distribution and spin polarization for the photonic excitation remain unchanged, it is worth to further consider excitation/ionization by electronic or ionic impact. Though the latter is not within the scope of this investigation, the general equations include the case of a primary excitation with ions. Hence, the following results do apply for such type of experiments, too.

Where the usual parameters of orientation and alignment are commonly referred to the first and second rank tensors $\mathcal{O}_{1Q}$ and $\mathcal{A}_{2Q}$ a new orientation parameter of third rank can occur if a primary hole with $J = 3/2$ is generated. This is physically important since these parameters are not accessible in ionization processes with photons due to the dipole approximation (see Sect. 2.5.5).

The general expression for the angular distribution depends on even rank tensors, only. Thus, it remains unchanged and no additional information can be gained from angle resolved experiments.

The expressions for the spin polarization vector show however a dependency on the third rank tensor. Using the results of the previous section we get additional terms

$$p_x(\theta)^e_{J=3/2} = p_x(\theta)^e_{J=1} + \frac{\sqrt{3}}{2N^e_{J=1}} \, \xi_3 \left(\sqrt{3} \mathcal{O}_{30} \, \sin\theta \left(5 \cos^2\theta - 1 \right) \right.$$
$$\left. + Re \, \mathcal{O}_{31} \, \cos\theta \left(15 \cos^2\theta - 11 \right) \right),$$

(2.278)

$$p_y(\theta)^e_{J=3/2} = p_y(\theta)^e_{J=1} + \frac{\sqrt{3}}{2N^e_{J=1}} \, \xi_3 \, Im \, \mathcal{O}_{31} \left(1 - 5 \cos^2\theta \right),$$

(2.279)

and

$$p_z(\theta)^e_{J=3/2} = p_z(\theta)^e_{J=1} + \frac{1}{2N^e_{J=1}} \, \delta_3 \left(\mathcal{O}_{30} \, \cos\theta \left(5 \cos^2\theta - 3 \right) \right.$$
$$\left. - \sqrt{3} Re \, \mathcal{O}_{31} \, \sin\theta \left(5 \cos^2\theta - 1 \right) \right).$$

(2.280)

The third rank tensors might be seen as the higher order analogon to the magnetic dipole vector which is directly related to the orientation tensors of rank one, and are functions of the expectation values of different combinations of angular momentum operators. Explicit expressions for the reduced rotation matrices of higher rank may be found in the literature (Buckmaster 1964, 1966; Ying-Nan 1966).

2.7.4 The Case of an Unresolved Resonance

As another point of discussion we consider the angular distribution of Auger electrons for the case of an unresolved resonance. The fine structure of the resonantly excited or singly ionized intermediate state can often be hardly resolved. For some Auger transitions this is due to the fact that the fine structure splitting of the states is smaller than their natural line widths. Therefore, a coherent excitation takes place. Such effects have been first investigated by Mehlhorn and Taulbjerg (1980) and later by Kabachnik *et al.* (1994).

Supposing a spin independent primary ionization process and the target atoms randomly oriented then, for an unresolved multiplet $M(LS)$ of the singly ionized intermediate state, the angular distribution of the Auger electrons may be written as (Kabachnik *et al.* 1994),

$$I_{M \to J_f}(\theta) = \frac{I_0^{M \to J_f}}{4\pi} \left(1 + \alpha_2^{M \to J_f} \mathcal{A}_{20}(L) P_2(\cos\theta)\right). \qquad (2.281)$$

Though the anisotropy parameter can still be factorized, the alignment $\mathcal{A}_{20}(L)$ now depends on the angular momentum L only, and the anisotropy coefficient $\alpha_2^{M \to J_f}$ can be written as a coherent sum over the total angular momentum J

$$\alpha_2^{M \to J_f} = \frac{1}{N} \sum_{JJ'} A_2(J, J') \, (-1)^{J+L+S} \frac{\Gamma_{JJ'}}{\omega_{JJ'}^2 + \Gamma_{JJ'}^2}$$

$$\times \frac{\sqrt{(2J+1)(2J'+1)(2L+1)}}{2S+1} \begin{Bmatrix} J & L & S \\ L & J' & 2 \end{Bmatrix}, \qquad (2.282)$$

where $\omega_{JJ'} = E_J - E_{J'}$ is the energy splitting, $\Gamma_{JJ'} = (\Gamma_J + \Gamma_J')/2$, and Γ_J is the total width of the level J, and the absolute parameter $A_2(J, J')$ is given by

$$A_2(J, J') = (-1)^{J_f + J - 1/2} \sqrt{(2J+1)(2J'+1)}$$

$$\times \sum_{\ell\ell'jj'} \langle J_f \| V \| (Jj) J_f \rangle \langle J_f \| V \| (J'j') J_f \rangle^*$$

$$\times \sqrt{(2j+1)(2j'+1)(2\ell+1)(2\ell'+1)}$$

$$\times (\ell 0, \ell' 0 | 2 0) \begin{Bmatrix} j' & j & 2 \\ J & J' & J_f \end{Bmatrix} \begin{Bmatrix} j & \ell & 1/2 \\ \ell' & j' & 2 \end{Bmatrix}. \qquad (2.283)$$

The relative intensity is proportional to N

$$N = \sum_J \frac{2J+1}{(2L+1)(2S+1)} \frac{\Gamma_{J \to J_f}}{\Gamma_J}, \qquad (2.284)$$

where

$$\Gamma_{J \to J_f} = 2\pi \sum_{\ell j} |\langle J_f \| V \| (Jj) J_f \rangle|^2 \tag{2.285}$$

is the partial width. Note, that our expressions (2.282) and (2.284) differ from those by Kabachnik *et al.* (1994) by a factor of $\sqrt{2L+1}$. This is, because otherwise one would obtain different relative intensities for a singlet state by using either (2.132) or (2.284).

Eventually we note, that an intermediate singlet state can be generally used as a test case. Inserting $S = 0$ into (2.282)–(2.284), the $6j$-symbol yields $J = J' = L$ and after some short calculation one ends up with (2.139) for the case $K = 2$, i.e. the common α_2 parameter.

2.7.5 Asymmetry Parameters for Auger Emission After Electron Impact

As has been seen from the previous discussion, the experimental conditions may yield symmetry relations with respect to the angular dependence of the angular distribution and spin polarization parameters, respectively. In order to obtain quantities which allow for an easy interpretation of the experimental data it is therefore often useful to investigate the left–right asymmetry of the relevant parameters. A general asymmetry parameter can be defined as

$$A(\theta) = \frac{Q(\theta) - Q(-\theta)}{Q(\theta) + Q(-\theta)}, \tag{2.286}$$

where Q is an observable which depends on the emission angle θ. Such an asymmetry parameter is related to the symmetries of the reduced rotation matrices under the operation $\theta \longrightarrow -\theta$; see Appendix B.6.

For the case of electron impact we define the left–right asymmetry parameter A_s as a function of the angle dependent intensity of Auger emission as

$$A_s(\theta) = \frac{I(\theta)^e - I(-\theta)^e}{I(\theta)^e + I(-\theta)^e}. \tag{2.287}$$

Inserting the related equation of angular distribution (2.164) we obtain

$$A_s(\theta) = \frac{1}{N^{A_s}} \sum_{\substack{K \geq 2 \\ even}} \frac{2}{\sqrt{K(K+1)}} \alpha_K \, Re \, \mathcal{A}_{K1} \, P_K^1(\cos\theta), \tag{2.288}$$

where the parameter N^{A_s} has been defined as

$$N^{A_s} = 1 + \sum_{\substack{K \geq 2 \\ even}} \alpha_K \, \mathcal{A}_{K0} \, P_K(\cos\theta), \tag{2.289}$$

in full analogy to (2.165). Equation (2.288) is a function of the angular distribution anisotropy parameter α_K, and of the components $\mathcal{A}_{K0}$ and $Re \, \mathcal{A}_{K1}$ of the alignment tensor. Its angular dependence is given by the associated Legendre polynomial

$P_K^1(\cos\theta)$. For a certain Auger transition with a fixed total angular momentum J of the intermediate ionic state A^+ the summation over K in (2.288) is generally restricted to $2J - 1$ possible values

$$2 \leq K \leq 2J. \tag{2.290}$$

In particular, for an intermediate state with $J \leq 1/2$, no component $\mathcal{A}_{K1}$ of the alignment parameter can be generated and thus, no asymmetry can be observed. Hence, (2.288) equals zero

$$A_s(\theta, J \leq 1/2) = 0. \tag{2.291}$$

For an intermediate ionic state with $J \leq 3/2$ the summation over K can be omitted and we remain with the case $K = 2$ which yields the simple relation

$$A_s(\theta, J \leq 3/2) = \frac{-\sqrt{\frac{3}{2}}\,\alpha_2\,Re\,\mathcal{A}_{21}\,\sin 2\theta}{1 + \alpha_2\,\mathcal{A}_{20}\,P_2(\cos\theta)}, \tag{2.292}$$

where we used $P_2^1(\cos\theta) = -\frac{3}{2}\sin 2\theta$.

Observing the asymmetry parameter A_s at the *magic angle* $\theta_m = 54.7^o$ and, using $\sin(2\theta_m) = \sqrt{8}/3$, this relation can be further reduced

$$A_s(\theta_m, J \leq 3/2) = -\frac{2}{\sqrt{3}}\alpha_2\,Re\,\mathcal{A}_{21}. \tag{2.293}$$

Hence, measuring the angular asymmetry for the relevant Auger transitions yields information about the alignment component $Re\,\mathcal{A}_{21}$ and gives a direct measure for the rotation of the symmetry axis of the aligned electron cloud out of the electron beam axis.

Eventually, we consider the case $J \leq 5/2$. Here, an additional parameter occurs since multipoles of rank $K = 4$ can contribute to the angular distribution. From (2.288) we get

$$A_s(\theta, J \leq 5/2) = \frac{-\sqrt{\frac{3}{2}}\,\alpha_2\,Re\,\mathcal{A}_{21}\,\sin 2\theta + \frac{1}{\sqrt{5}}\alpha_4\,Re\,\mathcal{A}_{41}\,P_4^1(\cos\theta)}{1 + \alpha_2\,\mathcal{A}_{20}\,P_2(\cos\theta) + \alpha_4\,\mathcal{A}_{40}\,P_4(\cos\theta)}. \tag{2.294}$$

Inserting the *magic* angle we find $P_4(\cos\theta_m) = -7/18$ and $P_4^1(\cos\theta_m) = -5\sqrt{2}/9$ which yields

$$A_s(\theta_m, J \leq 5/2) = -2\frac{6\sqrt{3}\,\alpha_2\,Re\,\mathcal{A}_{21} + \sqrt{10}\,\alpha_4\,Re\,\mathcal{A}_{41}}{18 - 7\,\alpha_4\,\mathcal{A}_{40}}. \tag{2.295}$$

Though, to our knowledge, no experiments have been performed in order to determine the 4th rank tensor components $\mathcal{A}_{40}$ and $Re\,\mathcal{A}_{41}$, they are usually assumed as small. Supposing $\mathcal{A}_{40} \ll \mathcal{A}_{20}$ and $\mathcal{A}_{41} \ll \mathcal{A}_{21}$, we see that (2.295) asymptotically reduces to (2.293).

Neglect of Spin-Dependent Forces

Considering the general equation (2.288), it is worth noting that the parameter $Re\, \mathcal{A}_{K1}$ can be different from zero, only, if spin dependent forces are present during the Auger emission. I.e. the asymmetry parameter A_s vanishes otherwise. This can be shown by decoupling the alignment parameter $Re\, \mathcal{A}_{K1}(J)$, which is a function of the total angular momentum J, and express it in terms of state multipoles of the total orbital angular momentum L and the total spin S, respectively. We obtain

$$Re\, \mathcal{A}_{K1}(J) = \sqrt{\frac{(2K+1)(2J+1)}{(2L+1)(2S+1)}} \sum_{\substack{K'Q' \\ kq}} (-1)^{K'-k+1} \mathcal{A}_{K'Q'}(L)\, \mathcal{A}_{kq}(S)$$

$$\times \sqrt{(2K'+1)(2k+1)} \begin{pmatrix} K' & k & K \\ Q' & q & -1 \end{pmatrix} \begin{Bmatrix} K' & k & K \\ L & S & J \\ L & S & J \end{Bmatrix}. \qquad (2.296)$$

By neglecting spin-dependent forces during the excitation/ionization process the multipole tensors $\mathcal{A}_{K'Q'}(L)$ and $\mathcal{A}_{kq}(S)$ are uncorrelated and can be written as a simple product. Since $Re\, \mathcal{A}_{K1}$ is real the product $\mathcal{A}_{K'Q'}(L)\, \mathcal{A}_{kq}(S)$ must be a real number, too. I.e., both tensors are either solely real or imaginary.

Assuming a primary electronic excitation/ionization of an unpolarized target and not observing the scattered and ionized electrons, the combined electron–target system has axial symmetry in the angular momentum subspace. Therefore, only tensor components $\mathcal{A}_{K'Q'}(L)$ with zero magnetic component $Q' = 0$ and K' even can be generated. This yields further selection rules for the spin state multipoles $\mathcal{A}_{kq}(S)$. From the symmetry of the $3j$-symbol we get

$$q = 1. \qquad (2.297)$$

Since $\mathcal{A}_{K'0}(L)$ is always real, the spin multipole $\mathcal{A}_{k1}(S)$ must be a real parameter, too. We get

$$Re\, \mathcal{A}_{K1}(J) = \sqrt{\frac{(2K+1)(2J+1)}{(2L+1)(2S+1)}} \sum_{\substack{K'even \\ k}} (-1)^{1-k} \mathcal{A}_{K'0}(L)\, Re\, \mathcal{A}_{k1}(S)$$

$$\times \sqrt{(2K'+1)(2k+1)} \begin{pmatrix} K' & k & K \\ 0 & 1 & -1 \end{pmatrix} \begin{Bmatrix} K' & k & K \\ L & S & J \\ L & S & J \end{Bmatrix}. \qquad (2.298)$$

Interchanging the last two columns of the $9j$-symbol its symmetry relation yields

$$(-1)^{K'+k+K} = 1, \qquad (2.299)$$

and, since K' and K are both even numbers, we get

$$(-1)^k = 1. \qquad (2.300)$$

Thus, only spin multipoles $Re \, \mathcal{A}_{k1}(S)$ with even rank k can occur. However, for an unpolarized target the rank of the spin multipoles $\mathcal{A}_{kq}(S)$ must be the same as for the state multipoles describing the polarization state of the incoming electron beam which yields the general restriction

$$k \leq 1. \tag{2.301}$$

Thus, since $q = 1$, the rank of the spin tensors is fixed to $k = 1$ which violates (2.300) and in the absence of spin-dependent forces we therefore obtain

$$Re \, \mathcal{A}_{K1}(J) = 0. \tag{2.302}$$

This result remains true even if the projectile electron is polarized. Therefore, the multipole parameters $Re \, \mathcal{A}_{K1}(J)$, in particular the alignment parameter $Re \, \mathcal{A}_{21}(J)$, and thus the asymmetry parameter $A_s(\theta)$, can be seen as a direct measure for the presence of spin-dependent forces. However, one should note that a vanishing parameter $\mathcal{A}_{K1}(J)$ does not automatically imply the absence of spin-dependent forces.

The result (2.302) can be applied to the imaginary part $Im \, \mathcal{A}_{K1}(J)$ of the tensor parameters as well. Adapting the arguments used above we obtain

$$Im \, \mathcal{A}_{K1}(J) = 0. \tag{2.303}$$

Polarization Asymmetry Parameters

In analogy to the above asymmetry parameter A_s we may define the *weighted polarization asymmetry parameter* for the Cartesian components of the spin polarization vector.

$$A_{p_i}(\theta) = p_i(\theta)^e I(\theta)^e - p_i(-\theta)^e I(-\theta)^e, \tag{2.304}$$

where $i = x, y$, and z, respectively.

We obtain from (2.166)

$$A_{p_x}(\theta) = -\frac{I_0}{2\pi} \sum_{K \, odd} \xi_K \, \mathcal{O}_{K0} \, P_K^1(\cos\theta). \tag{2.305}$$

Analogously, (2.167) yields

$$A_{p_y}(\theta) = -\frac{I_0}{2\pi} \sum_{K \, even} \xi_K \, \mathcal{A}_{K0} \, P_K^1(\cos\theta), \tag{2.306}$$

and out of (2.168) we obtain

$$A_{p_z}(\theta) = \frac{I_0}{\pi} \sum_{K \, odd} \frac{1}{\sqrt{K(K+1)}} \delta_K \, Re \, \mathcal{O}_{K1} \, P_K^1(\cos\theta). \tag{2.307}$$

Now, let us discuss some special cases in more detail. As a simple example we assume the total angular momentum of the intermediate ionic state to be $J = 1/2$. For this particular case (2.305)–(2.307) can be reduced to

$$A_{p_x}(\theta) = -\frac{I_0}{2\pi} \xi_1 \mathcal{O}_{10} P_1^1(\cos\theta), \tag{2.308}$$

$$A_{p_y}(\theta) = 0 \tag{2.309}$$

and

$$A_{p_z}(\theta) = \frac{I_0}{\sqrt{2}\pi} \delta_1 \, Re \, \mathcal{O}_{11} P_1^1(\cos\theta). \tag{2.310}$$

We note, that the asymmetry of the p_y-component of the spin polarization vector vanishes in this case. The p_x- and p_z-component, on the other hand, depend on orientation tensor components of the intermediate ionic state and can therefore be observed only if polarized electrons are used for the ionization process.

Performing the ratio of (2.308) and (2.310) yields an interesting relation

$$\frac{A_{p_x}}{A_{p_z}} = \frac{-1}{\sqrt{2}} \frac{\xi_1 \mathcal{O}_{10}}{\delta_1 \, Re \, \mathcal{O}_{11}}. \tag{2.311}$$

Hence, the ratio of the polarization asymmetry parameters becomes independent of the angle of Auger emission and is a function of the orientation and spin polarization parameters, only.

Considering the case of an intermediate ionic state with $J = 1$, we get the restriction $K \leq 2$. Thus, reduction of the general equations (2.305)–(2.307) yields

$$A_{p_x}(\theta) = -\frac{I_0}{2\pi} \xi_1 \mathcal{O}_{10} P_1^1(\cos\theta), \tag{2.312}$$

$$A_{p_y}(\theta) = -\frac{I_0}{2\pi} \xi_2 \mathcal{A}_{20} P_2^1(\cos\theta), \tag{2.313}$$

and

$$A_{p_z}(\theta) = \frac{I_0}{\sqrt{2}\pi} \delta_1 \, Re \, \mathcal{O}_{11} P_1^1(\cos\theta). \tag{2.314}$$

While the expressions for A_{p_x} and A_{p_z} remain unchanged to the previous case of $J = 1/2$ the component A_{p_y} can be non-zero. Note, that the ratio (2.311) is still valid.

Now, we consider the *normalized weighted polarization asymmetry parameters* as

$$A_{p_i}^n(\theta) = \frac{p_i(\theta)^e I(\theta)^e - p_i(-\theta)^e I(-\theta)^e}{p_i(\theta)^e I(\theta)^e + p_i(-\theta)^e I(-\theta)^e}, \tag{2.315}$$

with $i = x, y, z$, respectively. The general parameters yield a rather complex form.

We obtain

$$
A^n_{p_x}(\theta) = \left(\sum_{K\,even} \sqrt{K(K+1)}\, \xi_K\, Im\, \mathcal{A}_{K1} \left[d^{(K)}_{11}(\theta) + d^{(K)}_{-11}(\theta) \right] \right.
$$

$$
\left. - \sum_{K\,odd} \sqrt{K(K+1)}\, \xi_K\, Re\, \mathcal{O}_{K1} \left[d^{(K)}_{11}(\theta) - d^{(K)}_{-11}(\theta) \right] \right)^{-1}
$$

$$
\times \sum_{K\,odd} \xi_K\, \mathcal{O}_{K0}\, P^1_K(\cos\theta), \tag{2.316}
$$

$$
A^n_{p_y}(\theta) = \left(\sum_{K\,even} \sqrt{K(K+1)}\, \xi_K\, Re\, \mathcal{A}_{K1} \left[d^{(K)}_{11}(\theta) - d^{(K)}_{-11}(\theta) \right] \right.
$$

$$
\left. + \sum_{K\,odd} \sqrt{K(K+1)}\, \xi_K\, Im\, \mathcal{O}_{K1} \left[d^{(K)}_{11}(\theta) + d^{(K)}_{-11}(\theta) \right] \right)^{-1}
$$

$$
\times \sum_{K\,even} \xi_K\, \mathcal{A}_{K0}\, P^1_K(\cos\theta), \tag{2.317}
$$

and

$$
A^n_{p_z}(\theta) = \left(\sum_{K\,odd} \delta_K\, \mathcal{O}_{K0}\, P_K(\cos\theta) \right)^{-1}
$$

$$
\times \sum_{K\,odd} \frac{2}{\sqrt{K(K+1)}}\, \delta_K\, Re\, \mathcal{O}_{K1}\, P^1_K(\cos\theta). \tag{2.318}
$$

For illustrating the advantages of the general equations (2.316)–(2.318), we discuss two particular cases of total intermediate ionic angular momentum. Supposing an ionic state with total angular momentum $J = 1/2$, the general expressions (2.312)–(2.314) can be considerably reduced which yields

$$
A^n_{p_x}(\theta) = - \frac{\xi_1\, \mathcal{O}_{10}\, P^1_1(\cos\theta)}{\sqrt{2}\, \xi_1\, Re\, \mathcal{O}_{11} \left[d^{(1)}_{11}(\theta) - d^{(1)}_{-11}(\theta) \right]}, \tag{2.319}
$$

$$
A^n_{p_y}(\theta) = 0, \tag{2.320}
$$

and

$$
A^n_{p_z}(\theta) = \frac{\sqrt{2}\, \delta_1\, Re\, \mathcal{O}_{11}\, P^1_1(\cos\theta)}{\delta_1\, \mathcal{O}_{10}\, P_1(\cos\theta)}. \tag{2.321}
$$

Eventually, we find the p_y-component of the spin polarization vector to vanish and the two in-reaction plane components p_x and p_z independent of the Auger decay parameters δ_1 and ξ_1, respectively. Inserting expressions for the reduced rotation matrices and the associated Legendre polynomials we remain with

$$A_{p_x}^n(\theta) = \frac{\mathcal{O}_{10}}{\sqrt{2}\,Re\,\mathcal{O}_{11}}\,\tan\theta, \tag{2.322}$$

$$A_{p_y}^n(\theta) = 0, \tag{2.323}$$

and

$$A_{p_z}^n(\theta) = \frac{-\sqrt{2}\,Re\,\mathcal{O}_{11}}{\mathcal{O}_{10}}\,\tan\theta. \tag{2.324}$$

Equations (2.322) and (2.324) show some interesting relations. Multiplying both parameters yields

$$A_{p_x}^n(\theta)\,A_{p_z}^n(\theta) = -\tan^2\theta, \tag{2.325}$$

which leaves the product of the two asymmetry parameters independent of any parameters and a simple function of the scattering angle θ, only. On the other hand, the ratio yields

$$\frac{A_{p_x}^n(\theta)}{A_{p_z}^n(\theta)} = \frac{-(\mathcal{O}_{10})^2}{2\,(Re\,\mathcal{O}_{11})^2}, \tag{2.326}$$

which is independent of the scattering angle and solely depends on the orientation parameters. However, we need to recognize the fact, that both relations make sense only if the intermediate ionic or the incoming electron beam state is polarized in the reaction x–z plane.

Considering the case of an intermediate ionic state with $J = 1$ all parameters obey the restriction $K \leq 2$ and we obtain

$$A_{p_x}^n(\theta) = \frac{\xi_1\,\mathcal{O}_{10}\,P_1^1(\cos\theta)}{\left(\sqrt{6}\,\xi_2\,Im\,\mathcal{A}_{21}\left[d_{11}^{(2)}(\theta) + d_{-11}^{(2)}(\theta)\right] - \sqrt{2}\,\xi_1\,Re\,\mathcal{O}_{11}\left[d_{11}^{(1)}(\theta) - d_{-11}^{(1)}(\theta)\right]\right)}, \tag{2.327}$$

$$A_{p_y}^n(\theta) = \frac{\xi_2\,\mathcal{A}_{20}\,P_2^1(\cos\theta)}{\left(\sqrt{6}\,\xi_2\,Re\,\mathcal{A}_{21}\left[d_{11}^{(2)}(\theta) - d_{-11}^{(2)}(\theta)\right] + \sqrt{2}\,\xi_1\,Im\,\mathcal{O}_{11}\left[d_{11}^{(1)}(\theta) + d_{-11}^{(1)}(\theta)\right]\right)}, \tag{2.328}$$

and

$$A_{p_z}^n(\theta) = \frac{\sqrt{2}\,Re\,\mathcal{O}_{11}\,P_1^1(\cos\theta)}{\mathcal{O}_{10}\,P_1(\cos\theta)}. \tag{2.329}$$

Expressing the reduced rotation matrices in terms of angular functions yields

$$A_{p_x}^n(\theta) = \frac{\xi_1\,\mathcal{O}_{10}}{\sqrt{2}\,\xi_1\,Re\,\mathcal{O}_{11} - \sqrt{6}\,\xi_2\,Im\,\mathcal{A}_{21}}\,\tan\theta, \tag{2.330}$$

$$A_{p_y}^n(\theta) = \frac{-\frac{3}{2}\,\xi_2\,\mathcal{A}_{20}\,\sin(2\theta)}{\sqrt{2}\,\xi_1\,Im\,\mathcal{O}_{11} + \sqrt{6}\,\xi_2\,Re\,\mathcal{A}_{21}\,\cos(2\theta)}, \tag{2.331}$$

and

$$A^n_{p_z}(\theta) = \frac{-\sqrt{2}\, Re\, \mathcal{O}_{11}}{\mathcal{O}_{10}}\, \tan\theta. \tag{2.332}$$

Using (2.330) and (2.332) we compare the product and the ratio of the asymmetry parameters with equations (2.325) and (2.326). The product yields

$$A^n_{p_x}(\theta)\, A^n_{p_z}(\theta) = \frac{1}{\dfrac{\sqrt{3}\, \xi_2\, Im\, \mathcal{A}_{21}}{\xi_1\, Re\, \mathcal{O}_{11}} - 1}\, \tan^2\theta. \tag{2.333}$$

Thus, the angular dependency is identical to (2.325) besides a different amplitude. The ratio becomes independent of the Auger emission angle, too.

$$\frac{A^n_{p_x}(\theta)}{A^n_{p_z}(\theta)} = \frac{-(\mathcal{O}_{10})^2}{2\,(Re\, \mathcal{O}_{11})^2 \left[1 - \dfrac{\sqrt{3}\, \xi_2\, Im\, \mathcal{A}_{21}}{\xi_1\, Re\, \mathcal{O}_{11}} \right]}. \tag{2.334}$$

Thus, (2.333) and (2.334) for the case $J = 1$ show the same behaviour as (2.325) and (2.326) for the case $J = 1/2$, besides a modulation of their amplitudes.

If we are neglecting spin-dependent forces during the electronic excitation/ionization process, it is of importance to note the fact, that, applying (2.303), the relations (2.333) and (2.334) yield the same result as (2.325) and (2.326). This can be explained as follows. While explicitly spin-dependent forces cannot influence the asymmetry parameters for the case $J = 1/2$ since (2.325) and (2.326), both, depend on the orientation parameters, only, the asymmetry parameter $A^n_{p_x}(\theta, J = 1)$ of (2.330) depends on the alignment tensor $Im\, \mathcal{A}_{K1}$, too. As has been shown in the previous discussion, this parameter can be non-zero if spin-dependent forces are present, only. Thus, for a vanishing alignment tensor parameter $Im\, \mathcal{A}_{K1}$, we end up with the same result for, both, the cases of $J = 1/2$ and $J = 1$, respectively.

2.7.6 Linear Dichroism for Auger Electrons from Unpolarized Targets

Instead of investigating the left–right asymmetry parameter $A_s(\theta)$ of angular distribution, it is often useful to observe the asymmetry of angular distribution for a fixed angle θ of Auger emission but changing the polarization state of the ionizing species. For the case of a primary photoionization/excitation this is known as linear dichroism of angular distribution (LDAD).

We consider the angular distribution of Auger emission. The angular distribution depends on the alignment parameter $Re\, \mathcal{A}_{22}$ which is a function of the Stokes parameter η_3 and therefore depends on the degree of linear polarization of the incoming photon beam. We define or LDAD asymmetry parameter as

$$A_{LD}(\theta) = \frac{I(\theta, +)^\gamma - I(\theta, -)^\gamma}{I(\theta, +)^\gamma + I(\theta, -)^\gamma}, \tag{2.335}$$

where we have used the abbreviation $I(\theta, \pm) = I(\theta, \pm|\eta_3|)$.

Inserting the related equation of angular distribution (2.177) we obtain

$$A_{LD}(\theta) = \frac{\sqrt{\frac{3}{2}}\,\alpha_2\,Re\,\mathcal{A}_{22}\,\sin^2\theta}{1 + \alpha_2\,\mathcal{A}_{20}\,P_2(\cos\theta)}. \tag{2.336}$$

Using the relation $\mathcal{A}_{22} = \mathcal{A}_{20}\sqrt{3/2}(-\eta_3 + i\eta_1)$, see (2.182), we express the parameter $\mathcal{A}_{22}$ by the usual alignment parameter $\mathcal{A}_{20}$

$$A_{LD}(\theta) = \frac{-\frac{3}{2}\,|\eta_3|\,\alpha_2\,\mathcal{A}_{20}\,\sin^2\theta}{1 + \alpha_2\,\mathcal{A}_{20}\,P_2(\cos\theta)}. \tag{2.337}$$

Considering the case of a resonant Auger transition the alignment parameter $\mathcal{A}_{20}$ is fully analytic and becomes a constant number (see Table 2.9). This yields

$$A_{LD}(\theta) = \frac{-3\,\alpha_2\,|\eta_3|\,\sin^2\theta}{2\left(\sqrt{2} + \alpha_2\,P_2(\cos\theta)\right)}. \tag{2.338}$$

In this case the LDAD asymmetry parameter yields an interesting relation for the *magic angle*

$$A_{LD}(\theta_m) = \frac{-1}{\sqrt{2}}\,\alpha_2\,|\eta_3|. \tag{2.339}$$

Thus, for resonant Auger transitions we find the LDAD asymmetry parameter directly proportional to the angular distribution parameter α_2 and to the degree of linear polarization of the incoming synchrotron beam. That is, the LDAD vanishes for an unpolarized photon beam.

3 Numerical Methods

In this chapter the numerical methods developed to evaluate the relative intensities, angular distribution and spin polarization parameters are discussed. Common to all parameters is the numerical calculation of Auger transition amplitudes. From the previous chapter we learned that the investigation of angle and spin resolved Auger processes is connected with the determination of anisotropy parameters. The main numerical difficulty in the calculation of anisotropy parameters for Auger transitions is that they are not only functions of the transition amplitudes but of the scattering phases, too. Their explicit knowledge is therefore crucial for the calculation of numerical data.

Auger transitions in heavy atoms show large fine structure splittings. Thus, a relativistic approach is required for the calculation of the matrix elements. This has been taken into account by calculating the Auger transition matrix elements within a multiconfigurational Dirac–Fock (MCDF) approach.

The numerical calculations have been performed using the two program packages ANISO and RATR. They, both, have been developed in the context of scattering theory (cf. Åberg and Howat 1982), and apply selfconsistent field methods with configuration interaction (Δ-SCFCI method). Thus, intermediate coupling in the many-electron wavefunctions has been accounted for in both packages.

The ANISO package has been developed and frequently extended by ourselves (e.g. see Lohmann 1988, 1990, 1997). A number of useful approximations have been introduced and applied in the ANISO package. In particular, though applying a relativistic approach, we neglect information from the *small* component of the Dirac–Fock wavefunctions. Further, an energy dependent local exchange potential is used in the calculation of the continuum wavefunctions. This enables for a calculation of a decoupled set of differential equations and avoids the problem of solving the coupled set of integro-differential equations. This method is considerably improved compared to previous calculations (e.g. Aksela *et al.* 1984b; Chen *et al.* 1990; Lohmann 1990) which, for instance, completely neglect exchange with the continuum.

The RATR (relativistic Auger transition rates) package[1] is based on a previous version developed by Fritzsche (1991, 1992) for the calculation of Auger transition rates. It has been extended by Lohmann and Fritzsche (1994) to allow for the calcu-

---

[1] Meanwhile, a more extended version of RATR, called RATIP (relativistic atomic transition and ionization properties) is available (Fritzsche *et al.* 2000; Fritzsche 2001).

lation of anisotropy parameters. As pointed out, this requires explicit knowledge of the scattering phases. While a number of useful approximations have been applied in the ANISO package, RATR fully includes the relativistic framework. It takes the small component of the Dirac–Fock wavefunction into account by solving the coupled set of integro-differential equations for the evaluation of the continuum wavefunctions. Particularly, the right-hand side of the integro-differential equations contain an inhomogeneous part where orthogonality needs to be ensured by the introduction of Lagrange multipliers. Exchange interaction with the continuum is automatically accounted for in this approach. In addition, RATR provides a number of *switches* which allow for e.g. suppressing the Breit interaction or the continuum exchange. This is most useful for calculations using large basis sets of several 1000 configuration state functions (CSF). It also allows for a detailed investigation of the strength of relativistic or exchange effects. This will be discussed in detail in Sect. 3.2

Common to both packages is the calculation of the bound state wavefunctions which are obtained with the MCDF package of Grant and co-workers (Grant 1970; Grant *et al.* 1980; Dyall *et al.* 1989).

The general structure of the two packages and the applied numerical methods will be outlined in the following.

3.1 The ANISO Program

The ANISO program package basically consists of three major parts which is illustrated in Fig. 3.1.

The calculation of the bound state wavefunctions of the intermediate singly and the doubly ionized final state of the atom is done in the program part MCDF. The obtained wavefunctions are then interpolated to the appropriate mesh which is done in the program part POTRES. The model potential used for evaluating the continuum wavefunction is also calculated. In the ANISO program, relativistic corrections and exchange are included, and, eventually, the continuum wavefunction is generated which is used for calculating the Auger transition matrix elements. With this, the Auger rates and the relevant angular distribution and spin polarization parameters are calculated, respectively.

3.1.1 The Bound State Wavefunctions

The Auger transition matrix elements (see Sect. 3.1.7) are calculated with MCDF wavefunctions. In the MCDF model an atomic state function (ASF) is represented by a linear combination of configuration state functions (CSF)

$$|\psi_\alpha(PJM)\rangle = \sum_{r=1}^{n_c} c_r(\alpha) \, |\gamma_r PJM\rangle. \tag{3.1}$$

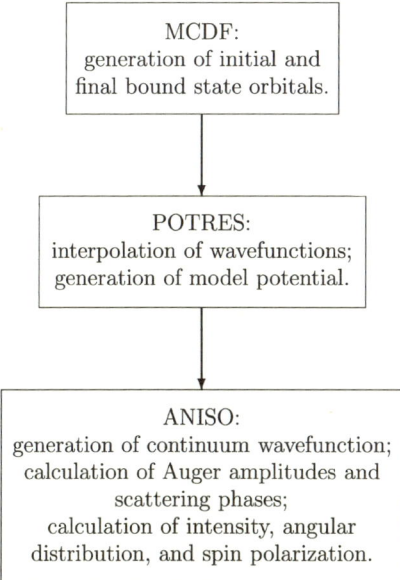

Fig. 3.1. The ANISO program package

The configuration states $|\gamma_r\,PJM\rangle$ are constructed from antisymmetrized products of Dirac orbitals which are eigenstates of the total (one-electron) angular momentum and parity. The label γ_r distinguish the occupation of the different subshells and the angular coupling schemes (see Grant 1970, for further details). n_c is the number of CSF included in the expansion and $c_r(\alpha)$, $r = 1,\ldots,n_c$, are the configuration mixing coefficients for the state α.

To generate the initial and final ionic bound states we used the atomic MCDF structure package of Grant *et al.* (1980). An explicit description may be found therein. In this code the radial orbitals are generated self-consistently with respect to the Dirac–Coulomb Hamiltonian

$$H_{DC} = \sum_{i=1}^{N} h_D(i) + \sum_{i,j}^{N} \frac{1}{r_{ij}} \tag{3.2}$$

and

$$h_D(i) = c\,\boldsymbol{\alpha}_i\,\mathbf{p}_i + (\beta_i - 1)\,c^2 + V_{nuc}(r_i), \tag{3.3}$$

where $h_D(i)$ is the one-electron Dirac Hamiltonian in the potential of the nucleus and $1/r_{ij}$ the static Coulomb repulsion between the pair i and j of electrons. In the standard MCDF scheme the transverse Breit interaction

$$b(i,j) = -\frac{\boldsymbol{\alpha}_i\,\boldsymbol{\alpha}_j}{r_{ij}} + (\boldsymbol{\alpha}_i\,\boldsymbol{\nabla}_i)\,(\boldsymbol{\alpha}_j\,\boldsymbol{\nabla}_j)\,\frac{\cos\left(\omega\,r_{ij} - 1\right)}{\omega^2 r_{ij}} \tag{3.4}$$

and the main radiative corrections are added to the Hamiltonian matrix as perturbation. Its final diagonalization gives the representation of the atomic states, i.e. the mixing coefficients $c_r(\alpha)$ in (3.1).

3.1.2 The Model Potential

For generating the continuum wavefunction of the emitted Auger electron, in the ANISO package, we need to generate a model potential which is calculated in the part POTRES.

Basically, POTRES calculates a relativistic Hartree potential, i.e. the static potential which is generated by the Coulomb interaction of the Auger electron with the charge distribution of the ion, where the electron shell and the charge distribution of the core have been taken into account. Intermediate coupling has been explicitly included. Thus, a specific intermediate coupling potential is generated for each final ionic state α,

$$U_{mod}(\mathbf{r}, \alpha) = \int \frac{\rho(\mathbf{r}', \alpha)}{|\mathbf{r} - \mathbf{r}'|} \, d\mathbf{r}'. \tag{3.5}$$

Generally, the charge density $\rho(\mathbf{r}', \alpha)$ may be anisotropic due to deformations of either the electronic charge cloud or the nucleus. Influences on the calculated data are, however, beyond the possible spectroscopic resolution, or even the natural Auger line widths. Therefore, both, the electronic as well as the nuclear charge densities can be assumed as isotropic. Hence, the charge density $\rho(\mathbf{r}', \alpha)$ is given as

$$\rho(\mathbf{r}', \alpha) = \rho(r', \alpha) = -\rho_H(r', \alpha) + \rho_{nuk}(r'). \tag{3.6}$$

Here, $\rho_H(r, \alpha)$ denotes the density of the electronic charge cloud for a certain ASF,

$$\rho_H(r, \alpha) = \frac{1}{4\pi r^2} \sum_{i,k} |c_k(\alpha)|^2 n_{ik} u_i^2(r). \tag{3.7}$$

Here, $u_i^2(r)$ is the square of the Dirac–Fock wavefunction

$$u_i^2(r) = p_i^2(r) + q_i^2(r), \tag{3.8}$$

where $p_i(r)$ and $q_i(r)$ are its *large* and *small* components. The configuration mixing coefficients $c_k(\alpha)$ have been introduced in (3.1). The occupation numbers n_{ik} give the number of electrons in a shell i for a configuration k and are simply associated to the total angular momentum j_{ik} via

$$n_{ik} = 2 j_{ik} + 1. \tag{3.9}$$

In particular, we define the real quantity

$$x_i(\alpha) = \sum_k |c_k(\alpha)|^2 n_{ik} \tag{3.10}$$

which can be interpreted as occupation number of the ith shell for the state α. Thus, numerically we are dealing with non-integer occupation numbers, e.g. the $2p_{3/2}$ shell might contain 2.734 electrons.

For completeness, we are including a finite core. For the charge density of the core we have chosen the Woods–Saxon form,

$$\rho_{nuk}(r) = \rho_0 \left(1 + \exp\left(\frac{r - R_{1/2}}{z}\right)\right)^{-1}, \tag{3.11}$$

where ρ_0 is the charge density of the core at the origin and $R_{1/2}$ is the point of half charge density; i.e. $\rho_{nuk}(R_{1/2}) = \rho_0/2$. The parameter z is a measure for the thickness of the core boundary. Numbers for these parameters may be found in the literature (e.g. Mayer-Kuckuk 1979). Typical values are

$$\rho_0 = 2.52 \times 10^{13} \frac{Z}{A} a_0^{-3},$$
$$z = 1.03 \times 10^{-5} a_0, \tag{3.12}$$
$$R_{1/2} = \left(2.13 \times 10^{-5} A^{1/3} - 1.68 \times 10^{-5} A^{-1/3}\right) a_0,$$

where A denotes the mass and Z the charge of the core. Of course, ρ_0 is normalized in POTRES such that

$$4\pi \int_0^\infty \rho_{nuk}(r) r^2 \, dr = Z \tag{3.13}$$

is fulfilled. Note, that using either a finite or point charge density of the core yields only small deviations in the calculations.

The numerical effort of calculating $U_{mod}(r, \alpha)$ can be reduced if we expand the denominator of (3.5) into a series of Legendre polynomials,

$$\frac{1}{|\mathbf{r} - \mathbf{r}'|} = \sum_{\ell=0}^\infty \gamma_\ell(r, r') P_\ell(\cos\theta), \tag{3.14}$$

where γ_ℓ is defined as

$$\gamma_\ell(r, r') = \frac{r_<^\ell}{r_>^{\ell+1}} \quad \text{with} \quad r_< = \begin{cases} r & \text{for } r < r', \\ r' & \text{for } r' < r. \end{cases} \tag{3.15}$$

Analogously we define $r_>$, and θ denotes the angle between the two vectors $\mathbf{r}$ and $\mathbf{r}'$.

With this, we may write

$$U_{mod}(\mathbf{r}, \alpha) = \iint \rho(r', \alpha) \left(\sum_{\ell=0}^\infty \gamma_\ell(r, r') P_\ell(\cos\theta)\right) d\Omega \, r'^2 dr'$$
$$= 2\pi \sum_{\ell=0}^\infty \int_0^\infty \rho(r', \alpha) \gamma_\ell(r, r') r'^2 dr' \int_1^{-1} P_\ell(\cos\theta) \, d(\cos\theta) . \tag{3.16}$$

Applying the orthogonality theorem of the Legendre polynomials and using $P_0(\cos\theta) = 1$ for any angle θ, (3.16) can be reduced to

$$U_{mod}(\mathbf{r}, \alpha) = -4\pi \int_0^\infty \rho(r', \alpha) \frac{1}{r_>} r'^2 \, dr'$$

$$= 4\pi \left(\frac{N_e - Z}{r} + \int_\infty^r \rho(r', \alpha) r' \, dr' - \frac{1}{r} \int_\infty^r \rho(r', \alpha) r'^2 \, dr' \right), \quad (3.17)$$

where we used the fact that

$$\int_0^\infty \rho(r', \alpha) r'^2 \, dr' = Z - N_e, \quad (3.18)$$

where N_e denotes the total number of electrons. Thus, using the symmetry of the charge density and applying orthogonality relations, the problem of calculating the 3-dimensional integral (3.5) has been reduced to the numerical integration of two one-dimensional integrals.

The calculation of the potentials is usually done in an area starting from ~ 70 a.u. inwards to $\sim 5 \cdot 10^{-8}$ a.u. Carrying out the inward integration yields the advantage that the potential is known after every integration step, i.e. both integrals need to be calculated only once.

The program structure of the POTRES part of the ANISO package is illustrated in Fig. 3.2. The subroutine XINIT initializes the used arrays. The routine NORM does necessary interpolations and re-normalizes the large component $u_i(r)$ of the MCDF wavefunctions. The subroutine NUCDEN calculates the core density for a finite nucleus. Eventually, the routine CALPOT calculates the model potential $U_{mod}(\mathbf{r}, \alpha)$ for a specific ASF. Since different grid systems have been used in the program parts, the subroutine INTERP does the appropriate interpolation of the results to the required mesh. The calculation of the energy independent Slater exchange potential has been included in this part, too. The exchange potentials are discussed in Sect. 3.1.4. As will be shown in the next section the relativistic correction potentials may require the derivative of the model potential which is done in subroutine DIFF1.

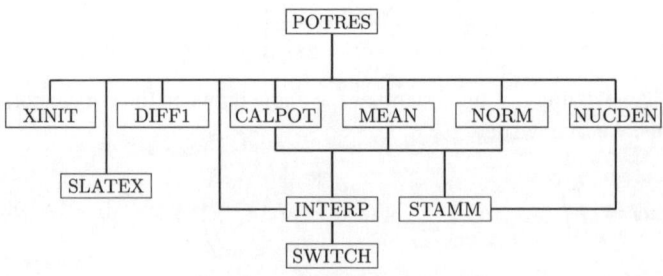

Fig. 3.2. The program part POTRES

As will be considered later, only the large component of the wavefunctions influences the numerical results significantly. A good measure to control the validity of this statement is the calculation of some expectation values like $\langle r \rangle$, $\langle r^2 \rangle$, or $\langle 1/r \rangle$, either by using the large and small components or solely the large component of the MCDF wavefunctions. This is done in subroutine MEAN. Only small deviations have been found for all investigated Auger transitions (order of magnitude $\sim 10^{-4}$).

A numerical integration is necessary in almost all subroutines. The integration routine is done in STAMM applying Simpson's rule and giving back the result at every second mesh point.

Though, $U_{mod}(\mathbf{r}, \alpha)^2$ is usually generated from the charge density of the doubly ionized final ionic state A^{++}, the structure of POTRES yields the possibility to calculate the model potential from the field of the intermediate singly ionized state A^+, too. Generally, POTRES allows for the calculation of $U_{mod}(\mathbf{r}, \alpha)$ for any level of ionization A^{n+}, including the neutral atomic state. Though, such type of calculations will not be considered here. Throughout this book, we are using the first possibility if not explicitly stating.

3.1.3 Exchange Potentials

For continuum states with an energy of several hundred eV the exchange potential becomes negligible. This can be understood simply by the fact that the electron is too fast for interacting with any of the electrons of the ionic charge cloud. However, with decreasing Auger energy the exchange potential becomes more important and does influence the calculation.

Generally, the inclusion of exchange leads to a set of coupled integro-differential equations with a non-local term. There are a number of methods available, approximating the non-local exchange term by a local potential, which allow to reduce the system of integro-differential equations to a set of differential equations of the Schrödinger type. This method has been applied in the ANISO program package for the inclusion of exchange with the continuum wavefunction.

Basically, electron exchange is a short-range effect which might be approximated by an exchange potential U_{ex}. For determining the sign of U_{ex}, that is whether the exchange potential is repulsive or attractive, it is useful to consider the correlation between only two electrons, i.e. the emitted Auger electron and its exchange partner. As pointed out before, considering the Auger emission as a *half scattering*, then, both electrons must have the same spin if we neglect excitation effects. Thus, the spin wavefunction must be symmetric. Requiring the total electron wavefunction to be antisymmetric, its spatial wavefunction must be antisymmetric and vanishes for the same spatial coordinates of the electrons. On average, the electrons are thus more apart then for a symmetric spatial wavefunction. Therefore, the repulsive Coulomb potential between the two electrons is reduced. From this follows that the exchange potential must be attractive and occurs for a Dirac–Fock orbital with an appropriate non-zero spin.

---

[2] The variable α is suppressed in the following if not causing ambiguities.

The package includes two different approximations of the exchange potential from which one can be chosen.

The energy independent form of the Slater approximation of the exchange potential (Slater 1951) is given by

$$U_{ex}^{SL}(r) = 3\left(\frac{3}{8\pi}\,\rho_H(r)\right)^{1/3}. \tag{3.19}$$

This approximation is based on the following idea. For a given quantization axis, a many-electron system consists of a number of electrons with spin up and down and electronic densities ρ_+ and ρ_-, respectively. Calculating the wavefunction for a specific electron, it can exchange with other electrons of the same spin, only. Thus, exchange effects lead to a potential, which would be generated by taking away electronic density of appropriate spin, such that the resulting charge is equal to the charge of a single electron. For a free electron gas this is known as the *Fermi hole*. The calculation has been implemented as subroutine SLATEX in the POTRES program part (see Fig. 3.2).

As a second possibility, an energy dependent form of the exchange potential, as has been introduced by Furness and Mc Carthy (1973) and Vanderpoorten (1975), has been implemented in the subroutine FURNEX (see Fig. 3.4). There are two types of potentials, known as the singlet and triplet version.

$$U_{ex}^{FM}(r, E) = \frac{1}{2}\left(\sqrt{(E - U_{mod}(r))^2 - 4\pi\,\tau\,\rho_H(r)} - (E - U_{mod}(r))\right), \tag{3.20}$$

where

$$\tau = \begin{cases} -1 & \text{for triplet}, \\ 1 & \text{for singlet}, \end{cases} \tag{3.21}$$

and the electronic charge density $\rho_H(r)$ has been given in (3.7). Throughout our calculations the first has been chosen which is caused by the fact that the singlet type is generally repulsive whereas the triplet version yields an attractive potential. In particular, we point out that the calculation of the singlet type is numerically not converging if applied to scattering problems.

A general advantage of the exchange potential of Furness and Mc Carthy (1973) is that it does not have any fitting parameters, i.e. the numerical results cannot be artificially improved. Generally, we note that the exchange potential vanishes for low densities or high energies, respectively.

With respect to Auger transitions, the Slater exchange potential leads to larger deviations from the experimental data then totally neglecting exchange. A similar behaviour has been found by Chen (1992). Thus, if not explicitly stating, we are using the triplet form of the exchange potential.

3.1.4 The Continuum Wavefunction

Generally, for calculating the continuum wavefunction we have to solve the Dirac–Fock equations of the emitted Auger electron in the stationary field of the ion. Considering the four-component relativistic wavefunction, it may be asymptotically expanded into a linear combination of plane and spherical waves

$$\lim_{r \to \infty} \psi_\lambda(\mathbf{r}) = a_\lambda e^{ikz} + \frac{1}{r} e^{ikr} F_\lambda(\theta, \phi), \tag{3.22}$$

where $\lambda = 1, \ldots, 4$. For most of the Auger transitions we are interested in, the kinetic energy of the outgoing Auger electron is much smaller than its rest energy ($E_{kin} < 1$ keV). This yields for the ratio of the *small* (a_3, a_4) to the *large* (a_1, a_2) components of the emitted plane wave

$$\left| \frac{a_3}{a_1} \right| = \left| \frac{a_4}{a_2} \right| = \frac{kc}{2c^2 + E_{kin}} \sim \frac{k}{2c}. \tag{3.23}$$

Since the emitted spherical wave can be expressed as a superposition of plane waves emitted into different directions, (3.23) is asymptotically valid for the angular functions F_λ, too. Thus, the calculation of the *small* components of the Dirac equation gives no more additional information, and, following some textbooks of relativistic quantum mechanics, we are able to reduce the four-component wavefunction to a two-component one.

$$\left(\frac{d^2}{dr^2} + k^2 - \frac{\ell(\ell+1)}{r^2} - 2U_\kappa^{fr}(r, E) \right) u_\kappa(r) = 0. \tag{3.24}$$

Since the used model potential is scalar and has spherical symmetry, this transformation is exact (see Mott and Massey 1965). Of course, the used local scalar potential is an approximation.

Here, we used $u_\kappa(r) = r R_\kappa(r)$, where $R_\kappa(r)$ denotes the radial part of the continuum wavefunction $\psi_{\varepsilon\ell j}(\mathbf{r})$ with κ abbreviating all quantum numbers necessary to uniquely define the state. The potential $U_\kappa^{fr}(r, E)$ is constructed from the model potential $U_{mod}(r)$ by calculating the spin-orbit, mass-velocity, Darwin term, and 4th-order corrections which yields a fully relativistic, energy dependent potential, including exchange, which has been discussed in the previous section

$$U_\kappa^{fr}(r, E) = U_{mod} + \kappa U_{SO}^{fr} + U_M^{fr} + U_D^{fr} + U_4^{fr} + U_{ex}, \tag{3.25}$$

where

$$\kappa = \begin{cases} \dfrac{1}{2}\ell & \text{if } j = \ell + 1/2, \\[2mm] -\dfrac{1}{2}(\ell+1) & \text{if } j = \ell - 1/2, \end{cases} \tag{3.26}$$

and the different potentials are given as

$$U_{SO}^{fr}(r, E) = -\frac{1}{2r\,\eta} \frac{d\eta}{dr}, \tag{3.27}$$

$$U_M^{fr}(r, E) = (\gamma - 1) U_{mod} - \frac{\alpha^2}{2} U_{mod}^2, \tag{3.28}$$

$$U_D^{fr}(r, E) = -\frac{1}{2r\,\eta}\frac{\mathrm{d}\eta}{\mathrm{d}r} - \frac{1}{4\eta}\frac{\mathrm{d}^2\eta}{\mathrm{d}r^2},\qquad(3.29)$$

and

$$U_4^{fr}(r, E) = \frac{3}{8}\left(\frac{1}{\eta}\frac{\mathrm{d}\eta}{\mathrm{d}r}\right)^2,\qquad(3.30)$$

where

$$\gamma(E) = \frac{1}{\sqrt{1 - \alpha^2 k^2}},\qquad(3.31)$$

and

$$\eta(r, E) = 2 + \alpha^2\left(\frac{k^2}{2} - U_{mod}(r)\right),\qquad(3.32)$$

where α denotes the fine structure constant ($\alpha = 1/137$) and in atomic units we have $k = \sqrt{2E}$. U_{ex} denotes the exchange potential.

On the other hand, we may apply the Foldy–Wouthuysen transformation (Foldy and Wouthuysen 1950), and obtain again a two-component Schrödinger equation. However, the model potential includes only terms up to the order of α^2 in this case. We will denote this kind of calculation as semi-relativistic. The calculation of the continuum wavefunction is then done by solving the radial two-component Schrödinger equation with the selected model potential, i.e.

$$\left(\frac{\mathrm{d}^2}{\mathrm{d}r^2} + k^2 - \frac{\ell(\ell+1)}{r^2} - 2U_\kappa^{sr}(r, E)\right)u_\kappa(r) = 0.\qquad(3.33)$$

Following Meister and Weiss (1968) we obtain the semi-relativistic potential as

$$U_\kappa^{sr}(r, E) = U_{mod} + \kappa U_{SO}^{sr} + U_M^{sr} + U_D^{sr} + U_{ex},\qquad(3.34)$$

where

$$U_{SO}^{sr}(r) = \alpha^2\frac{1}{4r}\frac{\mathrm{d}\,U_{mod}}{\mathrm{d}r},\qquad(3.35)$$

$$U_M^{sr}(r, E) = \frac{\alpha^2}{2}\left(k^2 U_{mod} - U_{mod}^2\right),\qquad(3.36)$$

and

$$U_D^{sr}(r) = \alpha^2\left(\frac{1}{4r}\frac{\mathrm{d}\,U_{mod}}{\mathrm{d}r} + \frac{1}{8}\frac{\mathrm{d}^2 U_{mod}}{\mathrm{d}r^2}\right).\qquad(3.37)$$

Generally, we note that the numerical results for a certain Auger transition remain almost unaffected by using either the full- or the semi-relativistic potential in the calculation.

The calculation of the relativistic potentials, the continuum wavefunctions and scattering phases, and of the required Slater integrals is controlled in the subroutine DWBA which is shown in Fig. 3.3. The subroutine MACHIN calculates some machine dependent parameters which are used to avoid an over- and underflow, respectively (Shampine and Gordon 1975). The routine SETPOT reads the calculated

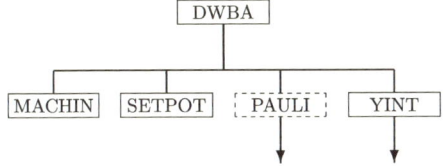

Fig. 3.3. The subroutine DWBA. The call of PAULI is optional

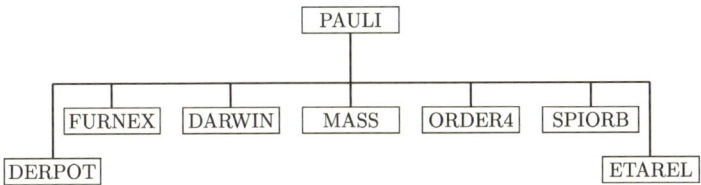

Fig. 3.4. The subroutine PAULI

model potential $U_{mod}(r)$, its derivative and the charge density of the residual ionic state from the program part POTRES. The calculation of the Slater integrals is controlled by the subroutine YINT which is discussed in Sect. 3.1.8.

The calculation of the potentials U_κ^{fr} or U_κ^{sr} is controlled by the subroutine PAULI which is illustrated in Fig. 3.4. The called subroutines SPIORB, DARWIN, MASS and ORDER4 allow for explicitly *switching on* any of the potentials according to (3.27)–(3.30) or (3.35)–(3.37), respectively. The calculation of the exchange (see Sect. 3.1.3) is included via the subroutine FURNEX. The subroutines ETAREL and DERPOT calculate $\eta(r, E)$ and its logarithmic derivative, respectively.

As a third possibility, the ANISO program may use non-relativistic wavefunctions with which a pure Hartree potential can be created generating a so-called non-relativistic continuum wavefunction, i.e. solving

$$\left(\frac{\mathrm{d}^2}{\mathrm{d}r^2} + k^2 - \frac{\ell(\ell+1)}{r^2} - 2U_{mod}(r) \right) u_\kappa(r) = 0. \tag{3.38}$$

This method has not been used very often since it, compared to the other two methods, leads to remarkable deviations in the results.

The calculation of the continuum wavefunction is controlled by the subroutine DWAVES for all possible values of ℓ and j. The program structure of DWAVES is shown in Fig. 3.5. In particular, the calculation of the Coulomb phase shift is done in subroutine COULGA. The routine COULFG calculates the regular and irregular asymptotic solutions of the continuum wavefunction (Barnett 1982). The matching procedure of the solution of the Schrödinger equation with the model potential to the asymptotic solution is done using a WKB method (Biedenharn *et al.* 1955). This is done in subroutine JWKB. The solution of the Schrödinger equation itself, with the model potential, is carried out in the routine BASFUN where the subroutines DERFUN, DEVGL and SIMEQN are used. Here, DERFUN provides the necessary derivatives of the continuum wavefunction for every second mesh point. The dif-

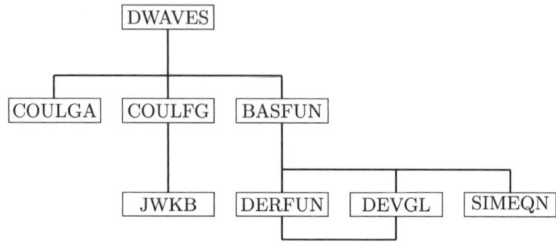

Fig. 3.5. The subroutine DWAVES

ferential equation is solved in subroutine DEVGL applying a de'Vogelaere method (Clenshaw *et al.* 1961). The routine SIMEQN can be generally used for solving a system of linearly coupled equations.

3.1.5 Calculation of the Scattering Phase

For calculating the scattering phase we make use of the fact that, independent of the model potential, the continuum wavefunctions show the same asymptotic behaviour. For small values of r we have

$$u_\kappa(r) \to 0 \quad \text{for} \quad r \to 0. \tag{3.39}$$

For large values of r the influence of the model potential can be neglected and the continuum wavefunction can be written as a linear combination of regular and irregular Coulomb functions

$$u_{\varepsilon j \ell}(r \to \infty) = \frac{1}{\sqrt{k}} \left(f_{\varepsilon j \ell}(r) - \tan \delta_\ell^j \, g_{\varepsilon j \ell}(r) \right). \tag{3.40}$$

The logarithmic derivation of this solution should match the solution of the Schrö-dinger equation in the inner region. Out of this we get a system of equations from which the phase shift can be evaluated. By adding the Coulomb phase σ_c, we receive the scattering phase as

$$\sigma_\ell^j = \sigma_c + \delta_\ell^j, \tag{3.41}$$

which is inserted into the relevant equations for the angular distribution and spin polarization parameters.

3.1.6 The Anisotropy Parameters

For calculating the Auger rates and the relevant anisotropy and spin polarization parameters, i.e. (2.132), (2.139), (2.144), (2.150), and (2.151), we need to evaluate the reduced Auger transition amplitudes and their scattering phases. This is done in the program part ANISO of the ANISO package. The overall structure of the ANISO part is shown in Fig. 3.6. Besides the calculation of frequently used numbers, which

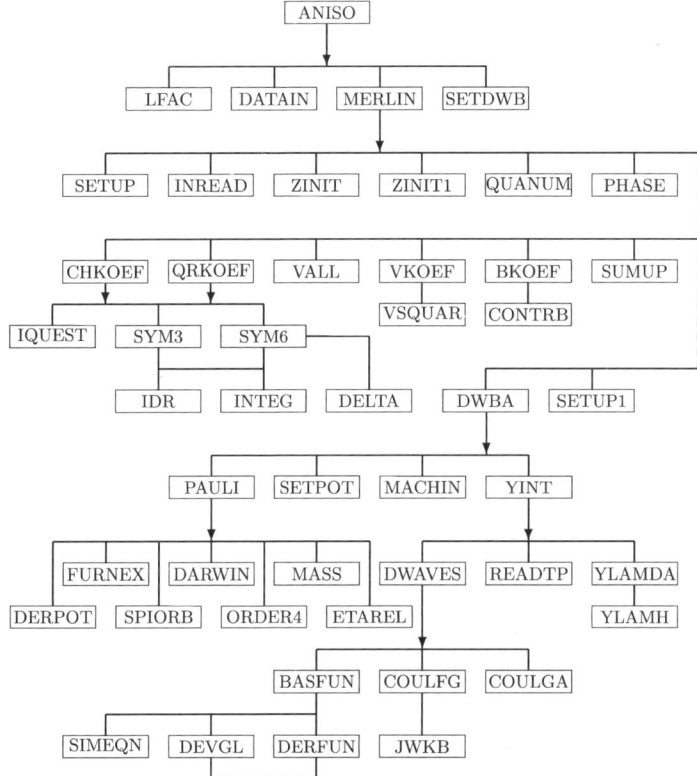

Fig. 3.6. The program part ANISO. Important branching points are marked by vectors

is done in subroutine LFAC, ANISO reads the input data via subroutines DATAIN and SETDWB. The further calculation is done in subroutine MERLIN which is the main control routine for all further calculations. Thus, besides the calculation of the bound state wavefunctions and the model potential all calculations are done in the ANISO part of the program package. Its main program steps are outlined in the following.

3.1.7 The Reduced Matrix Elements

For calculating the coefficients $A(KkQ)$, besides the calculation of the scattering phase, we need to evaluate the reduced matrix elements.

For simplicity, in the following we only consider cases with closed shells and sub-shells, respectively. I.e. $J_v = 0$, where J_v denotes the total angular momentum of the valence shell. Or, if $J_v \neq 0$, we assume that the valence shell electrons do not contribute to the Auger emission. I.e., considering the electronic configuration of the neutral atom, closed shells contribute to the Auger emission process, only. This is for instance perfectly fulfilled for the case of the rare gases.

With this, and adopting the notation of Sect. 3.1.1 we write the reduced matrix elements for a specific ASF as

$$\langle (j_1 j_2) J_f(\alpha) \| V \| (Jj) J_f(\alpha) \rangle = \sum_r c_r(\alpha)$$

$$\times \langle (n_1 \ell_1 j_1; n_2 \ell_2 j_2) J_f(\alpha) \| V \| (NLJ; n\ell j) J_f(\alpha) \rangle, \quad (3.42)$$

where (NLJ) denote the quantum numbers of the primary inner shell hole state. The doubly ionized final hole state is described by the total angular momentum J_f, and the two sets $(n_1 \ell_1 j_1)$ and $(n_2 \ell_2 j_2)$. Here, the index "1" is always associated with the innermost hole of the final state. The $c_r(\alpha)$ are the configuration mixing coefficients of the contributing CSF to the ASF, see (3.1), which have been obtained in the MCDF part. In order to simplify the discussion configuration mixing will be suppressed in the following.

The Auger electron is described as an initial hole in the continuum with quantum numbers $(n\ell j)$. In jj coupling this description is equivalent to an electronic one for the description of relativistic processes (see e.g. Grant 1970).

Due to the Fermi character of the electrons, and thus of the holes, we need to consider two types of transitions:

1. The *direct-process*: see Fig. 3.7. Here, the continuum hole goes down into the outer and the primary shell hole is shifted into the inner shell of the final state, respectively. I.e. the transitions

$$\left| NLJ \right\rangle \longrightarrow \left| n_1 \ell_1 j_1 \right\rangle \quad (3.43)$$

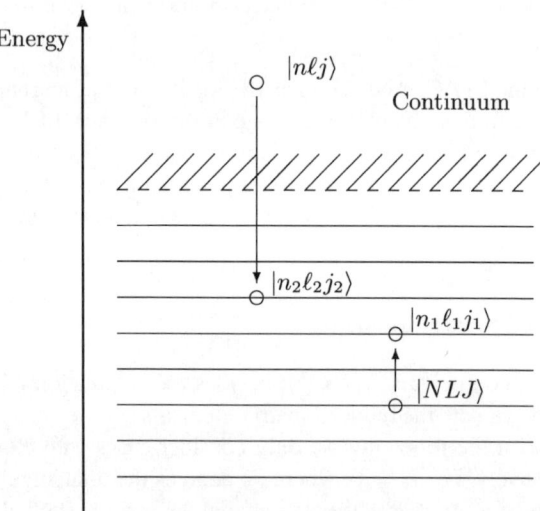

Fig. 3.7. The direct Auger emission process; the continuum hole state is transferred into the outer, the inner shell hole of the primary ion is shifted into the inner shell of the doubly ionized final state

and

$$|n\ell j\rangle \longrightarrow |n_2\ell_2 j_2\rangle \tag{3.44}$$

take place. The transition matrix element is denoted with the index "D".

2. The *exchange-process*: see Fig. 3.8. Here, the continuum hole goes down to the inner and the primary hole is shifted into the outer shell of the final state. I.e. the transitions

$$|NLJ\rangle \longrightarrow |n_2\ell_2 j_2\rangle \tag{3.45}$$

and

$$|n\ell j\rangle \longrightarrow |n_1\ell_1 j_1\rangle \tag{3.46}$$

occur. The transition matrix element is denoted with the index "E".

Both processes are indistinguishable and cannot be resolved within an experiment. Thus, we must anti-symmetrize to satisfy the Fermi statistics which yields

$$\langle (j_1 j_2) J_f \| V \| (Jj) J_f \rangle = \tau \Big(\langle (\phi(1)_{j_1}\phi(2)_{j_2}) J_f \| V \| (\phi(1)_J \phi(2)_j) J_f \rangle_D$$
$$- \langle (\phi(1)_{j_1}\phi(2)_{j_2}) J_f \| V \| (\phi(2)_J \phi(1)_j) J_f \rangle_E \Big), \tag{3.47}$$

where

$$\tau = \begin{cases} 1 & \text{for non-equivalent electrons,} \\ \dfrac{1}{\sqrt{2}} & \text{for equivalent electrons.} \end{cases} \tag{3.48}$$

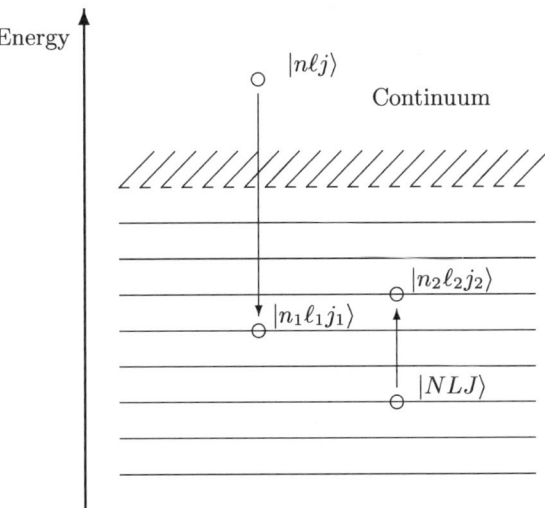

Fig. 3.8. The exchange Auger emission process: The primary inner shell hole is shifted into the outer, the continuum hole is transferred into the inner shell of the doubly ionized final state

Here, we introduced the parameter τ to include the case of a final state with two equivalent electrons (see e.g. Cowan 1981).

The function $\phi(1)_J$, for instance, denotes the CSF with the total angular momentum J; for simplicity, the other quantum numbers which are necessary to uniquely define the state have been suppressed. The notation 1, 2 abbreviates the coordinates $(r_1, \vartheta_1, \varphi_1)$ and $(r_2, \vartheta_2, \varphi_2)$ of the electronic and hole states, respectively.

For further evaluating the D- and E-terms of the transition matrix elements we expand the Coulomb operator into Legendre polynomials,

$$\frac{1}{r_{12}} = \sum_{\lambda=0}^{\infty} \gamma_\lambda(r_1, r_2) \, P_\lambda(\cos\theta_{12}), \tag{3.49}$$

where γ_λ is defined according to (3.15) and θ_{12} denotes the angle between (ϑ_1, φ_1) and (ϑ_2, φ_2). The Legendre polynomials can now be expressed in terms of a product of Racah tensors,

$$P_\lambda(\cos\theta_{12}) = \sum_\mu (-1)^\mu \, C_\mu^\lambda(\vartheta_1, \varphi_1) \, C_{-\mu}^\lambda(\vartheta_2, \varphi_2)$$

$$= \mathbf{C}_{(1)}^{(\lambda)} \, \mathbf{C}_{(2)}^{(\lambda)}. \tag{3.50}$$

The Racah tensors are proportional to the spherical harmonics

$$C_\mu^\lambda(\vartheta, \varphi) = \sqrt{\frac{4\pi}{2\lambda + 1}} \, Y_{\lambda\mu}(\vartheta, \varphi). \tag{3.51}$$

The Racah tensors show the advantage of irreducibility.

For the further reduction of direct and exchange terms we refer to the derivations by Asaad (1963a, 1963b) and Lohmann (1988, 1990). Eventually, we obtain the following expression for the reduced transition matrix elements

$$\langle (j_1 j_2) J_f \| V \| (Jj) J_f \rangle = \tau \left(\sum_\lambda (-1)^{J_f + j_2 + j} \mathcal{D}(\lambda) Q_s(\lambda, j) \right.$$

$$\left. - \sum_\mu (-1)^{j_2 - j} \mathcal{E}(\mu) R_s(\mu, j) \right). \tag{3.52}$$

Here, $\mathcal{D}$ and $\mathcal{E}$ denote the direct and exchange Slater integrals, respectively. They are defined as

$$\mathcal{D}(\lambda) = \mathcal{R}^{(\lambda)}(n_1 \ell_1 j_1; n_2 \ell_2 j_2; NLJ; n\ell j), \tag{3.53}$$

and

$$\mathcal{E}(\mu) = \mathcal{R}^{(\mu)}(n_1 \ell_1 j_1; n_2 \ell_2 j_2; n\ell j; NLJ), \tag{3.54}$$

where the Slater integrals are given as

$$\mathcal{R}^{(\lambda)}(a, b, c, d) = \iint R_a(r) R_b(s) \gamma_\lambda(r, s) R_c(r) R_d(s) \, r^2 \, s^2 \, dr \, ds. \tag{3.55}$$

Here, (a, b, c, d) abbreviates the quantum numbers of the radial wavefunctions and γ_λ has been defined in (3.15). The Slater integrals show some useful symmetries, which can be easily derived by interchanging r and s. In general we have

$$\mathcal{R}^{(\lambda)}(a, b, c, d) = \mathcal{R}^{(\lambda)}(b, a, d, c) \tag{3.56}$$

and if the two first or the two second radial wavefunctions are identical (e.g. $a = b$) we have

$$\mathcal{R}^{(\lambda)}(a, a, c, d) = \mathcal{R}^{(\lambda)}(a, a, d, c). \tag{3.57}$$

The coefficients Q_s and R_s basically consist of $3nj$-symbols. We have

$$Q_s(\lambda, j) = \Gamma(\hat{J})\sqrt{2J_f + 1} \begin{pmatrix} J & j_1 & \lambda \\ 1/2 & -1/2 & 0 \end{pmatrix} \begin{pmatrix} j & j_2 & \lambda \\ 1/2 & -1/2 & 0 \end{pmatrix} \begin{Bmatrix} j_2 & j & \lambda \\ J & j_1 & J_f \end{Bmatrix}, \tag{3.58}$$

and

$$R_s(\mu, j) = \Gamma(\hat{J})\sqrt{2J_f + 1} \begin{pmatrix} J & j_2 & \mu \\ 1/2 & -1/2 & 0 \end{pmatrix} \begin{pmatrix} j & j_1 & \mu \\ 1/2 & -1/2 & 0 \end{pmatrix} \begin{Bmatrix} j_1 & j & \mu \\ J & j_2 & J_f \end{Bmatrix}, \tag{3.59}$$

where the abbreviation

$$\Gamma(\hat{J}) = \sqrt{(2J + 1)(2j + 1)(2j_1 + 1)(2j_2 + 1)} \tag{3.60}$$

is used. Again, the coefficients Q_s and R_s have a useful symmetry. If we interchange j_1 and j_2 we find that

$$Q_s(j_1, j_2, \lambda, j) = R_s(j_2, j_1, \lambda, j). \tag{3.61}$$

An analogous expression can be found for interchanging J and j. Thus, if either $j_1 = j_2$ or $j = J$ is fulfilled, the coefficients Q_s and R_s are identical.

Note, that the reduced matrix elements, i.e. (3.52), are real numbers. This is, since we separated the expression for the complex scattering phase which is contained in the partial wave expansion coefficients c_ℓ^j.

Generally, we note that the *direct* term is only non-zero if

$$(-1)^{\ell_1 + \lambda + L} = 1 \quad \text{and} \quad (-1)^{\ell_2 + \lambda + \ell} = 1 \tag{3.62}$$

are simultaneously fulfilled. Analogously,

$$(-1)^{\ell_1 + \mu + \ell} = 1 \quad \text{and} \quad (-1)^{\ell_2 + \mu + L} = 1 \tag{3.63}$$

must be fulfilled for the *exchange* term. Thus, the summations over λ and μ in (3.52) are restricted to a finite number of elements, depending on the specific type of Auger transition.

If we are dealing with equivalent electrons, i.e.

$$n_1 = n_2, \quad \ell_1 = \ell_2, \quad \text{and} \quad j_1 = j_2 \tag{3.64}$$

for the final target state, the symmetries (3.61) and (3.57) apply, and using $\tau = 1/\sqrt{2}$, (3.52) can be simplified as follows

$$\langle (j_1 j_1) J_f \| V \| J_f (Jj) \rangle = \frac{1}{\sqrt{2}} \left(1 + (-1)^{J_f} \right)$$
$$\times \left(\sum_\lambda (-1)^{j_1 - j + 1} \mathcal{D}(\lambda) Q_s(\lambda, j) \right). \quad (3.65)$$

In this case the reduced matrix elements are only different form zero for an even total angular momentum J_f. This is however not a special result of our calculation but a general restriction in the jj coupling scheme based on the *Pauli principle*.

The calculation of the reduced matrix elements is controlled in the subroutine MERLIN which also controls the calculation of the relativistic intensities and the relevant angular distribution and spin polarization parameters. Its general structure is illustrated in Fig. 3.9. The subroutines SETUP and SETUP1 read additional input for the ANISO part where the call of the latter is optional. ZINIT initializes all arrays and QUANUM calculates all necessary upper and lower bounds for all summations to be carried out while ANISO is running. INREAD re-reads the calculated (stored) Slater integrals and the scattering phases. From the latter, the cosine and sine of the phase differences are obtained in subroutine PHASE whereas the Slater integrals are used in subroutines VKOEF and VSQUAR for the calculation of the reduced matrix elements. The subroutines BKOEF and SUMUP eventually calculate the relevant coefficients $A(KkQ)$ which are related to the relative intensities, angular distribution and spin polarization parameters.

The subroutines QRKOEF and CHKOEF calculate all occurring coefficients $Q_s(\lambda, j)$ or $R_s(\mu, j)$ of (3.58) and (3.59) and $C(K)_{jj'}$ or $H(K)_{jj'}$ of (2.140) and (2.152), respectively, in the limits which have been calculated in subroutine QUANUM. The structure of QRKOEF and CHKOEF is shown in Fig. 3.10. where both subroutines basically require the calculation of nj-symbols.

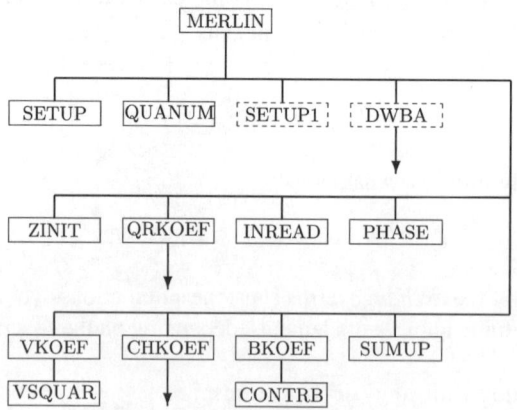

Fig. 3.9. The subroutine MERLIN

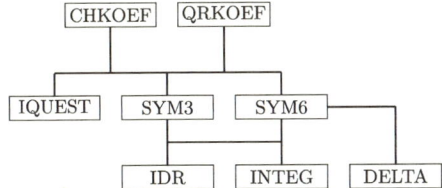

Fig. 3.10. The subroutines CHKOEF and QRKOEF

3.1.8 The Slater Integrals

The calculation of the Slater integrals is done in the program part YINT for all allowed combinations of quantum numbers which occur for a certain Auger transition. The integration is carried out for the large components $P_i(r) = rR_i(r)$ of the radial wavefunctions, only. Thus, we have to re-normalize

$$\int_0^\infty P_i(r)P_j(r)\,dr = \delta_{ij}. \tag{3.66}$$

For the calculation we start from (3.55). Using $P_i(r) = rR_i(r)$, we can write the Slater integrals as

$$\mathcal{R}^{(\lambda)}(a,b,c,d) = \int_0^\infty \int_0^\infty P_a(s)P_b(r)\gamma_\lambda(s,r)P_c(s)P_d(r)\,ds\,dr. \tag{3.67}$$

Here, the quantum numbers a and b are associated with the two hole states of the doubly ionized target, $|n_1\ell_1 j_1\rangle$ and $|n_2\ell_2 j_2\rangle$. The initial hole state $|NLJ\rangle$ is abbreviated as c and d denotes the continuum state $|n\ell j\rangle$. This choice of notation may influence the numerical calculation which will be discussed later in this section.

For carrying out the numerical integration we re-write the Slater integrals as

$$\mathcal{R}^{(\lambda)}(a,b,c,d) = \int_0^\infty P_b(r)P_d(r)\left(\frac{1}{r^{\lambda+1}}YL1(\lambda,r) + r^\lambda\,YL2(\lambda,r)\right)dr, \tag{3.68}$$

where we used the abbreviations

$$YL1(\lambda,r) = \int_0^r P_a(s)P_c(s)s^\lambda\,ds, \tag{3.69}$$

and

$$YL2(\lambda,r) = \int_r^\infty P_a(s)P_c(s)\frac{1}{s^{\lambda+1}}\,ds. \tag{3.70}$$

Calculating (3.69) and (3.70) using Simpson's formula eventually yields (3.68). The calculation of $YL1$ and $YL2$ is done at every second mesh point where the mesh goes from $0.5 \cdot 10^{-4}$ a.u. to 10.1808 a.u. Then, the calculation of $\mathcal{R}^{(\lambda)}$ is done for every forth mesh point.

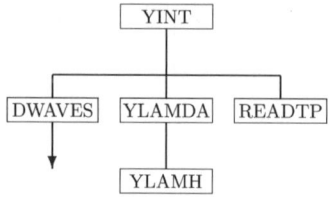

Fig. 3.11. The subroutine YINT

Considering the relation of the quantum numbers to the initial and final states, respectively, we denote the *direct integral* (3.53) as

$$\mathcal{D}(\lambda) = \mathcal{R}^{(\lambda)}(a, b, c, d), \tag{3.71}$$

and the *exchange integral* (3.54) can be written as

$$\mathcal{E}(\lambda) = \mathcal{R}^{(\lambda)}(a, b, d, c). \tag{3.72}$$

Applying this to (3.68), we find that, for the direct term, the integration over the continuum state is done in the second integration step, for the exchange term, however, in the first integration step.

As has been discussed, direct and exchange terms must be identically for equivalent electrons; i.e. for $a = b$, $\mathcal{D}(\lambda) = \mathcal{E}(\lambda)$ must be fulfilled. However, due to the fact that the integration over the continuum state is done either in the first or second integration step, we obtain slightly different results with a deviation $\leq 10^{-5}$ for the direct and exchange integrals, respectively.

The subroutine YINT controls the calculation of the Slater integrals $\mathcal{R}^{(\lambda)}(a, b, c, d)$ for all allowed values of λ. The range of λ has been already calculated in subroutine QUANUM. The structure of YINT is shown in Fig. 3.11. The continuum wavefunctions and the scattering phases are calculated in subroutine DWAVES. The subroutine READTP reads in the necessary orbitals of the bound states. These have been calculated with the MCDF code, and are then interpolated in the program POTNUC to the appropriate mesh. The routine YLAMDA calculates the Slater integral according to (3.68) using the results for the integrals $YL1$ and $YL2$ of (3.69) and (3.70) which are calculated in the subroutine YLAMH.

3.2 The Extended RATR Program

In the RATR package we calculate the Auger transition matrix elements with multiconfigurational Dirac–Fock (MCDF) wavefunctions. Thus, in the MCDF model an atomic state is represented by a linear combination of CSF, e.g. see (3.1). The generation of the initial and final ionic bound states is in analogy to the discussion in Sect. 3.1.1 and the final diagonalization of the Hamiltonian matrix, see (3.2)–(3.4), gives the representation of the atomic states.

The multiconfigurational computer code GRASP (McKenzie *et al.* 1980; Grant and Parpia 1992) has been extended for the calculation of autoionization amplitudes and phases of the emitted electrons (Fritzsche 1992). Our program RATR (relativistic Auger transition rates) (Lohmann and Fritzsche 1994) has been developed in the context of scattering theory (cf. Åberg and Howat 1982) and can be used for the numerical integration of the continuum orbitals and the scattering phases.

3.2.1 Properties of the RATR Package

The RATR package applies intermediate coupling in the many-electron wavefunctions. Electronic relaxation has been accounted for via the incomplete orthogonality of the initial and final state wavefunctions (Löwdin 1955). The RATR program is divided into four modules as displayed in Fig. 3.12.

The program RATR applies the Δ-SCFCI method. Therefore, it starts from two separate dump files of GRASP which contain the energies and wavefunctions of the initial and final states (multiplets) of the Auger transitions (the so-called I- and F-

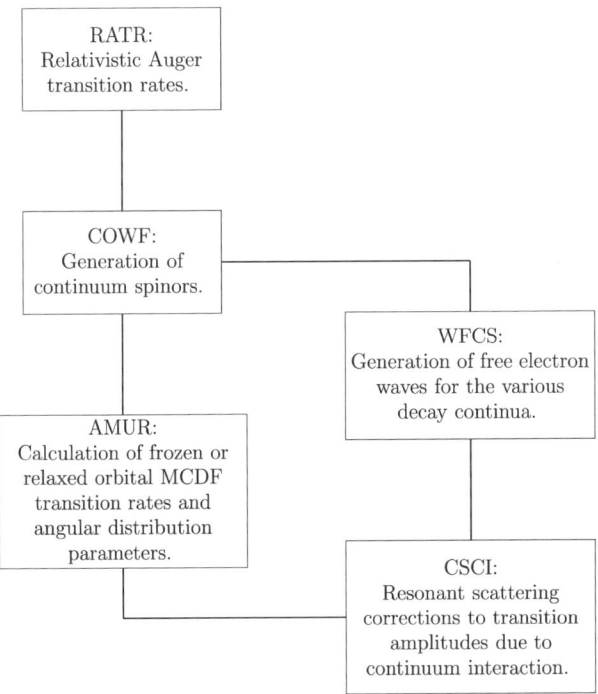

Fig. 3.12. Global structure of the RATR program package for the calculation of relativistic Auger transition rates and angular distributions (Fritzsche 1992; Lohmann and Fritzsche 1994). The program supposes two independent dump files of GRASP (Grant and Parpia 1992) which belong to the initial and final states, respectively. It is divided into four modules and controlled by an input data file which form is very similar to the input of the GRASP code

dumps, respectively). Here, we use the terms *initial* and *final* states with regard to a decay process. Initial is related to the initial vacancy state and final to the ionized outer hole states after the decay.

The continuum spinors in the construction of the final scattering states are generated by solving the Dirac–Fock equations for the emitted electron in the stationary field of the doubly ionized ion.

$$\left(\frac{d}{dr} + \frac{\kappa_a}{r}\right) P_{n_a \kappa_a}(r) - \left(2c - \frac{\epsilon_a}{c} + \frac{Y_a(r)}{cr}\right) Q_{n_a \kappa_a}(r) = -\frac{\xi_a^{(P)}(r)}{r} \tag{3.73a}$$

$$\left(\frac{d}{dr} - \frac{\kappa_a}{r}\right) Q_{n_a \kappa_a}(r) + \left(-\frac{\epsilon_a}{c} + \frac{Y_a(r)}{cr}\right) P_{n_a \kappa_a}(r) = \frac{\xi_a^{(Q)}(r)}{r}. \tag{3.73b}$$

Here we have used the notation of Dyall *et al.* (1989) with $\epsilon_a < 0$ for a free electron. In this coupled set of inhomogeneous differential equations $Y_a(r)$ denotes the *direct* potential, i.e.

$$Y_a(r) = -r V_{nuc}(r) - \sum_k \left(\sum_b y^k(ab) Y_k(bb; r) - \sum_{bd} y^k(abad) Y_k(bd; r)\right), \tag{3.74}$$

where the parameters $y^k(ab)$ and $y^k(abad)$ are expansion coefficients

$$y^k(ab) = \left(\frac{1 + \delta_{ab}}{\overline{q}_a}\right) \sum_r d_r^2 f_r^k(ab), \tag{3.75}$$

where $\overline{q}_a$ is the generalized occupation number

$$\overline{q}_a = \sum_r d_r^2 q_a(r), \tag{3.76}$$

and

$$y^k(abad) = \frac{1}{\overline{q}_a} \sum_{r,s} d_{rs} V_{rs}^k(abad). \tag{3.77}$$

The coefficients $d_r, r = 1, \ldots, n_c$, depend on the configuration mixing coefficients, and are chosen so that $\sum_r d_r^2 = 1$. In particular, for the average-level (AL) calculation mode (Grant *et al.* 1976), the need to obtain the coefficients d_{rs} iteratively is obviated by setting

$$d_{rs} = \begin{cases} (2J_r + 1)/\sum_t (2J_t + 1) & \text{if} \quad r = s, \\ 0 & \text{otherwise.} \end{cases} \tag{3.78}$$

For further details we refer to the cited literature (e.g. Dyall *et al.* 1989).

The relativistic Hartree Y-functions are defined as

$$Y_k(ab; r) = \int_0^\infty \gamma_k(r, s) \Big(P_a(s) P_b(s) + Q_a(s) Q_b(s)\Big) ds, \tag{3.79}$$

where $\gamma_k(r, s)$ has been given in (3.15). The right-hand side functions $\xi_a^{(P)}(r)$ and $\xi_a^{(Q)}(r)$ of the expressions (3.73a) and (3.73b) are the inhomogeneous terms. They contain terms for the exchange interaction between the continuum electron and the remaining bound state electrons and for ensuring their orthogonality where the latter has been achieved by introducing Lagrange multipliers. They may be expressed as

$$\xi_a^{(P)}(r) = X_a^{(P)}(r) + \frac{r}{c\bar{q}(a)} \sum_{b \neq a} \delta_{\kappa_a \kappa_b} \epsilon_{ab} Q_b(r), \tag{3.80a}$$

and

$$\xi_a^{(Q)}(r) = X_a^{(Q)}(r) + \frac{r}{c\bar{q}(a)} \sum_{b \neq a} \delta_{\kappa_a \kappa_b} \epsilon_{ab} P_b(r), \tag{3.80b}$$

where κ_a denotes the relativistic angular quantum number (seniority number), ϵ_{ab} are the Lagrange multipliers and the X-functions may be written as

$$X_a^{(P)}(r) = \frac{1}{c} \sum_k \left[\sum_{b \neq a} x^k(ab) Y_k(ab; r) Q_b(r) \right.$$
$$\left. - \sum_{\substack{b,c,d \\ c \neq a}} x^k(abcd) Y_k(bd; r) Q_c(r) \right], \tag{3.81a}$$

and

$$X_a^{(Q)}(r) = \frac{1}{c} \sum_k \left[\sum_{b \neq a} x^k(ab) Y_k(ab; r) P_b(r) \right.$$
$$\left. - \sum_{\substack{b,c,d \\ c \neq a}} x^k(abcd) Y_k(bd; r) P_c(r) \right], \tag{3.81b}$$

with the appropriate expansion coefficients $x^k(ab)$ and $x^k(abcd)$,

$$x^k(ab) = \frac{1}{\bar{q}_a} \sum_r d_r^2 g_r^k(ab), \tag{3.82}$$

and

$$x^k(abcd) = \frac{1}{\bar{q}_a} \sum_{r,s} d_{rs} V_{rs}^k(abcd). \tag{3.83}$$

Eventually, the Lagrange multipliers ϵ_{ab} are determined from either of the equations

$$\frac{\epsilon_{ab}}{\bar{q}(a)} = -I(ab) + c \int_0^\infty \frac{dr}{r} \left(P_b(r) X_a^{(Q)}(r) - Q_b(r) X_a^{(P)}(r) \right)$$
$$+ \int_0^\infty \frac{dr}{r} (Y_a(r) + r V_{nuc}(r)) (P_b(r) P_a(r) + Q_b(r) Q_a(r)), \tag{3.84a}$$

and

$$
\frac{\epsilon_{ab}}{\bar{q}(b)} = -I(ab) + c \int_0^\infty \frac{dr}{r} \left(P_a(r) X_b^{(Q)}(r) - Q_a(r) X_b^{(P)}(r) \right)
$$
$$
+ \int_0^\infty \frac{dr}{r} \left(Y_b(r) + r V_{nuc}(r) \right) \left(P_b(r) P_a(r) + Q_b(r) Q_a(r) \right), \quad (3.84b)
$$

or their difference and sum, respectively. The $I(ab)$ integrals stem from the one-body interaction and are given as

$$
I(ab) = \delta_{\kappa_a \kappa_b} \int_0^\infty dr \left[c \left(Q_a(r) P_b'(r) - P_a Q_b'(r) \right) - 2c^2 Q_a(r) Q_b(r) \right.
$$
$$
+ \frac{c\kappa_b}{r} \left(P_a(r) Q_b(r) + Q_a(r) P_b(r) \right)
$$
$$
\left. + V_{nuc}(r) \left(P_a(r) P_b(r) + Q_a(r) Q_b(r) \right) \right], \quad (3.85)
$$

where $f' \equiv df/dr$ denotes the first derivative. For a more detailed discussion we again refer to the cited paper by Dyall and co-workers and refs. therein.

To obtain the normalization and the phase of the continuum spinors a WKB method of Ong and Russek (1978) has been adapted.

The mixing coefficients between CSF of the same symmetry are usually obtained in the average-level scheme (Dyall *et al.* 1989). For the evaluation of the Auger amplitudes we used the bound orbitals of the initial resonant state and orthogonalized the continuum spinors with respect to the bound ones. Note, that we have to orthogonalize in any case since we are using different bound state orbitals for the description of the initial and final states.

The number of CSF to be included in the calculation is principally restricted by the available storage space and the actual CPU run-time, only. For the examples to be discussed in Chap. 4 we only used small basis sets which do not exceed 50 CSF. However, RATR has been designed for the use of large basis sets. E.g., transition rates for Rydberg Auger cascades from highly stripped ions have been calculated (Lohmann *et al.* 1998). For such calculations a large number of CSF need to be included. Common values are about 2000–3000 CSF.

3.2.2 Application Modes of RATR

With the RATR program, four different types of calculations can be performed. In a first calculation the continuum wavefunction is calculated by using only the pure Coulomb potential, and exchange interaction is neglected, that is $\xi_a^{(P)}(r) = \xi_a^{(Q)}(r) = 0$. (Note that spin-orbit terms etc. are automatically included since we solve the Dirac equation in full.) Secondly we can include the Breit interaction $b(i, j)$ of (3.4) in the calculation. As a third possibility the Dirac equation can be calculated including not only the Coulomb potential but also the exchange interaction in the generation of the continuum orbitals in (3.73a) and (3.73b). Finally, which is from our point of view the best approach, we can include all interactions in the Dirac equation that is Coulomb, Breit, and exchange interaction in the continuum. If not explicitly stating, this method has been used for all calculations with the RATR program. Examples will be discussed in the next chapter.

By obtaining the continuum and the bound state wavefunctions the Auger transition matrix elements can be evaluated of which the Auger rates and relative intensities can be obtained where the latter are normalized to the total rates. With respect to angular distribution and spin polarization parameters, at the present stage, the RATR program allows for the calculation of the angular distribution anisotropy parameter α_2, only. Calculation of α_2 requires explicit knowledge of the scattering phase. As a task for the future, the calculation of the spin polarization parameters shall be implemented in the program, too.

3.2.3 Influence of Relativistic and Exchange Effects

For concluding this section let us illustrate the influence of relativistic and exchange effects on the Auger transition rates, which have been investigated for the KLL Auger spectrum of the alkalis; see Sect. 4.8. The absolute Auger rates and relative intensities have been calculated in the four different approximations as discussed above. In calculation (a) we have only taken the static Coulomb repulsion in the evaluation of the Auger matrix elements into account. The analogous computation (b) includes additionally the (transverse) Breit interaction in the transition operator. Such relativistic corrections are particularly important for weak lines and/or heavier elements and may either increase or reduce the line intensities. By contrast, the exchange interaction between the emitted electron and the final ionic core in the

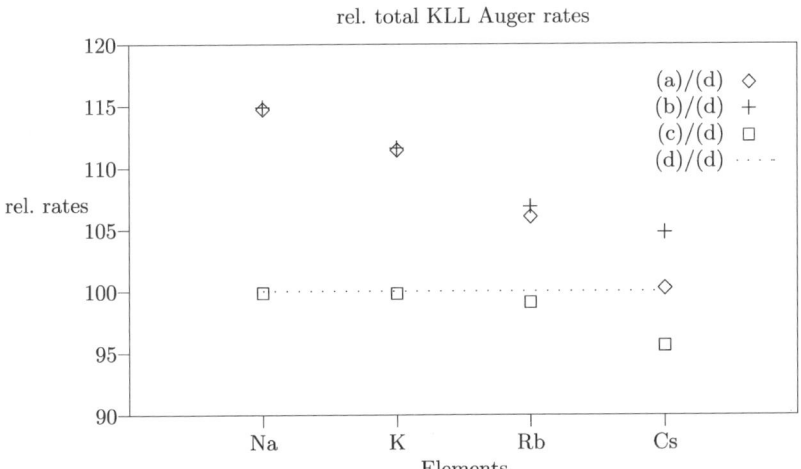

Fig. 3.13. The influence of the different theoretical approaches on the relative total Auger rates of the alkali KLL Auger spectra. The data are shown relative to the most accurate calculation (**d**) which is set equal 100. For the different types of calculations (**a**)–(**d**) see text. Data from Lohmann and Fritzsche (1994)

solution of the continuum wavefunctions reduces the absolute Auger rates in most cases. Calculation (c) shows the influence of the exchange interaction with respect to approach (a) and calculation (d) with respect to calculation (b). Since the effects of exchange depend on the individual transition one should note, that this may also result in a slightly higher intensity for some of the Auger lines.

To obtain a more general measure for the influence of Breit–QED and exchange interaction effects the total KLL Auger rates of the alkali elements for the different approaches are shown, relative to our most accurate calculation (d), in Fig. 3.13. It is interesting to note that our most exact calculation is close to the pure Coulomb-type result in the cesium case. This is due to the fact that Breit–QED and exchange interaction are almost of the same but opposite magnitude and cancel each other. However, the Coulomb-type calculation overestimates the total Auger rate for the light elements where the exchange interaction is dominant to the Breit–QED and generally reduces the total rates. One should note that this is only a general behaviour with respect to the total rates. The influence of the different interactions on certain Auger lines can show a considerably different behaviour.

4 Applications and Examples

In order to illustrate the applicability of the theoretical and numerical methods discussed in the previous chapters, we will now discuss a selected number of investigations of Auger transitions from free atoms. It is our aim, where possible, to compare our data with other theoretical results and with experimental data. Most data are available for the rare gases. This is, on the one hand, since rare gases are comparatively easy to handle in an experiment. On the other hand, their closed shell structure, i.e. $J = 0$, shows some numerical advantages. Thus, the discussion in this chapter will mainly consider our numerical results for Ar, Kr and Xe. Auger transitions on Ne have not been investigated since the ionization of its inner shell 1s hole does allow for neither orientation nor alignment, and thus, its KLL Auger spectrum is isotropic and no spin polarization for the Auger electrons can be observed. We did a few calculations on Rn (Lohmann 1988), since as the heavy most rare gas, it would be best to investigate relativistic effects. However, physicists are not keen on doing experiments with radon due to its radioactivity.

A detailed comparison between theoretical and experimental data of the angular anisotropy of Auger electrons emitted from noble gas atoms has been given by Kabachnik *et al.* (1991). In Sect. 4.1, we will discuss a few examples.

The spin polarization of Auger electrons from noble gases has been investigated in more detail by Lohmann *et al.* (1993), who provided full data sets, i.e. relative intensities, angular distribution and spin polarization parameters for a number of Auger transitions. In Sect. 4.2 the spin polarization of Auger electrons emitted after photoionization with circularly polarized light is discussed for Auger transitions in Ar, Kr and Xe. Since a circularly polarized photon beam generates an orientation in the intermediate ionic state, the effect of polarization transfer is considered in more detail. For this, the intrinsic in-reaction plane components of the spin polarization vector are investigated.

In Sect. 4.3, a special case of the previous section will be discussed. For a certain type of Auger transitions the intermediate ionic state can be oriented but not aligned. As a consequence the emitted Auger electrons can be spin polarized though remain isotropic with respect to their angular distribution. We will consider the intrinsic spin polarization of the isotropic Ar L_2MM Auger multiplet in detail (Lohmann and Larkins 1994). The fact that one of the intrinsic spin polarization parameters shows a constant value for most of the isotropic Auger multiplets has been explained by Lohmann (1996d).

The spin polarization of the Xe $M_{4,5}N_{4,5}N_{4,5}$ Auger lines has been measured in a high resolution experiment applying third generation synchrotron beam techniques (Hergenhahn 1996; Snell *et al.* 1996b), while we calculated the parameter set of relative intensities, angular anisotropy and spin polarization parameters (Kleiman *et al.* 1999a). Our results are discussed in Sect. 4.4.

Outside the rare gases, we investigated the Hg $N_{6,7}O_{4,5}O_{4,5}$ Auger transitions (Lohmann 1992, 1993), which can be observed in the gas phase, too. Here, numerical data stemming from an intermediate f-hole are provided. For mercury being a very heavy element, its relativistic effects are expected to be more pronounced than in the xenon case. This has been investigated considering different approaches for the initial and final state configuration interaction, and the parameter set for the line intensities, angular distribution, and spin polarization is discussed in Sect. 4.5.

A special field of research has been the Auger angular distribution of resonantly excited transitions. Applying some simple restrictions, a model has been developed to calculate the matrix elements for this type of Auger transitions (Lohmann 1991). This *strict spectator model* has been used for a detailed analysis of the angular distribution of resonant Auger transitions in Ar, Kr and Xe (Hergenhahn *et al.* 1991). In Sect. 4.6.1, we will discuss a more sophisticated version of the spectator model which has been applied by Hergenhahn *et al.* (1993).

Considering the dynamic spin polarization their values have been found as too small for experimental scrutiny for most of the diagram transitions. However, experiments (Snell *et al.* 1996a, 1996b) have given evidence of a large dynamic spin polarization for certain lines of the resonantly excited Xe* $(4d_{5/2} \rightarrow 6p)$ $N_5O_{2,3}O_{2,3}$ spectrum. In a more recent investigation (Lohmann 1999a), we derived simple propensity rules which allowed, for the first time, predictions for a large dynamic spin polarization of Auger electrons emitted in resonantly excited Auger transitions. Our predictions will be discussed in Sect. 4.6.2. It is shown, that they are in accordance with our numerical as well as with the experimental data. It is demonstrated that the effect of a large dynamic spin polarization is caused by a large shift of the scattering phase of the emitted $\varepsilon s_{1/2}$ partial wave whereas a small spin polarization is due to a cancellation between the Coulomb and scattering phases of the partial waves.

This behaviour is confirmed in Sect. 4.7, where we are discussing the show-case of resonantly excited Ar* $(2p \rightarrow 4s)$ $L_{2,3}M_{2,3}M_{2,3}$ Auger transitions, which has been investigated, both, experimentally and numerically, only recently (Lohmann *et al.* 2003b, 2005). In accordance with the results of Sect. 4.6.2, large transferred spin polarization has been found for most lines of the Auger spectrum, whereas the dynamic spin polarization vanishes for almost all Auger lines. The results are in accordance with our numerical calculation employing a small basis set for the final state configuration interaction. The experimental low resolution spectra, however, exhibit measurable dynamic spin polarization for, at least, one group of Auger lines. We have been able to explain this unexpected occurrence of dynamic spin polarization by strong configuration interaction in the final ionic state utilizing a large basis set for the calculation (Lohmann *et al.* 2005).

We also extended our investigations to open shell atoms. This is an open field of research where a number of theoretical predictions still need experimental proof. Our main aim has been the investigation of the angle resolved KLL Auger transitions. For closed shell atoms, like the rare gases, KLL Auger spectra have to be isotropic since no alignment can be generated. However, as has been shown by Dill *et al.* (1975) for the case of photoelectron emission, this is totally different for open shell atoms. We will discuss the KLL Auger transitions of the alkali atoms sodium, potassium, rubidium, and cesium, which have been investigated by Lohmann and Fritzsche (1994) in Sect. 4.8. Considering a primary ionization of the inner K-shell, its total angular momentum is coupled with the total angular momentum of the outer valence shell electron, which can result in a total angular momentum $J > 1/2$. Thereby, a non-zero alignment becomes possible for the intermediate ionic state which can result in an observable anisotropic angular distribution. In Sect. 4.9, we will consider the example of the open shell KLL Auger transitions of atomic oxygen. First data on the angle resolved O KLL spectrum have been published in an experiment by Krause *et al.* (1996), while we published theoretical data for the Auger rates and angular distribution parameters (Lohmann and Fritzsche 1996). Due to the coupling of the open $2p^4$ valence shell configuration to the inner shell K hole, the oxygen KLL spectrum is rather complex and allows for a variety of features in the Auger transitions.

4.1 Angular Anisotropy of Auger Electrons from Noble Gas Atoms

Let us now discuss the angular anisotropy of Auger electrons for the transitions Ar $L_3M_{2,3}M_{2,3}$, Kr $M_{4,5}N_{2,3}N_{2,3}$ and Xe $N_{4,5}O_{2,3}O_{2,3}$. Here, several experiments were carried out in which the angular distribution of Auger electrons in normal and resonant Auger processes of Ar, Kr and Xe were measured (Carlson *et al.* 1988, 1989; Kämmerling *et al.* 1989; Becker 1990a; Schmidt 1992). The measured anisotropy parameters α_2 of the normal Auger transitions were compared with theoretical predictions (Kabachnik and Sazhina 1988) based on simple, however rather crude approximations: LS coupling for the initial and final ionic states, Hartree–Slater single electron wavefunctions for the bound and Herman–Skillman potential for the continuum electron. Thus, it is not surprising that for some Auger lines a large discrepancy between theory and experiment was found.

Meanwhile, more refined theoretical models including electron-electron correlation and relativistic corrections were used in the calculations of the angular anisotropy and spin polarization of Auger electrons. It was demonstrated that for some Auger lines the improvement of the theoretical model leads to a considerable variation of the anisotropy parameter and spin polarization value. The discussion of the sensitivity of the calculated anisotropy and polarization on the theoretical model may be found in the cited literature as well as in the paper by Kämmerling *et al.* (1989).

Here, the relativistic Dirac–Fock model with intermediate coupling is used for the description of the initial and final state of the Auger transition. The results of our calculation are compared with earlier theoretical predictions and experimental data, as cited above. With a few exceptions good agreement with the experimental anisotropy is found.

4.1.1 Anisotropy Parameters

The calculated values of the anisotropy parameter α_2 are shown in Tables 4.1, 4.2 and 4.3, where they are compared with earlier calculations and experimental data. Before discussing the results in detail we should note the following. In all considered cases there are two states, $^1S_0$ and $^3P_0$, which have zero total angular momentum of the final ionic state. – Note, that the notation $^1S_0$ or $^3P_0$ in our case of intermediate coupling refers to the main component of the wavefunction. – It is well known (Cleff and Mehlhorn 1974b; Eichler and Fritsch 1976) that in this case the Auger transition proceeds through a single channel and therefore, the anisotropy parameter α_2 is independent of Auger amplitudes. Also see (2.142) and the discussion in Sect. 2.5.2. – This model independent value is usually used for the normalization of the experimental data. Two other transitions to the $^3P_{1,2}$ states are single channel transitions in the LS coupling approximation. Therefore, if the ion wavefunctions are close to LS coupling ones, the anisotropy should be practically model independent. The most general case is the $^1D_2$ transition which is a multi channel transition in any coupling scheme. One can expect that this transition will be the most sensitive to the details of the calculation.

Table 4.1. Theoretical (T) and experimental (E) data for the anisotropy coefficients α_2 of different Auger lines. The columns in Tables 4.1–4.3 are denoted as follows; (a): T Kabachnik et al. (1991); (b): T LS coupling by Kabachnik and Sazhina (1988); (c): E Sarkadi et al. (1990); (d): E Becker et al. (1989); (e): E Kämmerling et al. (1989); (f): E Carlson et al. (1989) evaluated by Kämmerling et al. (1989); (g): E Kämmerling et al. (1990); (h): E Becker (1990a, 1990b), data from Kabachnik et al. (1991)

Final state	Ar $L_3M_{2,3}M_{2,3}$ α_2			
	(a)	(b)	(c)	(d)
$^1S_0$	−1	−1	−1	−1
$^1D_2$	−0.441	0.056	−0.48(10)	−0.41
$^3P_0$	−1	−1		
$^3P_1$	−0.010	0		
$^3P_2$	0.795	0.8		
$^3P_\Sigma$	0.481	0.5	0.37(10)	0.49

Ar $L_3M_{2,3}M_{2,3}$

The Ar atom is well described by Russell–Saunders approximation. Therefore, one can expect that for the ^3P transitions the anisotropy should be close to the prediction of Kabachnik and Sazhina (1988) obtained in LS coupling approximation. This is confirmed by our results (see Table 4.1). A completely different situation occurs for the $^1D_2$ transition. Here, the results of the two calculations differ dramatically. The difference is mainly due to the following factors. First, Hartree–Slater wavefunctions were used by Kabachnik and Sazhina (1988) both, for bound and continuum electrons. However, the use of the Slater exchange potential for calculating the continuum wavefunction can lead to a large error in Auger amplitudes (Chen 1992). Second, the energies of Auger electrons were taken from the Hartree–Slater calculations and not from the experimental data like in the present work. Although the $^1D_2$ state may, in principle, be mixed with the $^3P_2$ state, the influence of the intermediate coupling is very weak in the Ar case. A very similar result for the $^1D_2$ transition was obtained by Lee (1990), $\alpha_2 = -0.49$ in pure LS coupling using Hartree–Fock wavefunctions.

There are two experimental data sets for the anisotropy of the discussed lines. The results by Carlson et al. (1989) are doubtful because a rather strong anisotropy was obtained for the line $2p_{1/2} \longrightarrow (3p^4)^3P$, which has to be isotropic due to symmetry considerations.[1] Sarkadi et al. (1990) have measured anisotropy coefficients in very good agreement with the present results.

There is also an indirect possibility to compare the calculated results with experimental data. The angular anisotropy of resonant Auger decay in Ar atoms, following a $2p_{3/2} \longrightarrow 4s$ excitation, was measured (Becker 1990a). As was pointed out first by Becker (1990b), in the strict spectator model of an s electron, there is a simple connection between the anisotropy of the resonant Auger electrons and the anisotropy of the corresponding normal Auger transition. This relation was also obtained theoretically (Hergenhahn et al. 1991; Lohmann 1991; Schmidt 1992). Using this relation and experimental data by Becker (1990a) one obtains the results shown in the last column of Table 4.1. It can be seen that there is good agreement between the experimental and the calculated values.

Kr $M_{4,5}N_{2,3}N_{2,3}$

The anisotropy coefficients of the $M_{4,5}N_{2,3}N_{2,3}$ ^3P lines from the two theoretical calculations differ only slightly. As it was expected (Kabachnik and Lee 1990b) the anisotropy coefficient of the $M_4N_{2,3}N_{2,3}$ $^3P_2$ line is nonzero due to intermediate coupling. It has the same sign as in the experiments (Kämmerling et al. 1989; Carlson et al. 1989). However, its value is much smaller than the experimental

---

[1] An anisotropic angular distribution requires an alignment which cannot be generated from a $J = 1/2$ state.

Table 4.2. The anisotropy coefficients α_2. For explanation see Table 4.1. Data from Kabachnik *et al.* (1991)

	Kr $M_4N_{2,3}N_{2,3}$			
Final	α_2			
state	(a)	(b)	(e)	(f)
$^1S_0$	-1	-1		
$^1D_2$	0.240	-0.093		
$^3P_0$	-1 ⎫	-1 ⎫		
$^3P_1$	-0.818 ⎬ -0.843	-0.8 ⎬ $-0.77(10)$		0
$^3P_2$	0.017 ⎭	0 ⎭	$0.21(9)$	1.45
	Kr $M_5N_{2,3}N_{2,3}$			
$^1S_0$	-1.07	-1.07		
$^1D_2$	0.330	-0.099	$0.18(4)$	0.50
$^3P_0$	-1.07	-1.07		
$^3P_1$	-0.739	-0.748		
$^3P_2$	-0.303	-0.382	$-0.31(6)$	0.14

one. (Note, that the last column in Table 4.2 contains the parameter α_2, calculated (Kämmerling *et al.* 1989) from the experimental anisotropy $\beta = \alpha_2 \mathcal{A}_{20}$ of Carlson *et al.* (1989), and the alignment parameter $\mathcal{A}_{20}$ obtained from the experimental anisotropy β of the transition to the $^1S_0$ state.) The presently calculated α_2 for the $M_5N_{2,3}N_{2,3}$ $^3P_2$ transition agrees very well with the experimental data (Kämmerling *et al.* 1989). On the other hand, the anisotropy coefficients α_2 for the transition to the $^1D_2$ state differ drastically in the two calculations. However, our result agrees much better with the experimental data than that obtained in LS coupling.

Xe $N_{4,5}O_{2,3}O_{2,3}$

The influence of intermediate coupling is largest in the Xe case. Our results are shown in Table 4.3. We obtained rather large positive anisotropy coefficients for the $N_4O_{2,3}O_{2,3}$ $^3P_2$ line which was predicted to be isotropic in the LS approximation. Our result is in a good agreement with experiments. As in the cases of Ar and Kr large differences between the two calculations were obtained for the $^1D_2$ transition. Our present calculation for the $N_{4,5}O_{2,3}O_{2,3}$ $^1D_2$ transitions agrees well with the experiments.

Concluding this section we note that the experimental anisotropy coefficients α_2 of Auger lines measured by different groups differ considerably. Taking into account the sensitivity of the anisotropy coefficient α_2 on the dynamics of Auger transition it seems worthwhile to measure the α_2 values as accurate as possible.

Table 4.3. The anisotropy coefficients α_2. For explanation see Table 4.1. Data from Kabachnik *et al.* (1991)

Final					
			Xe $N_4O_{2,3}O_{2,3}$		
			α_2		
state	(a)	(b)	(f)	(g)	(h)
$^1S_0$	-1	-1			
$^1D_2$	0.055	-0.554		0.05(6)	
$^3P_0$	-1 $\Big\}$ -0.839	-1			$\Big\}$ $-0.7(1)$
$^3P_1$	-0.835	-0.8	-0.43	$-0.73(11)$	
$^3P_2$	0.156	0	0.27	0.72(11)	1.2(2)
			Xe $N_5O_{2,3}O_{2,3}$		
$^1S_0$	-1.07	-1.07			
$^1D_2$	0.139	-0.592		0.23(4)	0.3
$^3P_0$	-1.07 $\Big\}$ -0.917	-1.07 $\Big\}$ -0.79		$-1.07(10)$	$-1.2(2)$
$^3P_1$	-0.734	-0.748		$-0.77(17)$	$-1.0(2)$
$^3P_2$	-0.227	-0.382		$-0.47(13)$	

The calculations of the Auger electron anisotropy with the Dirac-Fock intermediate coupling model give results which agree reasonably well with the experimental data. The only exception is the $M_4N_{2,3}N_{2,3}$ $^3P_2$ line in Kr for which the measured α_2 value is considerably larger than the calculated one. This line in Kr and the corresponding line in Xe are especially interesting because in pure LS coupling they should be isotropic. The anisotropy can arise due to relativistic effects and electron-electron correlation. Therefore, one can expect that the α_2 values of the $^3P_2$ lines should be very sensitive to these effects.

4.2 Complete Data Sets for Auger Electrons from Noble Gases

In this section we discuss numerical calculations for the spin polarization and angular anisotropy parameters of Auger decay for the Ar $L_3M_{2,3}M_{2,3}$, Kr $M_{4,5}N_{2,3}N_{2,3}$, Xe $N_{4,5}O_1O_{2,3}$ and Xe $N_{4,5}O_{2,3}O_{2,3}$ transitions. We give complete data sets, including relative intensities, angular distribution and all spin polarization parameters, for the transitions considered, which are calculated consistently within one approach. The relative intensities are given by contrast with experimental data.

We apply the two-step model and consider the Auger decay after photoionization with circularly polarized light (see Sect. 2.5.5). Then, two mechanisms for the production of an electron spin polarization exist (Klar 1980, see also the discussion in the first chapter).

- If the intermediate ionic state is aligned, the outgoing Auger electron can be spin polarized by a final state interaction with the core charge distribution. We will denote this as *dynamic spin polarization*.
- In the case of ionization by circularly polarized photons, an orientation of the intermediate state can be transferred partially to the Auger electron spin. This can be viewed as a consequence of the conservation of angular momentum and is due to the usually asymmetric population of the magnetic sublevels. We will therefore denote this as *transferred spin polarization*.

Theoretical calculations (Kabachnik and Sazhina 1988; Lohmann 1990, 1992) and experiments (Hahn *et al.* 1985; Merz and Semke 1990) on the dynamic spin polarization found evidence for this effect only in the form of very small polarization degrees. In contrast, first measurements of the polarization transfer showed this mechanism to be much more effective (Kuntze *et al.* 1993), also see Stoppmanns *et al.* (1992). In the following discussion we will focus our theoretical study on the latter phenomenon.

From the discussion in the previous chapters, it is clear that, in the usual two-step model of Auger decay, expressions for the degree of spin polarization can be formulated in terms of photon energy dependent parameters related to the primary ionization process, i.e. alignment and orientation, and fixed parameters dependent only on the dynamics of the Auger decay, the so-called *intrinsic* spin polarization parameters. In an experiment, analysis can aim at the latter or the former, provided that some knowledge or assumptions exist about the other quantity.

For the case of a circularly polarized photon beam the primary ionization process is sufficiently described by the parameters $\mathcal{A}_{20}$ and $\mathcal{O}_{10}$; see (2.136) and (2.137). The equations of angular distribution and spin polarization of the emitted Auger electrons for this case can be easily obtained from the general expressions in Sect. 2.5.5.

A first numerical study of the intrinsic spin polarization, i.e. the polarization transfer, has been performed by Kabachnik and Lee (1989) within the LS coupling scheme.

We are discussing our calculations of the spin polarization parameters for a number of Auger transitions in the noble gases, that are interesting for an experimental study. The calculations have been carried out within an MCDF approach, applying the ANISO program package using a relaxed orbital method. Thereby we take into account that the electron moves in the field of the residual ion. To probe the sensitivity of the parameters to the wavefunctions, we also carried out calculations in LSJ coupling for some of the transitions. The effects of exchange and channel coupling in the continuum have been neglected. The influence of exchange and channel coupling on the Auger amplitudes was considerable in the study by Tulkki *et al.* (1993) if each was taken separately, but tended to compensate when both were taken into account. – We refer to Fig. 3.13 and the discussion in Sect. 3.2, respectively. We are also discussing this effect for the case of open shell KLL Auger transitions in Sects. 4.8 and 4.9; also see Lohmann and Fritzsche (1994). – In many cases the α_2 parameter was shown to be not very sensitive on inclusion of the latter effects, too. As has been discussed in Sect. 2.5.2 the structural dependence of the δ_1 and ξ_1 parameters on the matrix elements of Auger decay is the same as for α_2. This is also true for the β_1 and γ_1 parameters which have been introduced by Kabachnik and Lee (1989); also see Sect. 2.5.2. Their connection to δ_1 and ξ_1 has been given in Table 2.7. In the following we will exploit the parameterization of the spin polarization vector with respect to the laboratory frame given by Kabachnik and Lee (1989). I.e. the quantization axis parallel to the beam axis which yields

$$P_X(\theta) = \frac{\frac{3}{4}\gamma_1 \mathcal{O}_{10} \sin(2\theta)}{1 + \alpha_2 \mathcal{A}_{20} P_2(\cos\theta)}, \tag{4.1}$$

$$P_Y(\theta) = \frac{\xi_2 \mathcal{A}_{20} \sin(2\theta)}{1 + \alpha_2 \mathcal{A}_{20} P_2(\cos\theta)}, \tag{4.2}$$

and

$$P_Z(\theta) = \frac{\mathcal{O}_{10}(\beta_1 + \gamma_1 P_2(\cos\theta))}{1 + \alpha_2 \mathcal{A}_{20} P_2(\cos\theta)}. \tag{4.3}$$

In our previous study (Kabachnik *et al.* 1991, also see the discussion in Sect. 4.1), as well as in an equivalent approach by Chen (1992) good agreement of theoretical results with the existing data on the angular anisotropy has been found. These facts let us expect, that our model has comparable power in explaining the transferred spin polarization. This has been confirmed by recent calculations (Kleiman *et al.* 1999a) and experimental data (Hergenhahn 1996) which are discussed in Sect. 4.4.

4.2.1 Spin Polarization Parameters β_1 and γ_1

We are giving theoretical results for the parameters β_1 and γ_1 of Auger transitions in Ar, Kr and Xe. Noble gases are experimentally easy to handle and can often be treated theoretically without regard to open shell effects. Therefore they provide an ideal testing ground for the investigation of new effects in the photophysics of atoms. However, the required ionization energies are higher than in the first experiment on barium by Kuntze *et al.* (1993).

For transitions to a final state with $J_f = 0$ the parameters β_1 and γ_1 as well as the angular anisotropy parameter α_2 are model independent and purely geometrical quantities (see Tables 2.4 and 2.8). Therefore, we will discuss the model dependent results for β_1 and γ_1 only, but give an overview of all intrinsic parameters and relative transition probabilities in Sect. 4.2.3. The results of our calculations in intermediate and LSJ coupling are presented in Tables 4.4 and 4.5. The LS designations of the final states should be understood as labels corresponding to the main component of the intermediate coupling wavefunction.

Ar $L_3M_{2,3}M_{2,3}$

The Ar atom is rather light and should be well described within the LS coupling approximation. As Table 4.4 shows, in fact the results in both coupling schemes

Table 4.4. Spin polarization parameters β_1 and γ_1, results calculated in intermediate and LSJ coupling. Data from Lohmann *et al.* (1993)

Transition		β_1	β_1 (LSJ)	γ_1	γ_1 (LSJ)
$Ar L_3 M_{2,3} M_{2,3}$	$^1D_2$	0.744	0.745	−0.136	−0.124
	$^3P_1$	−0.256	−0.248	0.990	0.994
	$^3P_2$	−0.275	−0.328	−0.358	−0.405
$Kr M_4 N_{2,3} N_{2,3}$	$^1D_2$	−0.441	−0.447	−0.346	−0.255
	$^3P_1$	0.607	0.626	−0.366	−0.358
	$^3P_2$	0.073	−0.089	−0.046	−0.358
$Kr M_5 N_{2,3} N_{2,3}$	$^1D_2$	0.670	0.683	0.171	0.117
	$^3P_1$	−0.400	−0.410	0.824	0.820
	$^3P_2$	−0.072	−0.254	−0.251	−0.206
$Xe N_4 O_{2,3} O_{2,3}$	$^1D_2$	−0.429	−0.447	−0.262	−0.105
	$^3P_1$	0.594	0.626	−0.371	−0.358
	$^3P_2$	0.110	−0.089	0.321	−0.105
$Xe N_5 O_{2,3} O_{2,3}$	$^1D_2$	0.643	0.683	0.103	0.057
	$^3P_1$	−0.392	−0.410	0.827	0.820
	$^3P_2$	0.070	−0.254	−0.220	−0.206

are nearly equal. Former comparison of values obtained for α_2 (see Sect. 4.1 or Kabachnik *et al.* 1991) within the same model showed good agreement with values obtained by angle resolved electron spectroscopy. Experimental measurements on this transition would therefore be a good test to assess the model sensitivity of spin polarization predictions relative to predictions of the angular distribution.

Kr $M_{4,5} N_{2,3} N_{2,3}$ and Xe $N_{4,5} O_{2,3} O_{2,3}$

These transitions in Kr and Xe occur between orbitals identical up to the main quantum number. As the atomic weight increases from Kr to Xe, we should expect a similar pattern for the β_1 and γ_1 parameters, but with increasing deviations from the LSJ coupling in the Xe case. In fact, the transitions to the $^3P_2$ states show this behaviour, with a pronounced model sensitivity of the Xe $N_4 O_{2,3} O_{2,3}$ ($^3P_2$) decay. This is caused by an additional selection rule suppressing εg partial waves in the case of LSJ coupling. As we turn to intermediate coupling, which is the case for Kr and Xe, the $\varepsilon g_{7/2}$ partial amplitude takes on non-zero values and influences, via a strong interference term with the $\varepsilon d_{5/2}$ partial wave, the value of the γ_1 parameter remarkably. The same has been already found for the angular distribution parameter by Chen (1992). In contrast, transitions to a $^3P_1$ state show no conspicuous model dependencies, thus indicating, that in this case the εd partial amplitude, allowed exclusively in LSJ coupling, dominates the transition even in intermediate coupling. As the transitions to ^3P_J states proceed in LSJ coupling only via a single channel, their values are identical for Kr and Xe in this approximation. The $^1D_2$ transitions

Table 4.5. Spin polarization parameters β_1 and γ_1 for the Xe $N_{4,5}O_1O_{2,3}$ decays, comparison between relativistic and non-relativistic results. For further explanation, see text. Data from Lohmann *et al.* (1993)

Transition	β_1 IC	β_1 HF	β_1 MBPT	γ_1 IC	γ_1 HF	γ_1 MBPT
$N_4O_1O_{2,3}$ $^1P_1$	−0.445	−0.45	−0.45	0.577	0.73	0.76
$^3P_1$	−0.009	−0.11	−0.04	−0.048	−0.06	0.03
$^3P_2$	0.262	0.18	−0.06	−0.414	−0.38	−0.55
$N_5O_1O_{2,3}$ $^1P_1$	0.681	0.68	0.68	−0.222	−0.32	−0.33
$^3P_1$	0.044	−0.01	0.42	−0.198	−0.05	0.05
$^3P_2$	−0.075	−0.30	−0.39	−0.173	0.09	0.07

IC: MCDF calculation with intermediate coupling, Lohmann *et al.* (1993).
HF: Hartree–Fock calculation, Kabachnik and Lee (1989).
MBPT: Many body perturbation theory, Kabachnik and Lee (1989).

are multichannel transitions in any coupling scheme. The β_1 parameter is slightly affected by utilizing intermediate coupling, whereas the γ_1 parameter shows larger differences.

Xe $N_{4,5}O_1O_{2,3}$

A non-relativistic calculation of β_1 and γ_1 for this case has been undertaken before by Kabachnik and Lee (1989). They used one approximation on the Hartree–Fock level and one taking into account correlations within the many-body perturbation theory (MBPT), but nevertheless in LSJ coupling. Their results are compared with our intermediate coupling calculations in Table 4.5. For most transitions all calculations are in rather good agreement with each other, however for β_1 in $N_5O_1O_{2,3}$ ($^3P_1$) the MBPT result deviates drastically from the two other calculations, while for γ_1 in $N_5O_1O_{2,3}$ ($^3P_2$) our value is significantly different from the two non-relativistic ones. This might be interpreted as caused by a different trend of correlation and relativistic effects. Nevertheless we think, a model being able of taking into account both (and other) effects consistently is necessary to confirm this.

4.2.2 Observable Degree of Spin Polarization

The observable degree of spin polarization is dependent on products of intrinsic Auger decay parameters with the values of orientation and alignment. Determining the latter parameters – dependent on photon energy – requires knowledge of the dipole matrix elements of photoionization. This, especially photoionization of deep inner shells, is still an active area of research. Only recently, a detailed Hartree–Fock calculation of deep inner shell alignment and orientation after photoionization for a variety of closed-shell atoms and cations has been published (Kleiman and Lohmann 2003; Kleiman and Becker 2005). However, a complete discussion of the alignment and orientation parameters would be beyond the scope of this book. Nevertheless, we can draw some conclusions based on published results.

As a general remark we want to point out, that measurement of the Cartesian components of the spin polarization vector under one angle is sufficient for the determination of the spin polarization parameters, provided that alignment and orientation are known. We suggest to measure at the *magic* angle $\theta_m = 54.74^o$. Performing the experiment under this angle simplifies (4.1)–(4.3) enormously. As $P_2(\cos \theta_m) = 0$ the denominator equals one and using $\sin(2\theta_m) = \sqrt{8}/3$ we obtain the simple relations

$$P_X(\theta_m) = \sqrt{\frac{1}{2}} \, \gamma_1 \, \mathcal{O}_{10}, \tag{4.4}$$

$$P_Y(\theta_m) = \frac{1}{3}\sqrt{8} \, \xi_2 \, \mathcal{A}_{20}, \tag{4.5}$$

and

$$P_Z(\theta_m) = \beta_1 \, \mathcal{O}_{10}. \tag{4.6}$$

If only a measurement of the P_Z component is possible, three different angles should be enough to determine β_1, γ_1 and the angular anisotropy $\alpha_2 \, \mathcal{A}_{20}$, if a value for $\mathcal{O}_{10}$ can be exploited.

If the ionic state parameters $\mathcal{O}_{10}$ and $\mathcal{A}_{20}$ are not known in advance, β_1 and γ_1 can be determined only up to a common factor, as only products $\beta_1 \, \mathcal{O}_{10}$ and $\gamma_1 \, \mathcal{O}_{10}$ enter (4.1) and (4.3). This situation is similar to the analysis of Auger angular distributions, where usually at first $\mathcal{A}_{20}$ is determined from the observable angular anisotropy parameter $\alpha_2 \, \mathcal{A}_{20}$ of a model independent, i.e. $J_f = 0$, transition from the considered core hole.

Expressions for the $\mathcal{O}_{10}$ and $\mathcal{A}_{20}$ parameters in LSJ coupling have been given by Kabachnik and Lee (1989) and Kleiman and Lohmann (2003). In this approach only two different partial amplitudes for photoionization of closed shell atoms are possible and only their ratio enters the expressions. Usually, out of the region of a Cooper minimum and well above threshold, the $\varepsilon(\ell+1)$ partial amplitude dominates photoionization from the $n\ell$ shell. Limiting values for alignment and orientation can be calculated for the case of total neglect of the other partial amplitude. We get $\mathcal{O}_{10} = -0.671$ and $\mathcal{A}_{20} = 0.1$ for photoionization of a $d_{3/2}$ electron and $\mathcal{O}_{10} = -0.683$ and $\mathcal{A}_{20} = 0.107$ for a $d_{5/2}$ electron. Experimental and theoretical results (Southworth *et al.* 1983; Berezhko *et al.* 1978b) indicate that for Kr 3d and Xe 4d photoionization these values are in fact realized. In Fig. 4.1 we therefore give the θ dependence of the spin polarization component P_Z for the decays Xe $N_4O_{2,3}O_{2,3}$ ($^1S_0$) and Xe $N_5O_{2,3}O_{2,3}$ ($^1S_0$, $^1D_2$) inserting the limiting values for $\mathcal{O}_{10}$ and $\mathcal{A}_{20}$. We would like to point out, that these limiting values for the ionic parameters are met in a region of photon energy, where the partial cross section is high, e.g. for Xe 4d ionization around 100 eV.

For photoionization of the Xe 4d shell Berezhko *et al.* (1978a) and Kleiman and Lohmann (2003) performed a Hartree–Fock calculation of the alignment as a function of the photon energy and Kabachnik and Lee (1989) and Kleiman and Lohmann

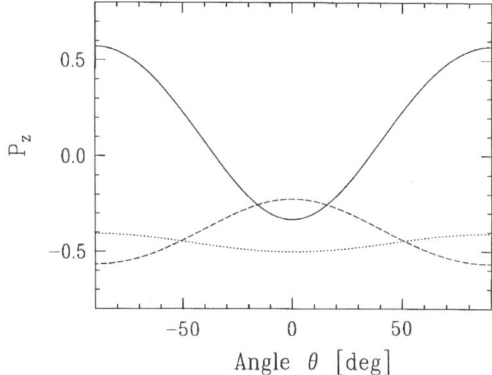

Fig. 4.1. P_Z dependent on the Auger electron emission angle: full curve, Xe $N_4O_{2,3}O_{2,3}$ ($^1S_0$); dotted curve, Xe $N_5O_{2,3}O_{2,3}$ ($^1D_2$); broken curve, Xe $N_5O_{2,3}O_{2,3}$ ($^1S_0$). For details see text. Data from Lohmann *et al.* (1993)

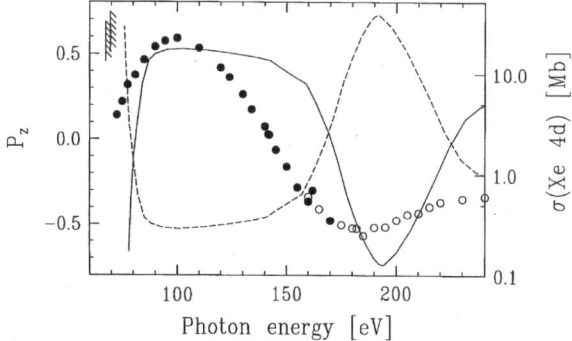

Fig. 4.2. P_Z dependent on photon energy, solid line: Xe $N_4O_{2,3}O_{2,3}\,^1S_0$, dashed line: Xe $N_5O_{2,3}O_{2,3}\,^1S_0$ under $\theta = 90^o$ (Lohmann *et al.* 1993). The points indicate experimental values for the Xe $4d$ partial cross section, filled dots: Becker *et al.* (1989), open dots: Lindle *et al.* (1988)

(2003) gave an orientation curve for this case. Both results were based on the same model. As the calculations were performed in LSJ coupling, $\mathcal{O}_{10}$ and $\mathcal{A}_{20}$ values for subshells with different spin coupling are distinguished only by a constant, geometrical factor: $\mathcal{O}_{10}(N_4) = \sqrt{\frac{27}{28}}\,\mathcal{O}_{10}(N_5)$ and $\mathcal{A}_{20}(N_4) = \sqrt{\frac{7}{8}}\,\mathcal{A}_{20}(N_5)$. Additionally, the curves are shifted in energy according to the values of the different ionization thresholds. For illustrative purposes we give in Fig. 4.2 the photon energy dependence of the P_Z component for the Xe $N_{4,5}O_{2,3}O_{2,3}$ ($^1S_0$) decays based on that simple model. Additionally, experimental values for the Xe $4d$ partial cross section are drawn (Becker *et al.* 1989; Lindle *et al.* 1988). We conclude, that the highest absolute values for the spin polarization are observed in the region of the Cooper minimum, but comparable polarization degrees are also found in other regions of photon energy, where photoionization cross sections are two orders of magnitude larger. This should constitute a favourable situation for the experimental proof of the transferred spin polarization.

4.2.3 Complete Data Sets for the Considered Auger Transitions

In Tables 4.6 to 4.8 we present complete results of our calculations, relative intensities and results for the parameters α_2, β_1, γ_1 and ξ_2, to provide a consistent data set for the analysis of experiments. The relative intensities have been obtained by normalization on the decay to the $^1S_0$ state for each initial vacancy. Some results for α_2 differ in the per cent range from the ones of Kabachnik *et al.* (1991) because we used a slightly different procedure of calculating the scattering phase difference. For the ξ_2 parameter the relative differences are somewhat larger, however the absolute values stay small. – Note, that the definition of ξ_2 in this section differs from that of Kabachnik *et al.* (1991) by inclusion of a factor $-3/2$; also see Table 2.7. Moreover, their angle ϑ was defined from the Z-axis, i.e. the beam axis, to the z-axis, i.e. the direction of Auger emission, opposite to our convention. Therefore all values for ξ_2 have to be reversed in sign for a comparison. – The tables contain also the results for the $J_f = 0$ transitions, which are independent of the model employed.

The tables show the range of possible behaviour of the spin polarization within the reaction plane. A high integral degree of spin polarization, related to a large β_1, can occur together with a low or even comparably high differential spin polarization, related to the γ_1 parameter (examples: Xe $N_5O_{2,3}O_{2,3}$ ($^1D_2$), Xe $N_4O_{2,3}O_{2,3}$ ($J_f = 0$), see also Fig. 4.1). Vice versa, the spin polarization can also be dominated by the differential term with a comparably low angle independent contribution, as for Ar $L_3M_{2,3}M_{2,3}$ ($^3P_1$). All intermediate stages between the extrema are also met. Only a minority of lines seems not to be spin polarized at all.

Concerning an analysis of experiments aiming at the ionic state parameters $\mathcal{O}_{10}$ and $\mathcal{A}_{20}$, we would like to point out the high spin polarization values and high inten-

Table 4.6. The relative intensities, angular anisotropy, and spin polarization parameters for the Ar $L_3M_{2,3}M_{2,3}$ Auger transitions. Data from Lohmann *et al.* (1993). Parameters of $J_f = 0$ transitions are independent of the matrix elements

	Rel. intensity		Parameter			
Transition	Theo.	Exp.†	α_2	β_1	γ_1	ξ_2
$L_3M_{2,3}M_{2,3}$ $^1S_0$	1.0	1.0	−1.0	0.745	−0.298	0
$^1D_2$	3.91	4.15	−0.441	0.744	−0.136	0.001
$^3P_0$	0.40	0.26	−1.0	0.745	−0.298	0
$^3P_1$	1.26	0.94	−0.010	−0.256	0.990	0.002
$^3P_2$	3.86	2.94	0.795	−0.275	−0.358	0.002

†: Väyrynen and Aksela (1979).

Table 4.7. The relative intensities, angular anisotropy, and spin polarization parameters for the Kr $M_{4,5}N_{2,3}N_{2,3}$ Auger transitions. Data from Lohmann *et al.* (1993). Parameters of $J_f = 0$ transitions are independent of the matrix elements

Transition	Rel. intensity		α_2	β_1	γ_1	ξ_2
	Theo.	Exp.†				
$M_4N_{2,3}N_{2,3}$ $^1S_0$	1.0	1.0	−1.0	−0.447	0.894	0
$^1D_2$	1.29	1.45‡	0.239	−0.441	−0.346	−0.001
$^3P_0$	0.06		−1.0	−0.447	0.894	0
$^3P_1$	0.35	0.53§	−0.818	0.607	−0.366	−0.005
$^3P_2$	0.19	0.28	0.019	0.073	−0.046	−0.060
$M_5N_{2,3}N_{2,3}$ $^1S_0$	1.0	1.0	−1.069	0.683	−0.390	0
$^1D_2$	1.35	1.23	0.330	0.670	0.171	0.035
$^3P_0$	0.17	‡	−1.069	0.683	−0.390	0
$^3P_1$	0.26	‡	−0.739	−0.400	0.824	0.001
$^3P_2$	0.70	1.12	−0.303	−0.072	−0.251	−0.033

†: Aksela *et al.* (1984a, 1984b).
‡: $N_4O_{2,3}O_{2,3}$ $^1D_2$ + $N_5O_{2,3}O_{2,3}$ $^3P_{0,1}$.
§: Not observed.

sities of the Kr $M_{4,5}N_{2,3}N_{2,3}$ ($^1S_0$) and Xe $N_{4,5}O_{2,3}O_{2,3}$ ($^1S_0$) transitions, that were well resolved in the high resolution Auger spectra by Werme *et al.* (1972). As the intrinsic angular anisotropy and spin polarization parameters for these transitions are model independent, the orientation and alignment of the intermediate vacancy state can be inferred from these lines. Once $\mathcal{O}_{10}$ and $\mathcal{A}_{20}$ are known, intrinsic parameters for decays to other final states can be determined.

Probably, it will not be possible to resolve all lines calculated by us, but though different unresolved contributions add up incoherently, a non-vanishing spin polarization might still remain. We would expect this, for example, for the Xe $N_4O_{2,3}$ $O_{2,3}(^1D_2)$ and Xe $N_5O_{2,3}O_{2,3}$ ($^3P_2$) transitions not resolved by Werme *et al.* (1972). In other cases a rather strong cancellation is likely to occur, e.g. for Xe $N_5O_{2,3}$ $O_{2,3}(^3P_{0,1})$. The decays $M_4N_{2,3}N_{2,3}$ ($^3P_2$) in Kr and $N_4O_{2,3}O_{2,3}$ ($^3P_2$) in Xe, that are interesting from our point of view, are also well resolved, but have somewhat smaller intensities than other lines. Note, that some designations in Fig. 7 of Werme *et al.* (1972) have to be corrected according to Persson *et al.* (1988).

Measurements of relative transition probabilities have been performed by electron impact excitation for all the decays considered (Väyrynen and Aksela 1979; Aksela *et al.* 1984c). The results have been included in Tables 4.6–4.8. In general they compare favourably with our theoretical values. Experimental values for the α_2 parameter were compiled and compared with our results (Kabachnik *et al.* 1991), where good agreement for most transitions has been found.

Aksela *et al.* (1984b) showed the importance of configuration interaction in the final state for the Xe $N_{4,5}O_1O_{2,3}$ spectrum, namely between Xe $5s^15p^5$ and Xe

Table 4.8. The relative intensities, angular anisotropy, and spin polarization parameters for the Xe $N_{4,5}O_1O_{2,3}$ and $N_{4,5}O_{2,3}O_{2,3}$ Auger transitions; data from Lohmann *et al.* (1993). Parameters of $J_f = 0$ transitions are independent of the matrix elements

Transition	Rel. intensity		Parameter			
	Theo.	Exp.†	α_2	β_1	γ_1	ξ_2
$N_4O_{2,3}O_{2,3}$ $^1S_0$	1.0	1.0	−1.0	−0.447	0.894	0
$^1D_2$	1.60	1.46	0.052	−0.429	−0.262	−0.073
$^3P_0$	0.01	0.03	−1.0	−0.447	0.894	0
$^3P_1$	0.45	0.35	−0.835	0.594	−0.371	−0.014
$^3P_2$	0.18	0.11	0.165	0.110	0.321	0.216
$N_4O_1O_{2,3}$ $^1P_1$	6.97	0.60	−0.679	−0.445	0.577	−0.051
$^3P_0$	0.06	‡	−1.0	0.745	−0.298	0
$^3P_1$	0.08	0.14	−0.716	−0.009	0.048	−0.453
$^3P_2$	0.78	0.13	−0.824	0.262	−0.414	0.065
$N_5O_{2,3}O_{2,3}$ $^1S_0$	1.0	1.0	−1.069	0.683	−0.390	0
$^1D_2$	2.32	1.42	0.137	0.643	0.103	0.137
$^3P_0$	0.52	0.51	−1.069	0.683	−0.390	0
$^3P_1$	0.44	0.51	−0.734	−0.392	0.827	0.001
$^3P_2$	1.37	0.63	−0.227	0.070	−0.220	−0.183
$N_5O_1O_{2,3}$ $^1P_1$	14.48	0.96	−0.676	0.681	−0.222	0.017
$^3P_0$	0.50	0.10	−1.069	−0.488	0.781	0
$^3P_1$	1.66	‡	−1.039	0.044	−0.198	−0.197
$^3P_2$	1.04	0.18	−0.757	−0.075	−0.173	0.125

†: Aksela *et al.* (1984a, 1984b).
‡: Not observed.

$5s^2 5p^3 5d^1$. This was neglected in our calculation, probably leading to the spurious relative intensities of the $^1P_1$ lines. Finally, note the unusually large value of the ξ_2 parameter for the $5s^1 5p^5$ ($^3P_1$) final states.

In this section we investigated the spin polarization of Auger electrons after photoionization of atoms with circularly polarized light. Our results indicate, that a large degree of spin polarization within the reaction plane should be observable. This has been confirmed by related experiments and further theoretical calculations to be discussed in the next sections. This is true for model independent Auger transitions to $J_f = 0$ final states as well as for the other transitions. The former are, therefore, especially well suited to extract information about the initial vacancy from spin polarization measurements. Spin polarization perpendicular to the reaction plane created by dynamical effects is, with two exceptions, found to be small. We observe a strong dependence of the spin polarization parameters on the coupling scheme of the final state wavefunction for some transitions in the heavier noble gases. The comparison between our and former non-relativistic calculations of the Xe $N_{4,5}O_1O_{2,3}$ transitions shows good agreement for most transitions. Some considerable deviations might be caused by neglecting configuration interaction and correlation effects.

4.3 Intrinsic Spin Polarization for Isotropic Auger Multiplets

Auger transitions having a non-vanishing spin polarization are usually anisotropic with respect to their angular distribution. The spin polarization vector components depend also on the angular distribution of the Auger electrons. Thus, their experimental determination is somewhat tedious in the most cases. An anisotropic angular distribution depends on the existence of an alignment whereas the in-reaction plane components of the spin polarization vector are connected with both the orientation and the alignment of the primary ion. It is possible however to identify special cases where the ion is oriented but not aligned. For this special type of Auger transitions the general equations describing the spin polarization of the Auger electrons can be simplified which should give the opportunity of a simpler experiment. In this section we discuss the calculated spin polarization parameters for some of these special Auger transitions for the case of the rare gases. Applying the extreme LS coupling approximation upper and lower bounds for the initial orientation can be given which allow predictions for the in-reaction plane components of the spin polarization vector for these transitions.

Let us consider an intermediate ionic hole state with $J = 1/2$. This case has been already discussed in Sect. 2.6.1. Here, the angular distribution becomes isotropic because no alignment can be generated, and, for the same reason, the spin polarization component p_y, perpendicular to the reaction plane, vanishes. Thus, we remain with the in-plane components of the spin polarization vector see (2.268) and (2.272). Adopting the parameterization of Kabachnik and Lee (1989) we may write

$$P_x = \frac{3}{4} \gamma_1 \mathcal{O}_{10} \sin 2\theta, \tag{4.7}$$

and

$$P_z = \mathcal{O}_{10}\big(\beta_1 + \gamma_1 \, P_2(\cos\theta)\big), \tag{4.8}$$

where we expressed the spin polarization vector in terms of β_1 and γ_1 instead of ξ_1 and δ_1; see Table 2.7.

From a physical point of view, we now have the most interesting feature that the emitted Auger electrons can be spin polarized although still remain isotropic with respect to their angular distribution. This aspect simplifies the experiment enormously because only the three parameters $\mathcal{O}_{10}$, β_1, and γ_1 have to be determined, whereas the spin polarization vector components P_x and P_z in such experiments generally depend on α_2 and $\mathcal{A}_{20}$, too. As an example, we consider the Ar L_2MM Auger multiplet in detail. The calculations have been done with the ANISO program where the relaxed orbital method as has been discussed in Sect. 3.1 has been applied.

Since the Ar LMM Auger spectrum has been investigated both, experimentally and theoretically, by several authors the calculation of the relative intensities of the

Table 4.9. The Ar L_2MM Auger multiplet relative intensities in comparison with the LS coupling predictions and other theoretical and experimental results. Data from Lohmann and Larkins (1994). The Ar $L_2M_1M_{2,3}$ $^3P_2$ Auger transition has zero intensity in extreme LS coupling. The Auger energies have been taken from Mehlhorn and Stalherm (1968) and are given in a.u. T and E denote theoretical and experimental data sets, respectively. (a): T Lohmann and Larkins (1994); (b): T Asaad and Mehlhorn (1968); (c): E Väyrynen and Aksela (1979); (d): E Mehlhorn and Stalherm (1968); (e): E Werme *et al.* (1973); (f): E Ridder *et al.* (1976); (g): T The triplet ratios in LS coupling

Final state	Auger energies	(a)	(b)	(c)	(d)	(e)	(f)	(g)
				Ar $L_2M_{2,3}M_{2,3}$ I_0				
$^1S_0$	7.4605	14.03	9.49	12.1	12.4	14.7		
$^1D_2$	7.5483	38.83	49.18	48.9	48.0	48.5		
$^3P_0$	7.6053	10.97	9.19	9.5			7.8	4
$^3P_1$	7.6075	24.32	20.67	18.9	39.6	36.8	23.2	9
$^3P_2$	7.6123	11.85	11.47	10.6			8.0	5
Σ		100	100	100	100	100		
				Ar $L_2M_1M_{2,3}$				
$^1P_1$	6.9566	12.64	17.00		8.4			
$^3P_0$	7.0867	6.17	5.53					1
$^3P_1$	7.0889	12.40	11.09		13.3			2
$^3P_2$	7.0940†	0.03	0.16					0
Σ		31.24	33.78		21.7			
				Ar $L_2M_1M_1$				
$^1S_0$	6.6123	0.26	1.94		7.1			

†: The energy has been obtained by averaging the experimental Auger energies and the theoretical values of the MCDF program (Grant *et al.* 1980) for the considered multiplet. The difference has then been subtracted from the theoretical value.

Ar L_2MM Auger transitions provides a good testing ground for our model. The values obtained are shown in comparison with other experimental and theoretical data in Table 4.9. For the convenience of the reader the Auger energies have been included. Relatively good agreement has been found for the intensities. Applying the LS coupling scheme, ratios for the intensities of the triplet states can be given (also see Mehlhorn and Stalherm 1968). Our relative intensities are close to these predictions demonstrating that LS coupling is still applicable for argon. However, the total intensities for the $L_2M_1M_{2,3}$ and $L_2M_1M_1$ multiplets are too high and too low, respectively. As pointed out by Mehlhorn and Stalherm (1968) this is due to the fact that excited states have been neglected. We have only included the 10 possible

Table 4.10. The intrinsic spin polarization parameters β_1 and γ_1 for the Ar L_2MM Auger transitions; data from Lohmann and Larkins (1994). α_2 and $\xi_2 = 0$, see text

Ar $L_2M_{2,3}M_{2,3}$			Ar $L_2M_1M_{2,3}$		
Final state	β_1	γ_1	Final state	β_1	γ_1
$^1S_0$	−0.3333	1.3333	$^1P_1$	−0.3333	−0.1226
$^1D_2$	−0.3333	0.5754	$^3P_0$	1	0
$^3P_0$	−0.3333	1.3333	$^3P_1$	−0.3333	0.0130
$^3P_1$	0.5555	−0.2222	$^3P_2$	0.2009	−0.3995
$^3P_2$	−0.3333	0.0898	Ar $L_2M_1M_1$		
			$^1S_0$	−0.3333	1.3333

final ground states in our calculation. Considering the ratio $R = L_3MM/L_2MM$, Mehlhorn and Stalherm (1968) reported $R = 2.0 \pm 0.1$ where we have obtained $R = 2.07$. This shows that our model can serve as a good starting point for the prediction and analysis of forthcoming experiments investigating the intrinsic spin polarization of Auger electrons.

We have calculated the intrinsic spin polarization parameters β_1 and γ_1 for the L_2MM Auger transitions of Ar. The results are presented in Table 4.10. Our calculations show that non-zero intrinsic spin polarization parameters should be observed even for isotropic Auger lines.

As has been discussed in Sect. 2.5.2 and by Lohmann et al. (1993), for a $J_f = 0$ final ionic state, the Auger decay proceeds via one single channel and the spin polarization parameters are therefore independent of the matrix elements.

At a first glance, the constant value of $\beta_1 = -1/3$ for most of the Auger transitions might be surprising. This can be explained by inspecting the general expression (2.159) for the parameter β_1 in more detail. For this we consider the example of the $L_2M_{2,3}M_{2,3}$ $^3P_2$ transition. With the angular momentum coupling rules we find that only two partial waves $\varepsilon p_{3/2}$ and $\varepsilon f_{5/2}$ occur during the Auger emission. However, in contrast to the parameter γ_1, the summation in the general expression for β_1 allows no interference terms between matrix elements with different angular momenta ℓ. Out of this one ends up with the special selection rule $j = j'$ for the considered case. Adopting the notation of Lohmann et al. (1993), and denoting the matrix element as $\langle J_f(\ell\frac{1}{2})j : J \| V \| J \rangle$, (2.159) yields

$$\beta_1 = -\left(\sum_{\ell j} |\langle 2(\ell\frac{1}{2})j : 1/2 \| V \| 1/2 \rangle|^2 \right)^{-1}$$
$$\times 2 \sum_{\ell j} (2j+1) \begin{Bmatrix} \frac{1}{2} & \frac{1}{2} & 1 \\ j & j & 2 \end{Bmatrix} \begin{Bmatrix} \frac{1}{2} & \frac{1}{2} & 1 \\ j & j & \ell \end{Bmatrix} |\langle 2(\ell\frac{1}{2})j : 1/2 \| V \| 1/2 \rangle|^2. \quad (4.9)$$

Carrying out the sum over the two possible partial waves, the product of Wigner 6-j coefficients in the numerator turns out to be $1/3$ for both matrix elements. Thus the parameter β_1 becomes independent of the matrix elements which finally yields the result. The same argument holds for the $L_2M_{2,3}M_{2,3}$ $^1D_2$ and $L_2M_1M_{2,3}$ $^1P_1$ and $^3P_1$ transitions. This is physically important since it shows an interesting feature of the Ar L_2MM spectrum. Although more than one partial wave is emitted, i.e. the Auger decay proceeds via two channels, the intrinsic spin polarization parameter β_1 remains independent of the matrix elements for most of the Auger transitions. This is due to the fact that the usually extended summation over the possible interfering partial waves (e.g. see (2.159) and (2.160) or Lohmann et al. 1993) of the emitted Auger electron is enormously reduced for a $J = 1/2$ initial ionic state by selection rules.

Even the β_1 parameter for the $L_2M_{2,3}M_{2,3}$ $^3P_1$ transition turns out to be independent on the Auger matrix elements, however, by different reasons. Again, only the emission of two partial waves $\varepsilon p_{1/2}$ and $\varepsilon p_{3/2}$ is possible. In contrast to the previous discussion, interference terms between different total angular momenta j of the Auger electron are possible. However, due to selection rules the matrix element $\langle 1\, \varepsilon p_{1/2} : 1/2 \| V \| 1/2 \rangle$ turns out to be zero.[2]

Only the β_1 parameter of the $L_2M_1M_{2,3}$ $^3P_2$ transitions shows a more complex structure. However, the transition exhibits almost vanishing intensity in our calculation and is forbidden in extreme LS coupling.

As has been shown (Lohmann 1996d) this behaviour of the β_1 parameter is not restricted to the case of the Ar L_2MM spectrum but is in fact a more general one. We have demonstrated this in a recent calculation of the $L_2M_{2,3}M_{4,5}$ Auger transitions of krypton (Lohmann 1996b).

The in-plane components P_x and P_z of the spin polarization vector are functions of the intrinsic spin polarization parameters β_1 and γ_1 and of the orientation parameter $\mathcal{O}_{10}$. Let us assume that the primary ionization process is by photoionization from the ground state, which is $^1S_0$ for the rare gases. Following the work of Bussert and Klar (1983) we obtain for the orientation of an intermediate ion state with a $np_{1/2}$ hole the general expression

$$\mathcal{O}_{10}(np_{1/2}) = \frac{D(s_{1/2})^2 - \frac{1}{2}D(d_{3/2})^2}{D(s_{1/2})^2 + D(d_{3/2})^2}, \tag{4.10}$$

where $D(\ell_j) = \langle (\ell_j, J = \frac{1}{2})J_t = 1 \| r \| J_0 = 0 \rangle$ denotes the dipole matrix element. Here, ℓ and j describe the angular and total angular momenta of the photoelectron, J_t is the total angular momentum of the final state, and J and J_0 denote the total angular momenta of the singly ionized and neutral atom state, respectively.

For further predictions it is useful to consider the extreme LS coupling scheme, investigated for photoionization for instance by Berezhko et al. (1978b) and more recently by Kleiman and Lohmann (2003). One obtains a similar expression as (4.10) where only the indices "1/2" and "3/2" have to be omitted. This yields upper and lower bounds for the orientation

$$-0.5 \leq \mathcal{O}_{10}(2p_{1/2}) \leq 1.0. \tag{4.11}$$

---

[2] For further information we refer to Lohmann (1990), Appendix A.

In LS coupling only two partial waves εs and εd are emitted. Usually, out of the region of the Cooper minimum and well above threshold the $\varepsilon(\ell+1)$-partial amplitude dominates photoionization from the $n\ell$ shell, i.e. $\varepsilon d \gg \varepsilon s$. This yields

$$\mathcal{O}_{10}(2p_{1/2}) = -0.5. \tag{4.12}$$

As pointed out by Lohmann *et al.* (1993), this value of orientation usually occurs in a region of photon energy where the cross section is high. Applying this result predictions for the spin polarization vector can be given.

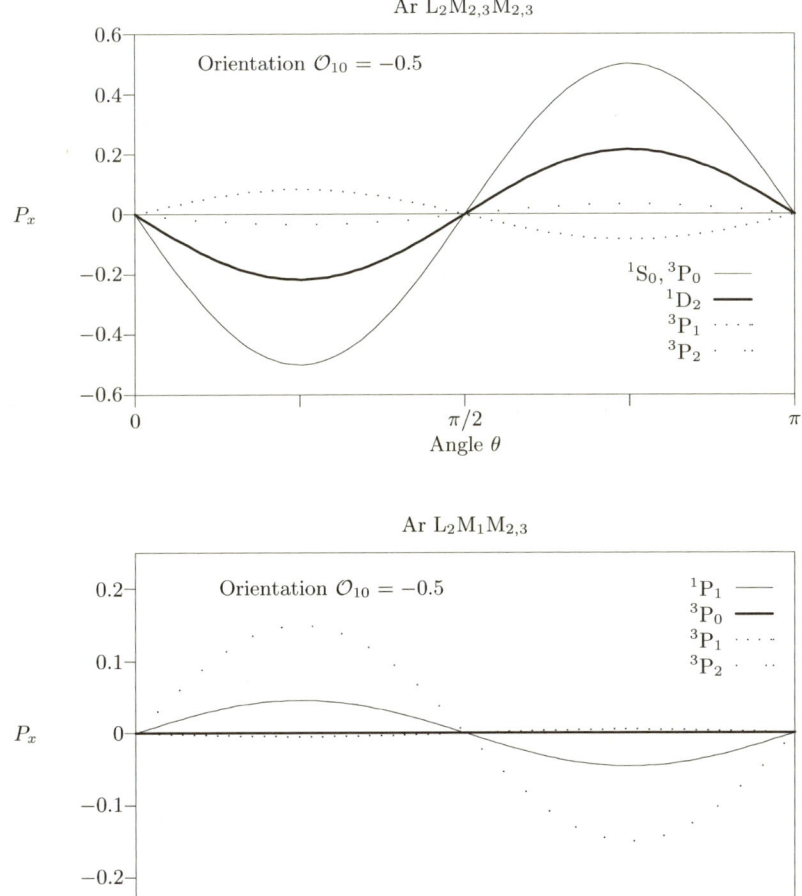

Fig. 4.3. P_x dependent on the Auger electron emission angle θ. Data from Lohmann and Larkins (1994)

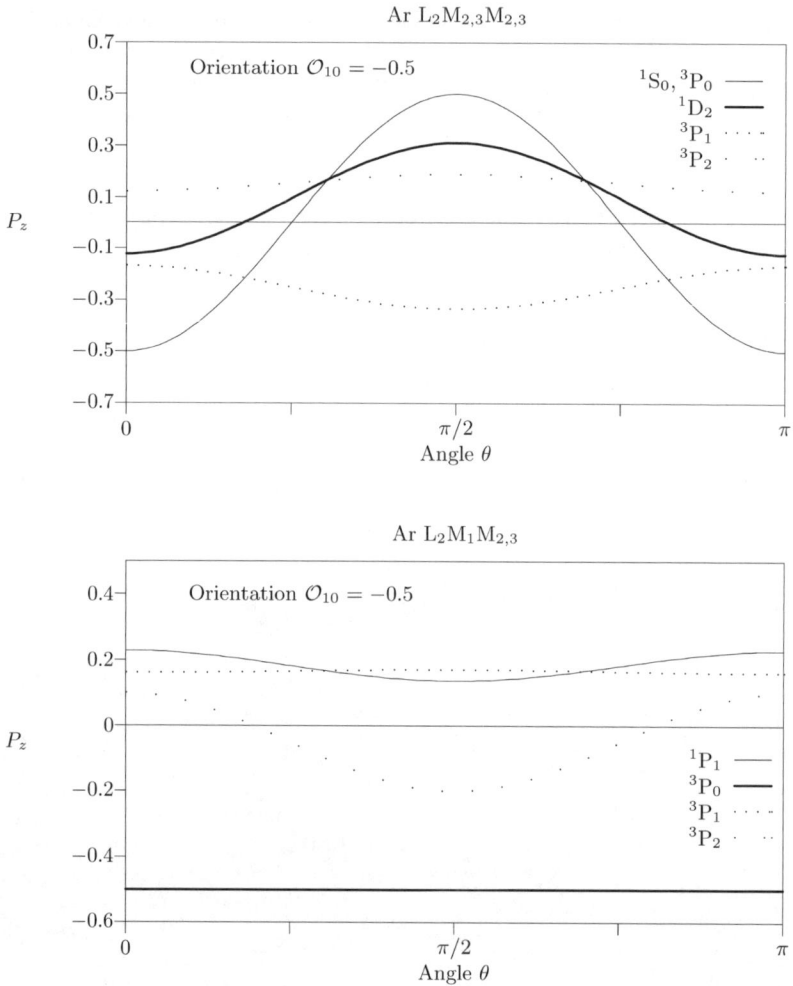

Fig. 4.4. P_z dependent on the Auger electron emission angle θ. Data from Lohmann and Larkins (1994)

The components P_x and P_z of the spin polarization vector are plotted against the angle of emission θ in Figs. 4.3 and 4.4 for the Ar L_2MM multiplet. The data show that an experiment should observe a non-zero spin polarization even for isotropic Auger lines. As can be seen from the two figures, the best angle of observation for the x-component of the spin polarization vector is at 45° where the z-component is better observed at 90°. Generally, the $L_2M_{2,3}M_{2,3}$ multiplet shows a higher degree of spin polarization than the $L_2M_1M_{2,3}$ transitions. The highest degree of 50% spin polarization can be found for the single channel transitions to $J_f = 0$ final states. Spin polarizations up to 35% can be predicted for most of the other Auger lines.

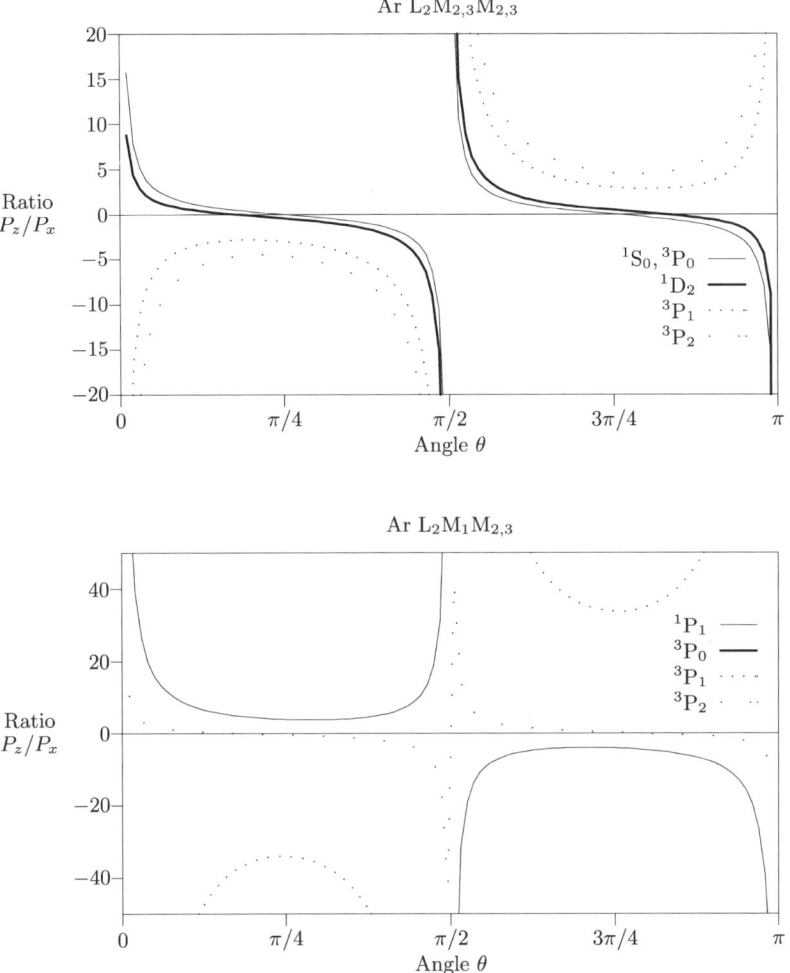

Fig. 4.5. The ratio P_z/P_x dependent on the Auger electron emission angle θ. The cut-off on the left edge of the figure is caused by the plot program. Data from Lohmann and Larkins (1994)

The x-component of the spin polarization vector is larger for the $L_2M_{2,3}M_{2,3}$ singlet states. Its observation will be more difficult for the $^3P_{1,2}$ states and the $L_2M_1M_{2,3}$ Auger lines. The observation of the z-component of the spin polarization vector should be possible for all Auger lines of the L_2MM multiplet. Its angular dependence shows a constant behaviour for the $L_2M_1M_{2,3}$ $^3P_0$ transition, and is almost constant for the $L_2M_1M_{2,3}$ $^3P_1$ Auger decay. While the $L_2M_1M_{2,3}$ $^1P_1$ and $L_2M_{2,3}M_{2,3}$ $^3P_{1,2}$ Auger transitions show only slightly variations of P_z, the z-component can even change its sign for the $L_2M_{2,3}M_{2,3}$ $^1S_0$, $^3P_0$, $^1D_2$, and $L_2M_1M_{2,3}$ $^3P_2$ transitions.

The $L_2M_1M_1$ $^1S_0$ Auger transition has not been plotted. It is identical to the $L_2M_{2,3}M_{2,3}$ $^1S_0$ and $^3P_0$ transitions due to the single channel character and the identical parity for these Auger decays.

By dividing (4.7) and (4.8), the ratio of the components P_x and P_z of the spin polarization vector turns out to be independent of the orientation $\mathcal{O}_{10}$. (However, $\mathcal{O}_{10}$ should be different from zero.) The ratio P_z/P_x is plotted against the angle of emission θ in Fig. 4.5. While the ratios of the $L_2M_{2,3}M_{2,3}$ $^1S_0$, $^3P_0$, and $^1D_2$ Auger transitions show a behaviour similar to a $\cot\theta$ function, the $L_2M_{2,3}M_{2,3}$ $^3P_1$, $^3P_2$ and $L_2M_1M_{2,3}$ $^1P_1$, $^3P_1$ transitions are closer to $\pm 1/\sin\theta$. The $L_2M_1M_{2,3}$ $^3P_2$ Auger transition just seems to connect these two extremes.

We have pointed out that for certain Auger transitions the in-reaction plane components of the spin polarization vector can be expected non-zero even though the angular distribution of the Auger electrons is isotropic. We have shown that the number of independent variables in a spin polarization experiment is considerably reduced for this type of Auger transitions. We have demonstrated this using the example of the Ar L_2MM multiplet. The spin polarization parameters β_1 and γ_1 have been calculated for these Auger transitions. Applying the extreme LS coupling limit for the primary photoionization with circularly polarized light, estimates for the orientation parameter $\mathcal{O}_{10}$ are possible. With this, predictions for the components P_x and p_z of the spin polarization vector are possible. An observable intrinsic spin polarization has been predicted. The ratio P_z/P_x has been found independent of the orientation, provided the orientation is non-zero. As has been shown in a calculation by Lohmann (1996b), a further candidate for such kind of experiments is the Kr $L_2M_{2,3}M_{4,5}$ Auger multiplet.

4.4 Spin Polarization for the Xe $M_{4,5}N_{4,5}N_{4,5}$ Auger Spectrum

The Xe $M_{4,5}N_{4,5}N_{4,5}$ Auger spectrum has been investigated in an experiment using synchrotron radiation (Hergenhahn 1996; Snell et al. 1996b) where high resolution measurements of the intrinsic spin polarization parameter ξ_1 have been done. Encouraged by the experiment we calculated the parameter set of relative intensities, angular anisotropy and spin polarization parameters, respectively, applying an MCDF approach (see Sect. 3.1).

4.4.1 Complete Data Sets

Our results are presented in Tables 4.11 and 4.12. Though we are using a relativistic approach, the spectroscopic notation has been adopted to identify the final states. The corresponding Auger energies have been obtained from the two independent computations for the initial and final state configuration as $E_{Auger} = |E_f - E_i|$. In order to get a measure for the validity of our calculation we evaluated the ratio of intensities for the total multiplet. We obtained $I(M_4N_{4,5}N_{4,5})/I(M_5N_{4,5}N_{4,5}) = 0.657$ which is close to the statistical ratio of $2/3$. This shows that our calculational

Table 4.11. The Auger energies, relative line intensities and intrinsic and dynamic spin polarization parameters for the Xe $M_4N_{4,5}N_{4,5}$ Auger transitions. The Auger energies are given in eV. The total line intensities add up to 100%. Data after Kleiman *et al.* (1999a)

final Auger state	energy	relative intensity	α_2	δ_1	ξ_1	ξ_2
$^1S_0$	528.74	5.96	−1.000	0.447	−0.894	0
$^1D_2$	534.97	23.70	0.085	−0.421	−0.439	−0.002
$^1G_4$	535.09	27.55	−0.655	0.102	−0.722	−0.001
$^3P_1$	535.80	8.89	−0.799	0.269	0.805	−0.000
$^3P_0$	536.03	1.50	−1.000	0.447	−0.894	0
$^3P_2$	536.98	12.14	0.358	0.778	0.681	−0.004
$^3F_2$	538.65	3.21	0.955	1.118	0.365	0.025
$^3F_3$	538.66	15.40	0.432	−0.097	0.111	−0.000
$^3F_4$	540.47	1.65	−0.824	−0.165	−0.426	0.010

Table 4.12. The Auger energies, relative line intensities and intrinsic and dynamic spin polarization parameters for the Xe $M_5N_{4,5}N_{4,5}$ Auger transitions. The Auger energies are given in eV. Total line intensities add up to 100%. Data after Kleiman *et al.* (1999a)

final Auger state	energy	relative intensity	α_2	δ_1	ξ_1	ξ_2
$^1S_0$	515.74	2.09	−1.069	0.293	0.878	0
$^1D_2$	521.98	3.79	0.126	0.495	−0.294	−0.001
$^1G_4$	522.10	23.03	−0.720	0.313	0.823	0.002
$^3P_1$	522.81	6.05	−0.748	0.409	−0.820	0.000
$^3P_0$	523.04	3.21	−1.069	0.293	0.878	0
$^3P_2$	523.99	9.85	−0.403	−0.402	−0.371	0.006
$^3F_2$	525.66	15.60	−0.169	0.402	0.749	−0.005
$^3F_3$	525.67	10.60	0.330	0.004	−0.053	−0.003
$^3F_4$	527.47	25.78	0.377	−0.343	−0.255	−0.002

approach has enough power to allow for a detailed comparison between experimental and theoretical data.

As we learned from the discussion of Sect. 2.5.2 the angular distribution parameter α_2, as well as the spin polarization parameters δ_1, ξ_1 and ξ_2 are constant values for all transitions to a final state with a vanishing total angular momentum $J_f = 0$. In this case the parameters are pure geometrical factors, that is they are independent of the transition matrix elements (see Sect. 2.5.2 and Lohmann 1990) and therefore, in particular, do not depend on the applied theoretical model. Moreover, the dynamic spin polarization parameter ξ_2 equals zero. Thus, for this type of transitions the Auger electrons can be polarized within the reaction plane only. As has been already pointed out the x-component of the spin polarization vector can be different from zero even if the photons are completely linearly polarized, e.g. see (2.179). However, the parameter ξ_2 entirely takes small values and thus, an experimental determination of p_x using linearly polarized radiation becomes difficult. Therefore, a partially circularly polarized photon beam must be used. In that case the value of the component p_x is also influenced by the intrinsic spin polarization parameter ξ_1 which is at least one magnitude larger than the dynamic parameter ξ_2.

4.4.2 Comparison with Experimental Data

Although some investigations concerning the Xe $M_{4,5}N_{4,5}N_{4,5}$ Auger spectrum have been performed (e.g. see Aksela *et al.* 1979; Tulkki *et al.* 1993) we will confine ourselves to a comparison with an experiment (Hergenhahn 1996; Snell *et al.* 1996b) where highly circularly polarized synchrotron radiation with a linearly polarized admixture has been used. In particular, apart from the usual measurements of the Auger energies and the relative line intensities the intrinsic spin polarization parameter ξ_1 has been extracted from the transverse in-reaction plane component p_x of the spin polarization vector.

Our theoretical data for the Auger transitions from the intermediate $3d_{3/2}$ and $3d_{5/2}$ hole states are compared to the experimental results in Tables 4.13 and 4.14, respectively. Note, that some lines cannot be resolved experimentally. Here, the theoretical data have been averaged over the contributing final states. Analyzing Tables 4.13 and 4.14 it turns out that there is good agreement between the theoretical and experimental values of the Auger energies as well as the relative line intensities. Within the experimental uncertainties this also holds for the intrinsic spin polarization parameter ξ_1 except for those transitions that result in the final states $^3F_4$ and $^3P_{0,1}$ belonging to the Xe $M_4N_{4,5}N_{4,5}$ Auger spectrum. Concerning the averaged $^3P_{0,1}$ state, we note that the $^3P_0$ state is strongly mixed with the final $^1S_0$ state

Table 4.13. Numerical data, (a): Kleiman *et al.* (1999a), for the Xe $M_4N_{4,5}N_{4,5}$ Auger spectrum compared with experimental results; (b): Hergenhahn (1996); (c): Snell *et al.* (1996b)

	Xe $M_4N_{4,5}N_{4,5}$					
final state	Auger energy		rel. intensity		ξ_1	
	(a)	(b)	(a)	(b)	(a)	(c)
$^1S_0$	528.7	527.0	6.0	5.8	−0.89	—
$^1D_2$, $^1G_4$	535.0	532.3	51.3	46.6	−0.59	−0.69±0.09
$^3P_1$, $^3P_0$	535.9	533.3	10.4	15.9	0.56	0.23±0.08
$^3P_2$	537.0	534.2	12.1	12.9	0.68	0.85±0.19
$^3F_2$, $^3F_3$	538.7	535.4	18.6	16.2	0.15	0.25±0.05
$^3F_4$	540.5	537.0	1.7	2.4	−0.43	0.18±0.10

Table 4.14. Numerical data, (a): Kleiman *et al.* (1999b), for the Xe $M_5N_{4,5}N_{4,5}$ Auger spectrum compared with experimental results; (b): Hergenhahn (1996); (c): Snell *et al.* (1996b)

	Xe $M_5N_{4,5}N_{4,5}$					
final state	Auger energy		rel. intensity		ξ_1	
	(a)	(b)	(a)	(b)	(a)	(c)
$^1S_0$	515.7	—	2.1	—	0.88	—
$^1D_2$, $^1G_4$	522.0	520.1	26.8	23.6	0.67	0.71±0.10
$^3P_1$, $^3P_0$	522.9	521.2	9.3	9.2	−0.23	−0.39±0.12
$^3P_2$	524.0	521.8	9.8	8.9	−0.37	−0.30±0.07
$^3F_2$, $^3F_3$	525.7	523.3	26.2	29.2	0.42	0.62±0.08
$^3F_4$	527.5	525.0	25.8	29.2	−0.26	−0.30±0.06

which is energetically closer to threshold. Such states are generally more influenced by correlation effects than the relatively strong bound states (e.g. Lohmann 1993). We neglected excited configuration state functions in our approach, especially the Xe $M_{4,5}N_1N_{4,5}$ Auger transitions, which may interact with the considered lines. In addition, no electron exchange with the continuum has been accounted for though this effect may be negligible since the Auger electrons possess a comparatively high kinetic energy (≈ 515–540 eV).

Besides the values presented in the two previous tables, we give predictions for the transverse in-reaction plane component p_x of the spin polarization vector selecting a 70-degree angle of detection θ and taking fully into account the polarization properties of the synchrotron beam ($\eta_1 = -0.32$, $|\eta_2| = 0.89$, $\eta_3 = 0$) which has been used in the experiment (Hergenhahn 1996; Snell *et al.* 1996b). Applying the LS coupling scheme, we can get upper and lower bounds for the photoionization parameter $B(K')$, see (2.51), of which estimates for the orientation and alignment parameters can be obtained. Our actual values are supposed to agree fairly well with the lower bounds of the intervals. We get $\mathcal{O}_{10}^{d_{3/2}} = -0.671\eta_2$, and $\mathcal{O}_{10}^{d_{5/2}} = -0.683\eta_2$ for the orientation and $\mathcal{A}_{20}^{d_{3/2}} = 0.100$ and $\mathcal{A}_{20}^{d_{5/2}} = 0.107$ for the alignment. Our predictions are expected to be sensible, since the energy of the incident photons is 834.5 eV which is far beyond ionization threshold of the $3d_{3/2,5/2}$ shells (Siegbahn *et al.* 1969) and the Cooper minimum. Inserting these estimated values and the calculated parameters α_2, ξ_1 and ξ_2 into (2.179) and using (2.182) then leads to the predicted values which are given in Table 4.15.

With the exception of the states $^1S_0$ and $^3P_{0,1}$ as well as $^3F_4$ of the Xe$M_4N_{4,5}N_{4,5}$ Auger spectrum good agreement between our results and the experimental data within the error bars can be found for all remaining transitions. This demonstrates, that the calculation can serve as a good starting point for more detailed investigations including correlation effects with higher excited states.

Table 4.15. Numerical data, (a): Kleiman *et al.* (1999a), for the component p_x of the spin polarization vector compared with experimental results; (b): Hergenhahn (1996); (c): Snell *et al.* (1996b)

final state	Xe $M_4N_{4,5}N_{4,5}$ $p_x(70°, 45°)$		Xe $M_5N_{4,5}N_{4,5}$ $p_x(70°, 45°)$	
	(a)	(b, c)	(a)	(b, c)
$^1S_0$	−0.49	−0.39±0.06	0.48	—
$^1D_2$, $^1G_4$	−0.33	−0.30±0.04	0.38	0.31±0.05
$^3P_1$, $^3P_0$	0.31	0.10±0.03	−0.13	−0.17±0.05
$^3P_2$	0.39	0.37±0.06	−0.21	−0.13±0.05
$^3F_2$, $^3F_3$	0.09	0.11±0.03	0.24	0.27±0.04
$^3F_4$	−0.23	0.08±0.07	−0.15	−0.13±0.03

4.5 Correlation Effects for the $N_{6,7}O_{4,5}O_{4,5}$ Auger Transitions of Mercury

As has been shown in the previous sections most numerical and experimental investigations concentrated on the rare gases applying different theoretical models. Correlation effects on the angular distribution and spin polarization of Auger electrons in Kr and Xe have been investigated by Kabachnik *et al.* (1988).

Theoretically and experimentally, there are only few investigations on mercury (Aksela *et al.* 1977; Oenning 1989; Lohmann 1992). Mercury is a very heavy element. Thus, the intermediate coupling scheme becomes necessary for an accurate description.

In the following we are discussing the influence of electron-electron-correlation effects on the relative intensities and anisotropy parameters α_2 and ξ_2 of angular distribution and spin polarization of the Hg $N_{6,7}O_{4,5}O_{4,5}$ Auger transitions. In particular, initial and final state configuration interaction (ISCI and FISCI) are taken into account.

4.5.1 Calculational Details

The wavefunctions of the initial and final state of the ion were constructed using the multiconfigurational Dirac–Fock computer code of Grant *et al.* (1980). For the investigation of a heavy element like mercury we used the intermediate coupling scheme for the description of the states. The mixing coefficients have been calculated with the code of Grant *et al.* applying the average level calculation mode. The bound electron wavefunctions were calculated in the field of the singly ionized atom. The continuum wavefunction of the Auger electron was evaluated by solving the Dirac-equation with a model potential, which was constructed from the single-electron wavefunctions of the doubly ionized atom. For all calculations the experimental energies of Aksela *et al.* (1977) and an extended version of the ANISO computer code by Lohmann (1988) have been used; also see Sect. 3.1 for further information.

To include correlation effects, all excited $6p^2$-configurations have been taken into account. Carrying out the calculation in the single ion field, this leads to 16 possible jj coupled $4f^{13}5s^25p^65d^{10}6n^2$ (n = s, p) configuration state functions (CSF; also see Sect. 3.1.1). We get the ionic state function $\psi_\alpha(J)$ as a linear combination of CSF

$$\psi_\alpha(J) = \sum_{r=1}^{n_c} c_r(\alpha)\, \phi_r(\gamma_r J), \tag{4.13}$$

where γ_r represents all information required to uniquely define the CSF. In the double ion field the number of possible $5d^86s^2$ and $5d^86p^2$ configurations varies from 13 up to 32 CSF depending on the total angular momentum of the final state. Of course all configurations have the same parity. (Note, that the inclusion of excited $6s6d$ and $6d^2$ states might be useful, too. Due to computer limitations this was not possible in the present study.)

Two kinds of calculations have been performed. In a first calculation, correlation effects were included only in the construction of the final state wavefunctions (FISCI), which were calculated in the intermediate coupling scheme, and the strongest admixtures of the final state CSF have been included. – Following a first run of the Grant-code, we suppress all configurations with a mixing coefficient $< 10^{-2}$. Then, we re-run the Grant program with the remaining CSF to obtain the mixing coefficients. – Depending on the total angular momentum of the final state the number of included configurations varies from 3 to 9. The initial state wavefunction is calculated in the $4f^{13}5s^25p^65d^{10}6s^2$ single hole ground state.

In the second calculation the 16 possible CSF of the initial state are included for the construction of the initial state wavefunction, too. That means both, FISCI and ISCI have been taken into account.

4.5.2 Configuration Mixing

As was pointed out in the last section only the strongest admixtures of the excited $5d^86p^2$ configurations were included in the construction of the final state. The mixing coefficients of the single-configuration wavefunctions are shown in Table 4.16. It is obvious, that correlation effects with the excited CSF cannot be neglected due to the strong admixture (In the case of the $^1D_2$ final state the admixture of some excited CSF is larger than one of the ground state CSF). It is interesting that all large admixtures are from the $\left|5d^8(6p^2)_0J\right\rangle$ and $\left|5d^8(6\bar{p}^2)_0J\right\rangle$ excited CSF. (Here, $\bar{p}$ denotes $p_{1/2}$, and p is $p_{3/2}$.) All other admixtures of $6p^2$-configurations are at least by a factor of 10 smaller. However, there might be further strong admixtures by $6s6d$ or $6d^2$ configurations.

Table 4.16. The mixing coefficients for the included final state configurations of the Hg $N_{6,7}O_{4,5}O_{4,5}$ Auger transitions of mercury; data from Lohmann (1993)

Final states	Mixing coefficients								
Configurations	$^3F_4$	$^3P_2$	$^3F_3$	$^3F_2$	$^3P_0$	$^3P_1$	$^1G_4$	$^1D_2$	$^1S_0$
$\lvert 5\bar{d}^2 5d^6 6s^2 J\rangle$	0	-0.161028	0	0.382740	0.449836	0	0	0.894530	-0.877681
$\lvert 5\bar{d}^3 5d^5 6s^2 J\rangle$	-0.254567	-0.233497	0.985893	0.864712	0	0.985905	0.952486	-0.412011	0
$\lvert 5\bar{d}^4 5d^4 6s^2 J\rangle$	-0.952164	0.943913	0	0.278969	-0.877100	0	-0.254471	0.050550	-0.449489
$\lvert 5\bar{d}^2 5d^6 (6p^2)_0 J\rangle$	0	-0.019025	0	0.045259	0.053066	0	0	0.105817	-0.103827
$\lvert 5\bar{d}^2 5d^6 6p^2 J\rangle$	0	-0.019299	0	0.045645	0.054079	0	0	0.106453	-0.103933
$\lvert 5\bar{d}^3 5d^5 (6p^2)_0 J\rangle$	-0.029901	-0.027519	0.116274	0.101949	0	0.116199	0.111980	-0.048546	0
$\lvert 5\bar{d}^3 5d^5 6p^2 J\rangle$	-0.031280	-0.028547	0.120392	0.105511	0	0.120367	0.116419	-0.050445	0
$\lvert 5\bar{d}^4 5d^4 (6p^2)_0 J\rangle$	-0.111552	0.110899	0	0.032754	-0.102936	0	-0.029767	0.005918	-0.052516
$\lvert 5\bar{d}^4 5d^4 6p^2 J\rangle$	-0.119425	0.118205	0	0.035059	-0.109571	0	-0.032178	0.0066447	-0.057408

4.5.3 Relative Intensities

The relative intensities of the different Auger lines are directly proportional to the square of the transition matrix elements. The results of the present calculations are given in Tables 4.17 and 4.18. Here, column (a) shows the results of the FISCI calculation including the strongest $6p^2$ CSF in the final states. In column (b) the results including also the excited initial CSF are shown (i.e. FISCI and ISCI included).

Comparing our two present calculations one finds small differences. Thus, in the framework of the used model, the influence of correlation effects due to ISCI can be assumed as small in the considered case. (Note, that this needs not be a general behaviour.)

The comparison with earlier results (Lohmann 1992), where only the $5d^8 6s^2$ ground state configurations have been included, shows small deviations. This is in agreement with Aksela *et al.* (1977) who pointed out, that the relative intensities of the Auger multiplet under discussion are rather insensitive to variations of the matrix elements. Though for the $N_7O_{4,5}O_{4,5}$ multiplet the differences are somewhat larger. The deviations become stronger for the energetically near-threshold $N_6O_{4,5}O_{4,5}^1 S_0$ and $N_7O_{4,5}O_{4,5}^1 G_4$, $^1D_2$, and $^1S_0$ states. This should be estimated because the influence of correlation effects should more affect the less bound, than the relatively strong bound states. For the convenience of the reader the energies have been included in Tables 4.17 and 4.18.

Table 4.17. The relative intensities for the $N_6O_{4,5}O_{4,5}$ Auger emission from Hg. The columns are denoted as follows: (a) FISCI, including $5d^8 6s^2$ CSF and the strongest admixtures of the excited $5d^8 6p^2$ CSF in the final state (Lohmann 1993); (b) FISCI as (a) and ISCI including the 16 possible $4f^{13}5s^2 5p^6 5d^{10} 6n^2$ ($n = s, p$) CSFs in the initial state (Lohmann 1993); (c) Lohmann (1992), intermediate coupling (IC) for the final state, including only the $6s^2$ ground state configuration; (d) LS coupling, with Slater integrals of McGuire (1975); (e) IC, with Slater integrals of Mc Guire; (f) Experiment; (d–f) Data by Aksela *et al.* (1977)

Final state	Energies a.u.	Relative intensities (a)	(b)	(c)	(d)	(e)	(f)
$^1S_0$	2.3332	16.0	15.7	14.8	6.4	10.5	16.7
$^1D_2$	2.4475	27.3	26.7	27.8	13.6	27.4	26.9
$^1G_4$	2.4798	17.5	18.9	17.7	13.0	15.5	20.2
$^3P_1$	2.4986	13.1	12.8	13.2	14.0	14.0	11.4
$^3P_0$	2.4986	0.7	0.7	0.7	5.3	1.2	
$^3F_2$	2.5129	12.6	12.4	13.0	12.7	15.1	12.9
$^3F_3$	2.5467	10.0	9.8	10.1	12.2	12.2	8.8
$^3P_2$	2.5761	0.3	0.3	0.3	16.7	0.5	0.8
$^3F_4$	2.6147	2.5	2.7	2.5	6.1	3.6	2.5

Table 4.18. The relative intensities for the $N_7O_{4,5}O_{4,5}$ Auger emission from Hg (Lohmann 1993). For explanations see Table 4.17

Hg $N_7O_{4,5}O_{4,5}$							
Final	Energies		Relative intensities				
state	a.u.	(a)	(b)	(c)	(d)	(e)	(f)
$^1S_0$	2.1884	3.0	3.2	5.0	6.4	2.6	6.2
$^1D_2$	2.3001	4.8	4.6	3.5	13.6	2.5	3.1
$^1G_4$	2.3332	10.4	11.0	12.3	13.0	11.1	16.4
$^3P_1$	2.3490	9.8	9.3	10.0	10.5	10.5	9.7
$^3P_0$	2.3490	9.2	9.8	7.9	3.0	6.8	
$^3F_2$	2.3688	13.7	13.1	12.7	3.4	13.4	18.8
$^3F_3$	2.3975	7.4	7.1	7.6	9.0	9.0	8.4
$^3P_2$	2.4251	23.3	22.3	23.8	22.5	23.6	21.4
$^3F_4$	2.4622	18.5	19.7	17.2	18.7	20.6	15.9

As in earlier calculations (Lohmann 1992) there is less agreement with the LS coupling results of Aksela *et al.* (1977). Their intermediate coupling results agree well with both of our calculations, while the $N_7O_{4,5}O_{4,5}$ data differ more slightly. There is a large discrepancy for the $N_6O_{4,5}O_{4,5}^1 S_0$ Auger line, where our value is in much better agreement with the experiment. Generally, the agreement of our data with the experiment is good. However, most of the $N_7O_{4,5}O_{4,5}$ Auger lines are fitted better by the earlier calculations (Lohmann 1992). This can be seen as a hint that further correlation effects should be included.

The relative numerical Auger line intensities of our described approach (b) have been plotted in Fig. 4.6 in comparison to the experimental results obtained by Aksela *et al.* (1977); column (f) in Tables 4.17 and 4.18. Here, Lorentz profiles have been assumed for the relative Auger line intensities which can be expressed as

$$I_0^{rel}(E) = \frac{\Gamma}{2\pi} \frac{1}{(E - E_0)^2 + \Gamma^2/4}, \qquad (4.14)$$

where E_0 denotes the Auger energy and Γ is the decay width (Frauenfelder and Henley 1979). The total spectrum has been normalized to 100. As can be seen, most of the final state fine structure is fully resolved in the spectrum where a decay width of $\Gamma_{Hg} = 0.12$ eV has been assumed. Both, the experimental and the numerical data have been plotted using the experimental Auger energies by Aksela *et al.* (1977).

Eventually, we note that Oenning (1989) measured the $N_{6,7}O_{4,5}O_{4,5}$ Auger spectra of mercury, applying a 100 keV proton beam. However, the large background and spectrometer dispersion make it difficult to compare their data with our results.

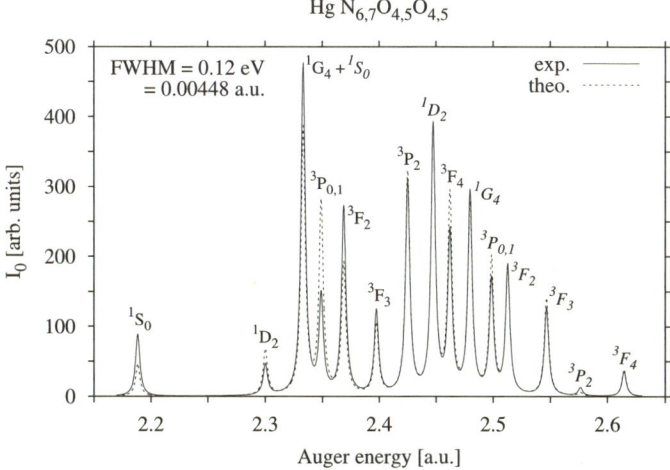

Fig. 4.6. The relative intensities of the $N_{6,7}O_{4,5}O_{4,5}$ Auger spectrum of mercury. The final fine structure states of the Auger lines have been given upright ($^{(2S+1)}L_J$) for transitions from an initial $4f_{7/2}$ and leaned ($^{(2S+1)}L_J$) from a $4f_{5/2}$ state, respectively. The numerical spectrum (Lohmann 1993) *(dashed curve)* has been plotted in comparison to the experimental results (Aksela *et al.* 1977) *(full curve)*. For both, the experimental energies have been used, and a FWHM $= 0.12$ eV have been assumed

4.5.4 Angular Distribution – Anisotropy Parameters

The results for the anisotropy parameters α_2 are shown in Tables 4.19 and 4.20. So far, there are no experimental data to compare with. The comparison of the two present calculations shows slight differences of about 1%. This is in agreement with our statement given above, that the initial state wavefunction is slightly affected by correlation effects. The only other theoretical data to compare with are the results of Lohmann (1992). For both calculations we find only small deviations of less than 5%.

From the discussion of the last section, we would expect deviations from our results (Lohmann 1992) only for the near-threshold transitions listed above. However, the situation is a bit different in the angular distribution case. The $N_{6,7}O_{4,5}O_{4,5}^1 S_0$ Auger transitions cannot be influenced by any correlation because they proceed via one single channel, and thus are independent of the transition matrix elements. Therefore, deviations in the data can be expected only for the $N_7O_{4,5}O_{4,5}^1 G_4$ and $^1D_2$ cases. A small difference appears for the $^1G_4$ Auger transition. An enormous effect however has been found for the $^1D_2$ transition. The measurement of the anisotropy of this Auger line would provide a good experimental test of the influence of correlation effects on the angular distribution of Auger electrons.

Table 4.19. The angular distribution parameters α_2 for the Hg $N_6O_{4,5}O_{4,5}$ Auger emission (Lohmann 1993). For explanations see Table 4.17

Final state	Hg $N_6O_{4,5}O_{4,5}$ α_2		
	(a)	(b)	(c)
$^1S_0$	−1.0690	−1.0690	−1.0690
$^1D_2$	−0.2037	−0.2022	−0.2038
$^1G_4$	1.0957	1.0972	1.1000
$^3P_1$	−0.9566	−0.9566	−0.9593
$^3P_0$	−1.0690	−1.0690	−1.0690
$^3F_2$	−0.2456	−0.2441	−0.2636
$^3F_3$	0.3777	0.3811	0.3799
$^3P_2$	−0.5246	−0.5229	−0.5304
$^3F_4$	0.5139	0.5098	0.5121

Table 4.20. The angular distribution parameters α_2 for the Hg $N_7O_{4,5}O_{4,5}$ Auger emission (Lohmann 1993). For explanations see Table 4.17

Final state	Hg $N_7O_{4,5}O_{4,5}$ α_2		
	(a)	(b)	(c)
$^1S_0$	−1.0911	−1.0911	−1.0911
$^1D_2$	−0.5688	−0.5675	−0.0749
$^1G_4$	1.0767	1.0788	1.1204
$^3P_1$	−0.9296	−0.9296	−0.9324
$^3P_0$	−1.0911	−1.0911	−1.0911
$^3F_2$	−0.6636	−0.6640	−0.6376
$^3F_3$	0.2693	0.2638	0.2713
$^3P_2$	−0.3987	−0.3975	−0.4014
$^3F_4$	0.5879	0.5930	0.5850

4.5.5 Spin Polarization Parameters

The results for the spin polarization parameters ξ_2 are shown in Tables 4.21 and 4.22.

The only other theoretical data obtained so far are those of Lohmann (1992). Moreover, Lohmann (1992) stated that an experimental check of the spin polarization parameters is difficult, because the theoretical values are small for all transitions. However, the behaviour discussed in the previous sections is confirmed by our present calculations. The differences between our present calculations (a) and (b) are less than 3%. The comparison with the former results of Lohmann (1992) where

Table 4.21. The coefficients ξ_2 for the Hg $N_6O_{4,5}O_{4,5}$ Auger emission (Lohmann 1993). For explanations see Table 4.17

Final state	Hg $N_6O_{4,5}O_{4,5}$ ξ_2		
	(a)	(b)	(c)
$^1S_0$	0	0	0
$^1D_2$	0.0192	0.0194	0.0185
$^1G_4$	−0.0131	−0.0130	−0.0145
$^3P_1$	0.0001	0.0001	0.0001
$^3P_0$	0	0	0
$^3F_2$	0.0132	0.0130	0.0106
$^3F_3$	−0.0056	−0.0059	−0.0063
$^3P_2$	0.0629	0.0634	0.0678
$^3F_4$	−0.0536	−0.0521	−0.0533

Table 4.22. The coefficients ξ_2 for the Hg $N_7O_{4,5}O_{4,5}$ Auger emission (Lohmann 1993). For explanations see Table 4.17

Final state	Hg $N_7O_{4,5}O_{4,5}$ ξ_2		
	(a)	(b)	(c)
$^1S_0$	0	0	0
$^1D_2$	0.0966	0.0971	0.0390
$^1G_4$	0.0735	0.0742	0.0394
$^3P_1$	−0.0002	−0.0002	−0.0002
$^3P_0$	0	0	0
$^3F_2$	0.0101	0.0100	0.0072
$^3F_3$	0.0377	0.0371	0.0379
$^3P_2$	0.0024	0.0022	0.0022
$^3F_4$	−0.0400	−0.0401	−0.0418

electron-electron correlation effects have been neglected, shows deviations of less than 10%. Considering again the near-threshold transitions, the two $^1S_0$ Auger transitions must be zero in all models due to their single-channel character. Thus, considerable deviations between the calculations with and without correlation should occur only for the $N_7O_{4,5}O_{4,5}^1 G_4$ and $^1D_2$ transitions. Indeed for both transitions deviations by a factor of two or three are found.

The investigation of relative intensities and anisotropy coefficients of angular distribution and spin polarization of the Auger electrons has shown small deviations compared to former calculations without correlation and with the experimental data. However, the near-threshold states of the investigated multiplet are strongly influenced by correlation effects. In particular, a large effect has been found for the anisotropy parameter of angular distribution of the $N_7O_{4,5}O^1_{4,5}D_2$ Auger transition. This may allow for an experimental test of the influence of correlation effects in Auger transitions.

4.6 Resonant Auger Transitions

Let us now consider the resonant Auger transitions. The theory has already been discussed in Sect. 2.6. In the following we discuss theoretical calculations for the angular anisotropy parameter $\beta = \alpha_2 \mathcal{A}_{20}$ of the resonantly excited Auger decays $Kr^* 3d^{-1}_{5/2}5p\, (J = 1) \rightarrow 4p^{-2}5p$ and $Xe^* 4d^{-1}_{5/2}6p\, (J = 1) \rightarrow 5p^{-2}6p$. The transition amplitudes are obtained within a model treating the outer electron as a spectator. Its influence on the final states is taken into account via intermediate coupling MCDF calculations. This model reproduces the unusually large negative β-values observed in the considered transitions.

In the second part of this section we consider the dynamic spin polarization parameter of resonant Auger transitions. While, in the so-called diagram transitions, the dynamic spin polarization has been found magnitudes smaller than the intrinsic, for certain Auger lines of the resonant $Xe^*(4d^{-1}_{5/2}6p)_{J=1}N_5O_{2,3}O_{2,3}$ spectrum we calculated large dynamic spin polarization parameters (Lohmann 1997, 1999a), and experimental proof has been given (Hergenhahn et al. 1999). Investigating these transitions, we have been able to derive simple propensity rules which allow, for the first time, to predict whether a large dynamic spin polarization can be expected. However, the propensity rules can be applied to resonant Auger transitions, only.

4.6.1 Angular Anisotropy in the Auger Decay of Resonantly Excited States

The decay of autoionizing core resonances of the rare gases has been intensively studied since the late 1980s. Spectator Auger decay, leaving the excited electron in its outer orbital, is in many cases the major decay mode. Experimental techniques used include high-resolution electron spectrometry (Aksela et al. 1986a, 1986b, 1988a, 1992; Meyer et al. 1991), electron-electron coincidence experiments (von Raven et al. 1990) and angle resolved electron spectroscopy (Carlson et al. 1988, 1989; Becker 1990a; Kämmerling et al. 1990; Caldwell 1990). In the latter cases unusually large degrees of the angular anisotropy parameter β with negative sign have been found for certain transitions. In a first theoretical study Cooper (1989) reproduced this phenomenon for the Ar case. For the heavier atoms sophisticated studies devoted to the relative line strengths have been performed (Aksela et al. 1992; Combet Farnoux 1992). Kämmerling et al. (1990) introduced a relationship between a normal Auger decay and the resonant Auger decay of the same hole configuration in presence of a spectator electron. They also probed this relation using their experimental data. Further investigations have been carried out by Lohmann (1991).

However, to our knowledge our earlier investigation (Hergenhahn *et al.* 1991) based on the *strict spectator model* has been the only attempt to explain angular distributions in the resonant Auger decays of heavier atoms on a purely theoretical basis. Although giving a satisfying agreement in general, our calculations were not able to reproduce the negative β values observed for transitions to the ground or lowest excited states of a singly ionized noble gas configuration.

In the following we will show that a refined version of this model can explain the large degree of negative anisotropy in resonant Auger decay.

We consider the angular anisotropy of the decays $Kr^* 3d_{5/2}^{-1}5p\,(J = 1) \rightarrow 4p^{-2}5p$ and $Xe^* 4d_{5/2}^{-1}6p\,(J = 1) \rightarrow 5p^{-2}6p$. In the description of the final states we take into account the interaction of the outer electron with the core explicitly via MCDF calculations. This interaction entails an additional mixing of core states which has been neglected by Hergenhahn *et al.* (1991). By inclusion of this we arrive at a substantially better agreement of the calculated anisotropy with experiment.

Following the discussion of Sect. 2.6 we apply the two-step picture of autoionization. Thus, describing the process as an excitation to a well defined intermediate state, followed by a decay caused by the Coulomb interaction. For photoexcitation, in the dipole approximation, the angular distribution can be expressed in terms of alignment and anisotropy parameters

$$I(\theta) = \frac{I_0}{4\pi}\big(1 + \alpha_2\,\mathcal{A}_{20}\,P_2(\cos\theta)\big). \tag{4.15}$$

In the case of the rare gases the ground state is $^1S_0$. For photoexcitation by linearly polarized photons the value of the alignment parameter is then fixed at a geometrical value of $\mathcal{A}_{20} = -\sqrt{2}$ (e.g. Kronast *et al.* 1986), where, in contrast to the discussion of Sect. 2.6.4, we have chosen the direction of the electric field vector as quantization axis. Thus, the observable anisotropy parameter $\beta = \alpha_2\mathcal{A}_{20}$ can be predicted from a calculation of the decay amplitudes.

The matrix elements of resonant Auger decay can be calculated with various degrees of sophistication. Hergenhahn *et al.* (1991) used the rather simple *strict spectator model*, which is based on the following assumptions:

1. Relaxation of bound state electronic orbitals during the transition is ignored. The single-electron wavefunctions are assumed to be the same in the initial and final state.
2. The wavefunctions of the core electrons remain unchanged in the presence of the spectator electrons compared to the case of the naked core.
   Assumptions 1 and 2 allows for relating the amplitudes of normal and resonant Auger transitions (e.g. see Lohmann 1991).
3. The initial and final states are described each by a single configuration. The final ionic core is characterized by good quantum numbers. There is no mixing of different core states by the outer electron. This implies application of a pure jK or jj coupling scheme for the final state.

For normal Auger decays application of intermediate coupling in the final state proved to be important for a quantitative description of the electron anisotropy

(Kabachnik and Lee 1990a; Kabachnik et al. 1991; Chen 1992). Also there is evidence that mixing of core states has an effect on the considered resonant Auger transition in Kr (Aksela et al. 1992). Thus, by giving up assumption 3 and using intermediate coupling MCDF-wavefunctions for describing the final state (double hole + spectator electron) configurations, we get a final state wavefunction $\psi_\alpha(J_f)$ as a linear combination of CSF

$$\psi_\alpha(J_f) = \sum_{r=1}^{n_c} c_r(\alpha)\, \phi_r\big([j_2, j_3]J_c(r), j_1 : J_f\big). \qquad (4.16)$$

For further explanation, we refer the reader to the discussion in Sect. 3.1 and to the review by Grant (1970).

The CSF are expressed in a jj-coupled basis with j_2, j_3 denoting the core holes. $J_c(r)$ is the core angular momentum of the CSF, j_1 refers to the spectator electron. By virtue of this, the matrix elements in the anisotropy parameter α_2, see (2.139), are replaced by linear combinations of matrix elements $\big\langle([j_2, j_3]J_c, j_1)J_f, j :$ $J \|V\|(j_0, j_1)J\big\rangle$. The initial state is still a single-configuration, pure jj-coupled state. For calculating the latter matrix elements we use assumption 1. Then changing the order of angular momentum coupling we employ the expression (Kämmerling et al. 1990; Lohmann 1991)

$$\big\langle(J_c, j_1)J_f, j : J\|V\|j_0, j_1 : J\big\rangle = (-1)^{J_f+j_1+j+j_0}\sqrt{(2J+1)(2J_f+1)}$$

$$\times \begin{Bmatrix} j & J & J_f \\ j_1 & J_c & j_0 \end{Bmatrix} \langle J_c, j : j_0\|V\|j_0\rangle, \qquad (4.17)$$

where j_0 is the total angular momentum of the initial hole. This equation implies, that j_1 might not change during the transition. The matrix element on the right side depends only on total angular quantum numbers of the core, though its radial part could be influenced by the outer electron. In the frame of our spectator model we nevertheless neglect this influence and insert amplitudes for the normal Auger decay in its place (assumption 2). For the details of the calculation we refer to the discussion in Sect. 3.1.

In Tables 4.23 and 4.24 we compare the results of our calculations with our earlier model and with recent measurements. Following Hansen and Persson (1987) we assigned jK coupling designations to the final states. However, as we use intermediate coupling in this work, they are merely to be understood as labels. For comparison with experimental results the contributions of unresolved states were averaged.

We see, that our model is well able to explain the large negative anisotropy coefficient of lines 1a, b for Kr and Xe. The proper description of the final state is crucial for this, as a comparison with Hergenhahn et al. (1991) shows. (There pure jK coupling according to the designations in the tables has been assumed.) Obviously this problem gets more important for the heavier atoms: In earlier investigations on the decay Ar $2p_{3/2}^{-1}4s \rightarrow 3p^{-2}4s$ a simple LS coupling scheme was sufficient to reproduce the experimental results (Cooper 1989; Hergenhahn et al. 1991). Regarding the

Table 4.23. Angular anisotropy parameter β for the decay Kr* $3d_{5/2}^{-1}5p \rightarrow 4p^{-2}5p$. Contributions from experimentally unresolved states were averaged. Data from (a): Hergenhahn *et al.* (1993); (b): Hergenhahn *et al.* (1991); (c): Carlson *et al.* (1988); (d): Caldwell (1990)

Line no	State labels	β_{th}^{a}	β_{th}^{b}	β_{exp}^{c}	β_{exp}^{d}
1a	$4p^4(^3P_2)5p[2]_{3/2,5/2}$	−0.990	0.007	−0.89	−0.76(2)
1b	$4p^4(^3P_2)5p[3]_{5/2,7/2}, [1]_{1/2}$	−0.823	−0.15	−0.98	−0.87(2)
1c	$4p^4(^3P_2)5p[1]_{3/2}, (^3P_0)5p[1]_{1/2}$	0.801	0.940	0.62	0.77(6)
1d	$4p^4(^3P_1)5p[0]_{1/2}, [2]_{5/2}, (^3P_0)5p[1]_{3/2}$	0.820	0.883	0.24	0.04(5)
1e	$4p^4(^3P_1)5p[1]_{1/2,3/2}, [2]_{3/2}$	0.467	0.100	0.19	0.31(6)
2a	$4p^4(^1D_2)5p[1]_{3/2}, [3]_{5/2,7/2}$	−0.066	−0.144	−0.06	0.27(3)
2b	$4p^4(^1D_2)5p[1]_{1/2}, [2]_{3/2,5/2}$	−0.248	−0.369	−0.12	0.05(3)
4	$4p^4(^1S_0)5p[1]_{3/2}$	0.759	0.8	0.73	

Table 4.24. Angular anisotropy parameter β for the decay Xe* $4d_{5/2}^{-1}6p \rightarrow 5p^{-2}6p$. Contributions from experimentally unresolved states were averaged. Data from (a): Hergenhahn *et al.* (1993); (b): Hergenhahn *et al.* (1991); (c): Carlson *et al.* (1989); (d): Kämmerling *et al.* (1990)

Line no	State labels	β_{th}^{a}	β_{th}^{b}	β_{exp}^{c}	β_{exp}^{d}
1a	$5p^4(^3P_2)6p[2]_{3/2,5/2}$	−0.976	0.002	−0.88	−0.60(3)
1b	$5p^4(^3P_2)6p[3]_{5/2,7/2}, [1]_{1/2}$	−0.726	−0.478	−0.93	−0.90(2)
1c	$5p^4(^3P_2)6p[1]_{3/2}$	0.972	1.014	0.82	1.31(2)
2a	$5p^4(^3P_0)6p[1]_{1/2,3/2}, (^3P_1)6p[0]_{1/2}$	0.868	0.859	0.26	0.58(2)
2b	$5p^4(^3P_1)6p[1]_{1/2,3/2}, [2]_{3/2,5/2}$	0.394	0.100	0.16	0.54(3)
3a	$5p^4(^1D_2)6p[1]_{3/2}, [3]_{5/2,7/2}$	−0.048	−0.132	−0.02	0.23(2)
3b	$5p^4(^1D_2)6p[1]_{1/2}, [2]_{3/2,5/2}$	0.007	−0.081	−0.09	0.33(5)
5	$5p^4(^1S_0)6p[1]_{3/2}$	0.753	0.8	0.51	0.83(5)
6	$5p^4(^1D_2)7p[K_f]$	−0.01	−0.10	−0.03	0.12(6)
9	$5p^4(^1S_0)7p[1]_{1/2,3/2}$	0.753	0.8		0.76(3)

other lines at least the gross features of the angular distribution are well reproduced. As our model is still a relatively simple approximation to resonant Auger decay, we consider it to be in good accordance with experiment. We would like to point out two reasons for the remaining discrepancies.

1. For transitions to a pure state with vanishing total core angular momentum the anisotropy parameter would have the geometrical value $\beta = 0.8$ in our model

Table 4.25. Angular anisotropy parameter β for the decay $Xe^* 4d_{5/2}^{-1}6p \rightarrow 5s^{-1}5p^{-1}6p$. Data from Hergenhahn *et al.* (1993)

State no	E_{kin} [eV]	Parent states	β_{th}
1	25.82	$(^3P_2)6p(J = 3/2)$	-0.433
2, 3	25.54, 25.53	$(^3P_2)6p(J = 5/2, 7/2)$	-0.658
4–6	25.30, 25.27, 25.08	$(^3P_{0,1})6p$	0.784
7–11	24.70, 24.63, 24.56, 24.54, 24.41	$(^3P_{0,1})6p$	0.801
12, 13	23.98, 23.97	$(^3P_{0,1})6p$	0.821
14–16	19.48, 19.45, 19.32	$(^1P_1)6p$	0.956
17, 18	19.15, 19.10	$(^1P_1)6p$	0.468

(see Hergenhahn *et al.* 1991). Especially decays participating in lines 1d of Table 4.23 and 2a of Table 4.24 contain main components with $J_c = 0$ in their final states. So it can be understood, that β-values for these lines resemble the geometrical value in our calculation, but are somewhat smaller in nature.

2. For the transitions to states with dominant $^1D_2$ core we interpret the differences as consequence of dynamical effects introduced by the outer electron. Already in calculations for normal Auger decay these transitions showed up to be most model sensitive (Kabachnik *et al.* 1991).

By the same method we calculated β parameters for the decay $Xe^* 4d_{5/2}^{-1}6p$ ($J = 1$) $\rightarrow 5s^{-1}5p^{-1}6p$. Including only the latter (non-relativistic) configuration for generating the CSF we found 18 possible final states. In Table 4.25 they are represented by their numbers and a tentatively assigned parent state. The results given in Table 4.25 have been grouped assuming an energy resolution of 0.2 eV. Kinetic energies are taken from our Dirac–Fock calculations. We found, that the two lines with the highest kinetic energy have negative β values. Nevertheless, this transition requires more thorough theoretical investigations, particularly taking CI with the $5s^25p^35d\,6p$ configuration into account (see Aksela *et al.* 1986a). Besides experimental investigations are needed to confirm if this a general trend.

4.6.2 Propensity Rules for Dynamic Spin Polarization Parameters

Describing the spin polarization vector of Auger electrons by its Cartesian components the mechanism of generation of spin polarization for the in-reaction plane components is different to that for the component perpendicular to the reaction plane. In the first case an intrinsic orientation is necessary in the intermediate excited state, which for instance might be generated via photoexcitation with circularly polarized light. Thus, the spin polarization of the emitted Auger electron is generated via polarization transfer (see the discussion in Chap. 1). Therefore it is not surprising that comparatively large values for the so-called *intrinsic* spin polarization have been found theoretically and in the experiment (Lohmann *et al.* 1993; Müller *et al.* 1995, and refs. therein).

From a physical point of view the latter process is interesting. Here, in the intermediate state an alignment is generated during the ionization which, under certain physical conditions (e.g. see Klar 1980; Kabachnik 1981), can be transformed into a *dynamic* spin polarization of the Auger electron. Various experiments (Hahn *et al.* 1985; Merz and Semke 1990) and a number of theoretical calculations (Kabachnik and Sazhina 1988; Kabachnik *et al.* 1991; Lohmann 1993) reported almost vanishing dynamic spin polarization for the so-called diagram Auger transitions. Though, experiments (Snell *et al.* 1996a, 1996b) have given evidence of a large dynamic spin polarization for certain lines of the resonantly excited Xe^* $(4d_{5/2}6p_{3/2})_{J=1}$ $N_5O_{2,3}$ $O_{2,3}$ Auger spectrum; see (4.20) and (4.21).

For the planning of experiments it is important to have simple rules available from which conditions for the experiments can be derived. – E.g. for which atoms, and which transitions can significant spin polarizations be expected. – In the following we will derive such propensity rules for the dynamic spin polarization of resonant Auger transitions. The results can be used by experimentalists for determining whether a measurable dynamic spin polarization can be expected in an experiment. Due to the enormous experimental effort necessary in combined angle and spin resolved experiments there is need for propensity rules to guide experimentalists in planning, performing and the analysis of such experiments.

Describing the resonant Auger emission in the well observed two-step model (Mehlhorn 1990) the dynamic spin polarization factorizes into a product of two parameters, the alignment parameter $\mathcal{A}_{20}$ containing solely the dynamics of the primary excitation, and the parameter ξ_2 describing the dynamics of the resonant Auger decay. For a primary photoexcitation, using a fully linearly polarized photon beam, we get the advantage of a maximum alignment $\mathcal{A}_{20} = -\sqrt{2}$ (Kronast *et al.* 1986). This has been extensively discussed in Sect. 2.6. An expression for the component of the spin polarization vector perpendicular to the reaction plane has been given in (2.258). Usually, the spin polarization is observed at the *magic* angle, i.e. $\theta_m = 54.74^o$. Then, (2.258) can be reduced to

$$p_y(\theta_m) = \frac{\sqrt{8}\,\xi_2}{\alpha_2 - \sqrt{2}}. \qquad (4.18)$$

As pointed out before p_y is found to be small even for most of the resonant Auger transitions. This is because the α_2 parameter is commonly at least a magnitude larger than the dynamic spin polarization parameter ξ_2. Thus, the spin polarization is best observed as a function of ξ_2, only. We suggest to observe the spin polarization at the angle $\theta_\xi = 35.26^o$. For this angle the denominator of (2.258) becomes independent of α_2 and we obtain the simple relation

$$p_y(\theta_\xi) = -2\,\xi_2. \qquad (4.19)$$

Now, let us consider the resonant $Xe^* N_5O_{2,3}O_{2,3}$ Auger process. In the first step, we have a photoexcitation

$$\gamma + Xe(^1S_0) \longrightarrow Xe^*\left(4d_{5/2}^{-1}6p_{3/2}\right)_{J=1}. \tag{4.20}$$

Then, the resonantly excited intermediate state decays via Auger emission.

$$Xe^*\left(4d_{5/2}^{-1}, 6p_{3/2}\right)_{J=1} \longrightarrow Xe^{+*}\left(5p^{-2}, 6p_{3/2}\right)_{J_f} + e_{Auger}, \tag{4.21}$$

where the total angular momentum of the final state may take values $J_f = 1/2, \ldots,$ 7/2. We investigated these Auger transitions applying a relaxed orbital method within a MCDF approach where intermediate coupling between the different CSF forming the ASF of the final state, and exchange interaction with the continuum has been taken into account (Lohmann 1997). Within this approach we calculated the Auger rates as well as the relevant parameters of angular distribution and spin polarization of the emitted Auger electron. A complete discussion of our data and their comparison to recent experiments has been published (Lohmann and Kleiman 2001).

From our calculations we find the dynamic spin polarization parameter ξ_2 of the same magnitude as or even larger then the angular distribution anisotropy coefficient α_2 for all Auger transitions to final states with $J_f = 1/2$. The anisotropy and spin polarization parameters for these transitions are shown in Table 4.26 together with their Auger energies and relative intensities. Values for the dynamic spin polarization are also given at the angles θ_m and θ_ξ. As can be seen, a large dynamic spin polarization of up to 90% can be achieved at certain angles for some Auger lines of the resonant spectrum.

Considering the possibility of interference effects between different partial waves during the Auger emission it is obvious that destructive interference, and thus a small spin polarization parameter ξ_2 can be expected if a large number of partial waves is emitted. Thus, as a first and necessary propensity rule, see (2.151), we

Table 4.26. The Auger energies, relative intensities, anisotropy and dynamic spin polarization parameters and the spin polarization at angles θ_m and θ_ξ of our MCDF calculation for the $Xe^*(4d_{5/2}^{-1}6p_{3/2})_{J=1} N_5O_{2,3}O_{2,3}$ resonant Auger transitions to final states with $J_f = 1/2$. (a): The leading jj coupled CSF has been used to identify the state; (b): Line numbers according to Aksela *et al.* (1995); †: The whole multiplet has been normalized to 100. Data from Lohmann (1999a)

		$Xe^*(6p_{3/2}) N_5O_{2,3}O_{2,3}$					
final states (a)	(b)	[eV]	I_0†	α_2	ξ_2	$p_y(\theta_m)$	$p_y(\theta_\xi)$
$\|([5\bar{p}^25p^2]_26p^1)1/2\rangle$	23	39.54	1.430	−0.078	0.478	−0.906	−0.956
$\|([5\bar{p}^15p^3]_16p^1)1/2\rangle$	36	38.37	7.119	−0.750	−0.164	0.214	0.328
$\|([5\bar{p}^15p^3]_26p^1)1/2\rangle$	46	37.07	8.435	−0.291	−0.404	0.670	0.808

need to restrict the number of partial waves to its minimum, i.e. two, in order to minimize a decrease of dynamic spin polarization due to destructive interference.[3]

For photoexcitation of the rare gases we have an intermediate total angular momentum $J = 1$. Thus, the angular momentum coupling rules reduce the number of partial waves to be emitted during the Auger decay. As a matter of fact only two partial waves can be emitted for the considered case. I.e.

$$J_f + j = J \longrightarrow j = 1/2, 3/2, \tag{4.22}$$

where j denotes the total angular momentum of the Auger partial waves.

Inspecting the parity of the states we find negative parities π_i and π_f for the intermediate excited and final ionic states, respectively. Note, that for photoexcitation of the rare gases the initial state generally has an odd parity. Due to the fact that the Coulomb interaction conserves parity we have

$$\pi_i = \pi_f \, \pi_{Aug} = \pi_f \, (-1)^\ell, \tag{4.23}$$

where π_{Aug} denotes the parity of the emitted Auger partial waves which, in our case, must be even.

Combining (4.22) and (4.23) we find that only $\varepsilon s_{1/2}$ and $\varepsilon d_{3/2}$ partial waves are emitted. Thus, if the intermediate ionic alignment has been dynamically transformed into the spin polarization of the Auger electron it will be solely taken away by the $\varepsilon d_{3/2}$ partial wave. Thus, neglecting loss of spin polarization due to effects like electron other electron interaction in the electronic charge cloud during the Auger emission, the transferred dynamic spin polarization is conserved in the emitted $\varepsilon d_{3/2}$ partial wave. Even if interference with the $\varepsilon s_{1/2}$ wave takes place it cannot reduce the degree of spin polarization since the εs wave cannot carry away any polarization at all. Thus, though two partial waves are emitted there can be no decrease in spin polarization of the emitted Auger electron caused by destructive interference. On the other hand, if $J_f > 1/2$, more spin polarized partial waves with $\pi_{Aug} = 1$ can be emitted allowing for a destructive interference and thus ξ_2 will decrease.

As an important reason for a large dynamic spin polarization we consider the phase shift of the partial waves. The total phase σ can be split into the Coulomb phase σ_c and the pure scattering phase δ, i.e.

$$\sigma = \sigma_c + \delta. \tag{4.24}$$

The phase shifts for the considered Auger lines of the Xe transitions are shown in Table 4.27. From the general expression for ξ_2 e.g. see (2.151) or Lohmann (1991) it is clear that a large dynamic spin polarization can only be achieved if the relative phase shift between the two partial waves is $\sim \pi/2$. This is fulfilled for the considered transitions. Our calculation indicates that this is mainly due to the phase shift of the $\varepsilon s_{1/2}$ partial wave caused by its large scattering phase δ which exceeds the pure Coulomb phase by a factor of ~ 3.7. While for the $\varepsilon s_{1/2}$ partial wave the Coulomb

---

[3] Note, that ξ_2 vanishes if only one partial wave is emitted.

Table 4.27. The phase shifts of the partial waves of the $Xe^*(4d_{5/2}^{-1}6p_{3/2})_{J=1} N_5O_{2,3}O_{2,3}$ Auger transitions to $J_f = 1/2$. For identification of states see Table 4.23. Data from Lohmann (1999a)

	$\varepsilon s_{1/2}$		$\varepsilon d_{3/2}$	
no.	σ_c	δ	σ_c	δ
23	0.26940	0.99295	−0.54643	0.44391
36	0.27164	1.02048	−0.55485	0.46739
46	0.27415	1.04328	−0.56467	0.48350

and scattering phases have the same sign, they point in different directions for the $\varepsilon d_{3/2}$ wave. Here, σ_c and δ are approximately of the same magnitude and almost cancel each other. Thus, a large dynamic spin polarization parameter ξ_2 is mainly caused by a large phase shift of the $\varepsilon s_{1/2}$ partial wave.

As pointed out, such type of experiments are extremely severe to perform, and thus, we are able to compare our predictions to one data set, only. Recently, a high resolution experiment has been performed by Hergenhahn *et al.* (1999). They observed a large dynamic spin polarization $\xi_2 = -0.41$[4] for the unresolved lines 46–47, where we obtain a value of $\xi_2 = -0.404$ for the sole line 46. Aksela *et al.* (1995) reported the ratio of the relative line intensities as $46/47 = 102/6.7$. They identify line 47 as an excited $6p \rightarrow 7s$ shake-up state. Though we have not included shake-up processes in our calculation the ratio of the intensities clearly shows that a large ξ_2 of the unresolved line is mainly caused by a spin polarization of line 46 which is in good agreement with our prediction.

Now, discussing the opposite case, i.e. small ξ_2 parameters, we consider the resonant $Ar^*L_3M_{2,3}M_{2,3}$ Auger transitions,

$$\gamma + Ar(^1S_0) \longrightarrow Ar^*(2p_{3/2}^{-1}, 4s_{1/2})_{J=1}$$
$$\longmapsto Ar^{+*}(3p^{-2}, 4s_{1/2})_{J_f} + e_A. \qquad (4.25)$$

Again, only two partial waves, according to (4.22) are emitted. As pointed out, the parity of the excited intermediate Ar^* state is odd. However, the final ionic state shows an even parity. Thus, from (4.23) we obtain an odd parity $\pi_{Aug} = -1$ for the Auger partial waves. I.e. the emitted partial waves are $\varepsilon p_{1/2}$ and $\varepsilon p_{3/2}$, respectively. Both εp waves may carry parts of the overall spin polarization. Therefore, it is likely that they show destructive interference which should decrease the spin polarization for these Auger transitions. Thus, even for transitions to $J_f = 1/2$ only a small ξ_2 should be expected. This is confirmed by our calculations of the resonant $Ar^*(2p_{3/2}^{-1}4s_{1/2})_{J=1}L_3M_{2,3}M_{2,3}$ multiplet which is shown in Table 4.28. In particu-

---

[4] For comparison to our data the ξ_2 parameter of Hergenhahn *et al.* (1999) must be multiplied by $-2/3$.

Table 4.28. The Auger energies, relative intensities, anisotropy and dynamic spin polarization parameters and the spin polarization at angles θ_m and θ_ξ of our MCDF calculation for the $Ar^*(2p_{3/2}^{-1}4s_{1/2})_{J=1}$ $L_3M_{2,3}M_{2,3}$ resonant Auger transitions to final states with $J_f = 1/2$. †: The whole multiplet has been normalized to 100. Data from Lohmann (1999a)

$Ar^*(4s_{1/2})\,L_3M_{2,3}M_{2,3}$						
final states	[eV]	I_0†	α_2	ξ_2	$P_y(\theta_m)$	$P_y(\theta_\xi)$
$\|3p^{-2}(^3P)4s\,^4P_{1/2}\rangle$	213.56	6.889	−1.115	0.0052	−0.0057	−0.0104
$\|3p^{-2}(^3P)4s\,^2P_{1/2}\rangle$	213.07	8.842	0.617	−0.0041	0.0145	0.0082
$\|3p^{-2}(^1S)4s\,^2S_{1/2}\rangle$	208.45	0.114	−0.748	0.0002	−0.0003	0.0004

Table 4.29. The phase shifts of the partial waves of the $Ar^*(2p_{3/2}^{-1}4s_{1/2})_{J=1}L_3M_{2,3}M_{2,3}$ Auger transitions to $J_f = 1/2$. For identification of states see Table 4.25. Data from Lohmann (1999a)

	$\varepsilon p_{1/2}$		$\varepsilon p_{3/2}$	
tran.	σ_c	δ	σ_c	δ
$^4P_{1/2}$	−0.10779	0.14735	−0.10779	0.13603
$^2P_{1/2}$	−0.10791	0.14869	−0.10792	0.13737
$^2S_{1/2}$	−0.10913	0.16172	−0.10913	0.15041

lar, the spin polarization turns out to be approximately two magnitudes smaller than in the Xe case.

The Coulomb and scattering phases for the Ar transitions are shown in Table 4.29. Since the Coulomb phase solely depends on the angular momentum, σ_c has the same value for the $\varepsilon p_{1/2}$ and $\varepsilon p_{3/2}$ partial waves. In contrast to the Xe case, the Coulomb and scattering phases are of the same magnitude but of opposite sign. Thus, we have an almost vanishing phase shift between the two partial waves which eventually yields a small ξ_2.

As last example we consider the resonant $Ar^*L_3M_1M_{2,3}$ multiplet after $2p_{3/2} \longrightarrow 4s_{1/2}$ excitation. Again, we have $\pi_i = -1$ and $\pi_f = -1$. Thus, $\pi_{Aug} = 1$ which results in the emission of $\varepsilon s_{1/2}$ and $\varepsilon d_{3/2}$ partial waves. Thus, we end up again with the first case; i.e. we would expect large spin polarization parameters ξ_2 for Auger transitions to $J_f = 1/2$.

From the above discussion we are able to derive two propensity rules which must be fulfilled for resonant Auger transitions if a large dynamic spin polarization shall be observed.

Table 4.30. Predictions of ξ_2 for resonantly excited Auger transitions. Transitions where a large ξ_2 can be expected are indicated by a $+$ sign. Data from Lohmann (1999a)

exci.	tran.	ξ_2	exci.	tran.	ξ_2
p $\longrightarrow$ s, d	p $\longrightarrow$ sp	$+$	d $\longrightarrow$ p	d $\longrightarrow$ sp	$-$
	p $\longrightarrow$ sd	$-$		d $\longrightarrow$ sd	$+$
	p $\longrightarrow$ pp	$-$		d $\longrightarrow$ pp	$+$
	p $\longrightarrow$ pd	$+$		d $\longrightarrow$ pd	$-$
	p $\longrightarrow$ dd	$-$		d $\longrightarrow$ dd	$+$

- Ensure that only two partial waves are emitted. I.e. preferably investigate Auger transitions of the type

$$J = 1 \longrightarrow J_f = 1/2. \tag{4.26}$$

This is a necessary but not a sufficient condition and already reduces the number of possible candidates of Auger transitions with a large ξ_2. The second rule is based on parity arguments.

- Ensure that

$$\pi_i = \pi_f \quad \text{or} \quad \pi_{Aug} = 1. \tag{4.27}$$

This prevents destructive interference of spin polarization between the two partial waves.

With this we are able to give a number of predictions for the ξ_2 parameter which are shown in Table 4.30. However, one has to keep in mind that these are just propensity rules, i.e. physics might show a different behaviour for certain Auger transitions. In particular, even for the case of a large number of interfering partial waves, which is the common case for the diagram transitions, there might be the possibility of constructive interference which can in principle bring enhancements of the dynamic spin polarization. Besides a few exceptions, these Auger transitions show an almost vanishing dynamic spin polarization (Hahn *et al.* 1985; Merz and Semke 1990; Kabachnik and Sazhina 1988; Kabachnik *et al.* 1991; Lohmann 1993). Even from the present state it is hard to provide any rules or guidelines for this case.

Besides the considered examples, applying our propensity rules and using Table 4.30, we are able to predict large ξ_2 parameters, and thus a large dynamic spin polarization, e.g. for the resonantly excited $\text{Kr}^*(5p_{3/2})\,\text{M}_{4,5}\text{N}_{2,3}\text{N}_{2,3}$, $\text{Xe}^*(6p_{3/2})\,\text{M}_{4,5}\text{N}_{4,5}\text{N}_{4,5}$ or $\text{Xe}^*(6p_{3/2})\,\text{N}_4\text{O}_{2,3}\text{O}_{2,3}$ Auger transitions. On the other hand, we expect a vanishing dynamic spin polarization for instance for the resonantly excited $\text{Kr}^*(5p_{3/2})\,\text{M}_{4,5}\text{N}_1\text{N}_{2,3}$, $\text{Xe}^*(6p_{3/2})\,\text{M}_{4,5}\text{N}_{2,3}\text{N}_{4,5}$ or $\text{Xe}^*(6p_{3/2})\,\text{N}_{4,5}\text{O}_1\text{O}_{2,3}$ Auger transitions.

Deriving simple propensity rules, we introduced a method to identify Auger lines for which a large dynamic spin polarization can be expected. Our calculation indicates that a large dynamic spin polarization parameter is caused by a large scattering phase of the $\varepsilon s_{1/2}$ and an almost vanishing phase shift of the $\varepsilon d_{3/2}$ partial waves, respectively. On the other hand, a small spin polarization is due to a cancellation effect between the Coulomb and scattering phases of the partial waves. The derived rules allow to decide whether an Auger transition is likely to show a dynamic spin polarization before doing any experiment or calculation. The predictions from our propensity rules are in good accordance with recent experimental data.

4.7 Angle and Spin Resolved Analysis of the Resonantly Excited Ar*$(2p^{-1}_{1/2,3/2}4s_{1/2})_{J=1}$ Auger Decay

4.7.1 General Considerations

Spin polarization of photo- and Auger electrons, the most sensitive probe of the magnetic properties of matter, is directly related to the generally anisotropic electron–electron correlation during the electron emission. In particular, spin polarization manifests itself via two physically different effects. The electrons may be spin polarized due to polarization transfer by circularly polarized light, but also due to the interference between different partial waves emitted by linearly polarized light. The components of the spin polarization vector of the Auger emission in the reaction plane, defined by the incoming synchrotron beam axis and the direction of Auger emission (see Fig. 2.3), can show a non-zero spin polarization due to spin polarization transfer by circularly polarized light of the incoming synchrotron beam (Klar 1980; Kabachnik and Lee 1989; Blum *et al.* 1986; Lohmann *et al.* 1993). This transferred spin polarization (TSP) is generally large because of the asymmetric *m*-sublevel population (Lohmann *et al.* 1993; Müller *et al.* 1995; Snell *et al.* 1996b, 1999; Lohmann 1999b; Hergenhahn *et al.* 1999; Lohmann and Kleiman 2001) generated by circularly polarized light.

This is not the case for the component of the spin polarization vector perpendicular to the reaction plane, related to the dynamic spin polarization (DSP), which can have a non-vanishing spin polarization generated via dynamic effects during the Auger emission, i.e. interference between the different emitted partial waves, induced by linearly polarized light. In principle, a DSP should be possible even if the target and the photon beam are unpolarized (Klar 1980; Kabachnik 1981; Blum *et al.* 1986). The latter is related to the fact that photons are generally aligned due to their transversal character.

The DSP shows higher values only if certain conditions concerning the number of contributing partial waves and their relative phase shifts are fulfilled (Hergenhahn *et al.* 1999; Lohmann 1999a). *Propensity rules* for specific Auger transitions where a large dynamic spin polarization can be expected have been derived by Lohmann (1999a) and have been discussed in Sect. 4.6.2. Particularly, a fully resolved fine structure has been hitherto recognized as a prerequisite for a non-vanishing DSP in a non-relativistic approximation. In the following, we are discussing the first experimental analysis and theoretical interpretation of both types of spin polarization in one Auger spectrum considering the prominent LMM spectrum of Ar (Lohmann *et al.* 2005).

The resonant Auger decay, shown in Fig. 2.7 in contrast to the normal Auger emission, has been studied extensively in the past by several research groups (Tulkki *et al.* 1993; Aksela *et al.* 1997; Aksela and Mursu 1996; Hergenhahn *et al.* 1991; von Raven *et al.* 1990; de Gouw *et al.* 1995; Chen 1993; Farhat *et al.* 1997). We will consider the resonantly excited, angle and spin resolved argon Auger decay which has been recently investigated, both, experimentally and numerically (Lohmann *et al.* 2003b, 2005),

$$\gamma + Ar(^1S_0) \longrightarrow Ar^*\left(2p^{-1}_{1/2,3/2}, 4s_{1/2}\right)_{J=1}$$
$$\hookrightarrow Ar^{+*}\left(3p^{-2}_{1/2,3/2}, 4s_{1/2}\right.$$
$$\left.+3d_{3/2,5/2}\right)L^{2S+1=2,4}_{J=1/2,\ldots,9/2} + e_{Auger}. \quad (4.28)$$

The theory of angle and spin resolved Auger decay has been extensively discussed in Sect. 2.6 and the main aspects of the applied numerical codes have been considered in Chap. 3. In the following, we will consider the experimental set-up for the measurement and discuss the specific theoretical details, resulting from the experimental geometry, in order to determine the transferred (TSP) and dynamic (DSP) spin polarization, of the Auger lines. We will outline the particular numerical aspects of the calculation, and compare the numerical to the experimental data for the case of the transferred spin polarization. For the dynamic spin polarization we will give a detailed analysis in terms of phase shift differences and partial decay widths.

Resonant Auger decay from 2p-excited argon atoms is a show case for large TSP but vanishing DSP. The first may be understood from the close similarity with d-shell photoionization, particularly the case with f-wave suppression. The reason for the very small DSP is two-fold: on one hand it is in accordance with the discussed propensity rules of Sect. 4.6.2 (Lohmann 1999a), showing that for most resonant Auger transitions, due to parity arguments, their partial waves are emitted with equal orbital angular momentum, hence, having very small phase differences which in turn diminishes the DSP. This vanishing relativistic phase shift situation is also known from photoionization (Snell *et al.* 1999). The measurements performed at the *Advanced Light Source* (ALS) in Berkeley and the calculations obtained within a relativistic distorted wave approximation (RDWA) corroborate these assumptions – see the discussion in Sect. 4.7.5. – On the other hand, the different fine structure components tend to cancel their spin polarization with respect to each other due to the requirement of vanishing DSP for the whole multiplet in pure *LS* coupling. This is because all spin polarization effects are mediated via the spin orbit interaction (Klar 1980; Kabachnik 1981; Blum *et al.* 1986; Kabachnik *et al.* 1988; Drescher *et al.* 2003). As a consequence, non-resolved fine structure components should exhibit almost no measurable DSP. Therefore, virtually no DSP is expected for a light element such as argon, i.e. the resonant $L_{2,3}M_{2,3}M_{2,3}$ Auger spectrum following the Ar$^*(2p_{1/2,3/2}^{-1}4s_{1/2})_{J=1}$ excitation with linearly polarized light. However, the experimental low resolution spectra exhibit measurable DSP, at least for one unresolved group of lines. Lohmann et al. (2005) have been able to explain this unexpected occurrence of DSP by strong configuration interaction (CI) in the final ionic state; see Sect. 4.7.6. Their work represents the first combined experimental and theoretical demonstration of a new mechanism that gives rise to DSP of unresolved Auger lines.

4.7.2 Experimental Details and Set-Up

The spin resolved electron spectra of the Ar $2p_{3/2} \rightarrow 4s_{1/2}$ and Ar $2p_{1/2} \rightarrow 4s_{1/2}$ autoionization resonances have been measured (Lohmann et al. 2003b, 2005). The experiment is shown in Fig. 4.7, and has been performed at the ALS operating in two bunch mode utilizing the elliptically polarizing undulator (EPU) at Beamline 4.0.2 (Young et al. 2001). The EPU was set to deliver 100% circularly or linearly polarized light at all used photon energies. Electron energy analysis has been performed using a time-of-flight (TOF) spectrometer, collecting electrons emitted at 45° with respect to the storage ring plane[5] in the direction perpendicular to the photon propagation. A spherical Mott polarimeter of the Rice type, operated at 25 kV, mounted at the end of the TOF has been used to carry out the electron spin polarization analysis (Burnett et al. 1994; Snell et al. 2000, 2001). The geometry of the experiment has been chosen to measure the polarization of the spin components

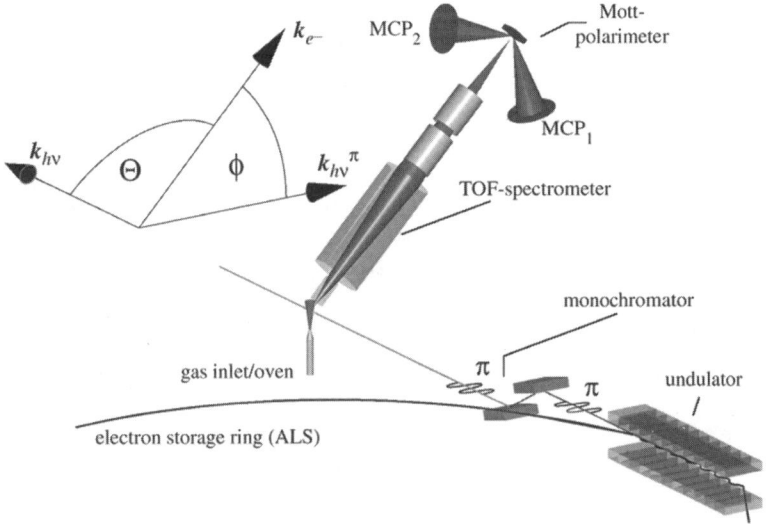

Fig. 4.7. Experimental set-up with undulator beamline, TOF electron analyzer and Mott detector (Lohmann et al. 2003b)

---

[5] This refers to $\phi_{exp} = 135^o$ in our chosen coordinate frame.

of the electrons along the photon propagation direction (Snell *et al.* 2002). The instrumental asymmetries of the Mott polarimeter have been eliminated by reversing the helicity in the case of circularly polarized light and switching the polarization from horizontal to vertical in the case of linearly polarized light. The transferred and dynamic electron spin polarizations, corresponding to $p_{trans}(\theta_{exp} = 90^o)$ and $p_{dyn}(\theta_{exp} = 90^o, \phi_{exp} = 135^o)$, respectively,[6] can be calculated from the four measured intensities I_1^+, I_1^-, I_2^+, and I_2^- as follows

$$p_{trans,dyn}(\theta_{exp}, \phi_{exp}) = \frac{1}{S_{eff}} \frac{\sqrt{I_1^+ I_2^-} - \sqrt{I_1^- I_2^+}}{\sqrt{I_1^+ I_2^-} + \sqrt{I_1^- I_2^+}}. \quad (4.29)$$

The lower index denotes the multi-channel plates (MCP) of the Mott detector and the upper index stands for the helicity, respectively the horizontal or vertical polarization of the light. S_{eff} describes the analyzing power of the polarimeter (effective Sherman function) which has been determined to be $S_{eff} = 0.13(3)$. Measurement of the DSP and TSP has been achieved arranging MCP_1 and MCP_2 perpendicular to the reaction plane using linearly and circularly polarized light for the DSP and TSP, respectively. Figure 4.7 shows the experimental set-up for measuring the DSP; i.e. MCP_1 and MCP_2 perpendicular to the reaction plane. Measurement of the TSP is achieved by rotating the MCPs by 90° around the TOF-axis. A more detailed description of the experiment and the analysis for the resonant Auger decay after photoexcitation, (4.28), has been given by Snell *et al.* (2002).

4.7.3 Determination of Transferred and Dynamic Spin Polarization

Defining the reaction plane by the incoming synchrotron beam axis and the direction of the Auger emission (see Fig. 4.7), both, DSP and TSP can be related to the in-reaction plane component of the spin polarization vector perpendicular to the direction of Auger emission (Kleiman *et al.* 1999b), which, in our chosen geometry, coincides with the x-component of the spin polarization vector; see (2.226). As we have outlined in Sect. 2.6 describing the resonant Auger transitions in the well observed two-step model (Mehlhorn 1990) allows for factorizing the DSP and TSP into a set of alignment and orientation parameters containing solely the dynamics of the primary excitation, and angular distribution and spin polarization parameters describing the dynamics of the resonant Auger decay. For resonant Auger decay the alignment and orientation parameters become constant numbers (Kronast *et al.* 1986) and are given in Table 2.9.

Measuring the DSP after photoexcitation with a linearly polarized photon beam and performing the experiment in the emission reference system under the fixed geometry $\theta_{exp} = 90^o$ and $\phi_{exp} = 135^o$, see Fig. 4.7, and inserting θ_{exp}, ϕ_{exp} into (2.226) we obtain the DSP as

$$p_{dyn}(\theta_{exp}, \phi_{exp}) = \frac{6\xi_2}{2\sqrt{2} - \alpha_2}. \quad (4.30)$$

---

[6] The angle ϕ_{exp} is redundant for measuring the TSP; see the next Sect. 4.7.3.

Thus, the DSP is directly proportional to the dynamic spin polarization parameter ξ_2 which can be evaluated from (2.151). The angular distribution parameter α_2 can be obtained from (2.139).

Similarly, for a fully circularly polarized photon beam, the TSP can be related to the x-component (2.226) of the spin polarization vector, too. The TSP becomes independent of the azimuthal angle ϕ because the combined photonic and target system, where the latter has been assumed as unpolarized, is axially symmetric with respect to the synchrotron beam axis. Thus, the TSP can be measured under the same solid angle as the DSP, i.e. $\theta_{exp} = 90^o$, while ϕ_{exp} becomes redundant, and inserting into (2.226) yields

$$p_{trans}(\theta_{exp}) = \frac{2\sqrt{3}\,\xi_1}{2\sqrt{2} - \alpha_2},$$
(4.31)

where ξ_1 refers to as the transferred spin polarization parameter given by (2.150).

4.7.4 Numerical Calculation Methods

An important point in obtaining numerical data is to apply a consistent model for the calculation of all parameters, and we briefly review the methods used for the calculation of the DSP and TSP, i.e. (4.30) and (4.31), for the resonant Auger decay. Further information may be found in Chap. 3, or in Lohmann (1999b), Lohmann and Kleiman (2001).

The numerical data have been obtained employing a relativistic distorted wave approximation (RDWA). Here, the bound state wavefunctions of the excited intermediate and the ionized final state of the atom are constructed using the MCDF computer code of Grant et al. (1980). Intermediate coupling has been taken into account where the mixing coefficients have been calculated applying the average level calculation mode. The calculation of the Auger transition matrix elements has been done applying a relaxed orbital method. Thus, the bound electron wavefunctions of the intermediate state are calculated in the field of the excited atom, whereas the bound electron wavefunctions of the final state are calculated in the field of the singly ionized atom.

The ASF of the intermediate excited and the singly ionized final state have been constructed as linear combinations of CSF

$$\left|\psi_\alpha(PJM)\right\rangle = \sum_{r=1}^{n_c} c_r(\alpha)\left|\gamma_r\,PJM\right\rangle,$$
(4.32)

see Sect. 3.1, or Grant (1970) for further details.

The intermediate excited state has been generated as linear combination from the five possible jj-coupled $Ar^*(2p_{1/2,3/2}^{-1}4s_{1/2})_{J=1}$ and $Ar^*(2p_{1/2,3/2}^{-1}3d_{3/2,5/2})_{J=1}$ CSF. We find the $Ar^*(2p_{3/2}^{-1}4s_{1/2})_{J=1}$ and $Ar^*(2p_{1/2}^{-1}4s_{1/2})_{J=1}$ ASF as almost pure states.

Two calculations have been performed for the final ionic state. In calculation (a) a basis set of 8 CSF has been used to generate the final state ASF from the possible linear combinations of the $Ar^+(3p_{1/2,3/2}^{-2}4s_{1/2})$ jj-coupled states (8 CSF-CI); see the discussion in Sect. 4.7.5. Since this approach has not been able to reveal all lines of the spectrum (Aksela and Mursu 1996; Mursu *et al.* 1996) our calculation (b) accounts for all jj-coupled $Ar^+(3p_{1/2,3/2}^{-2}3d_{3/2,5/2})$ CSF, too. Thus, a basis of 36 CSF has been used to generate the final state ASF (36 CSF-CI). The results will be discussed in Sect. 4.7.6.

Eventually, the continuum wavefunction of the Auger electron is evaluated by solving the Dirac-equation with an intermediate coupling potential where electron exchange with the continuum has been taken into account. The intermediate coupling potential is constructed from the mixed CSF of the final ionic state. Thereby we take into account, that the ejected electron moves within the field of the residual ion. With this, the Auger transition matrix elements are obtained for calculating the relevant angular anisotropy and spin polarization parameters, respectively. Note, that both are not functions of the transition matrix elements, only but explicitly depend on the scattering phases.

4.7.5 Analysis and Comparison of Theoretical and Experimental Data

The numerical results of Lohmann *et al.* (2003b) (8 CSF-CI) for the relative intensities, angular distribution and spin polarization parameters are shown in Table 4.31 for the resonant $Ar^*(2p_{3/2}^{-1}4s_{1/2})_{J=1}L_3M_{2,3}M_{2,3}$ Auger transitions together with the

Table 4.31. The relative intensities, angular distribution and spin polarization parameters for the $Ar^*(2p_{3/2}^{-1}4s_{1/2})_{J=1}L_3M_{2,3}M_{2,3}$ Auger multiplets (Lohmann *et al.* 2003b). Comparison of calculational to experimental results. The transferred spin polarization p_{trans} has been obtained for the angle of detection $\theta_{exp.} = 90^o$. †: The states have been identified in the LSJ coupling according to (Tulkki *et al.* 1993). ‡: The total intensity has been normalized to 100

					$Ar^*(4s_{1/2})L_3M_{2,3}M_{2,3}$				
Final		Energy	Int.		Ang. & spin pol. par.			p_{trans}
states†		(eV)	I_0‡	α_2	δ_1	ξ_1	ξ_2	Theo. Exp.
$^4P_{5/2}$		213.74	5.8	−0.18	−0.26	−0.48	−0.012	
$^4P_{3/2}$	^4P	213.63	11.3	0.63	−1.00	−0.09	−0.002	−0.38 −0.30
$^4P_{1/2}$		213.56	6.2	−1.12	0.15	−0.57	−0.005	
$^2P_{3/2}$	^2P	213.20	19.4	0.70	−0.46	0.06	0.000	−0.10 −0.22
$^2P_{1/2}$		213.07	7.9	0.62	1.17	−0.35	−0.004	
$^2D_{5/2}$	^2D	211.48	6.5	−0.93	−0.04	0.27	−0.228	0.82 0.78
$^2D_{3/2}$		211.47	32.7	−0.22	0.65	0.84	0.048	
$^2S_{1/2}$	^2S	208.44	10.2	−0.71	0.41	0.82	0.000	0.80 0.80

Ar*$(2p^{-1}_{3/2}4s_{1/2})_{J=1}L_3M_{2,3}M_{2,3}$

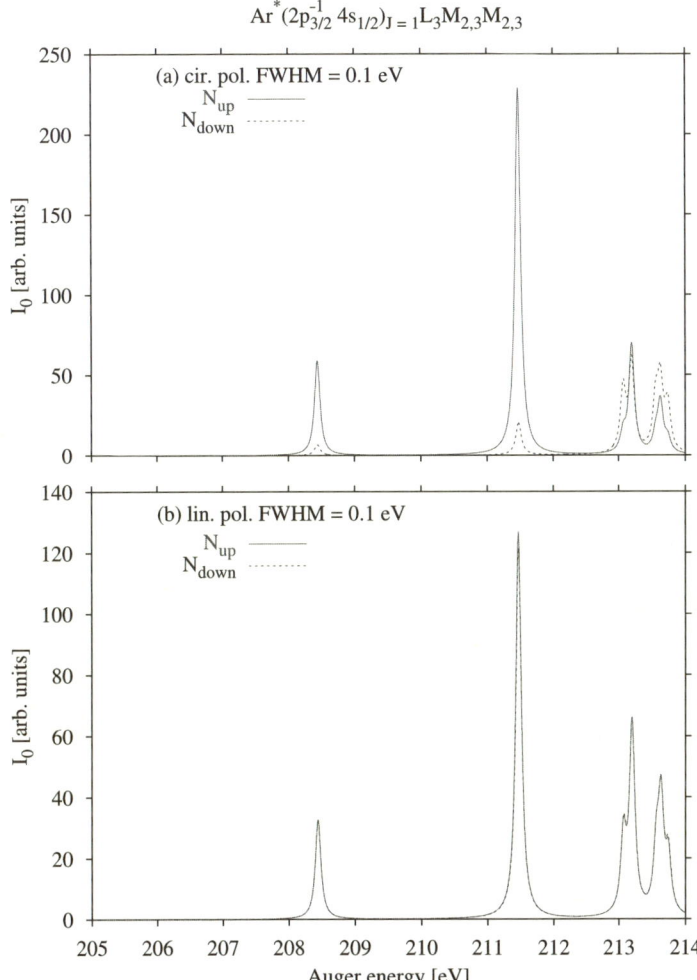

Fig. 4.8. The calculated 8 CSF-CI spin resolved intensities for excitation with (**a**) circularly and (**b**) linearly polarized light showing high degree of transferred spin polarization and almost no dynamic spin polarization for any of the Auger lines (Lohmann *et al.* 2003b). The data are calculated for the experimental geometry, $\theta_{exp.} = 90^o$, $\phi_{exp.} = 135^o$ (see Fig. 4.7); after Lohmann *et al.* (2003b)

numerical and experimental data for the TSP. As can be seen, a large TSP has been found in the experiment which is in good accordance with the numerical data.

The numerical predictions for the resolved spin-up and spin-down partial intensities for excitation with circularly polarized light are shown in Fig. 4.8a illustrating the high degree of transferred spin polarization for the eight lines of the calculated Auger spectrum (Lohmann *et al.* 2003b). Here, a Lorentz profile with

Table 4.32. The calculated relative phase shifts of the partial waves of the $Ar^*(2p_{3/2} \rightarrow 4s_{1/2})\, L_3M_{2,3}M_{2,3}$ Auger transitions (Lohmann *et al.* 2003b)

Final state	Relative phases			
	$\Delta_{\varepsilon p_{3/2}-\varepsilon p_{1/2}}$	$\Delta_{\varepsilon f_{5/2}-\varepsilon p_{3/2}}$	$\Delta_{\varepsilon f_{7/2}-\varepsilon p_{3/2}}$	$\Delta_{\varepsilon f_{7/2}-\varepsilon f_{5/2}}$
$^4P_{5/2}$	—	0.658	0.658	−0.0005
$^4P_{3/2}$	−0.011	0.658	—	—
$^4P_{1/2}$	−0.011	—	—	—
$^2P_{3/2}$	−0.011	0.656	—	—
$^2P_{1/2}$	−0.011	—	—	—
$^2D_{5/2}$	—	0.649	0.649	−0.0005
$^2D_{3/2}$	−0.011	0.649	—	—
$^2S_{1/2}$	−0.011	—	—	—

a FWHM $= 0.1\,\text{eV}$ has been assumed for all lines. On the other hand, only small DSP has been found in accordance with the theoretical predictions in terms of propensity rules; see Sect. 4.6.2. Therefore, the Auger decay of the Ar $(2p \rightarrow 4s)$ excited state can be considered as a show case for large transferred but almost vanishing dynamic spin polarization. The calculated spin-up and spin-down partial intensities are plotted in Fig. 4.8b illustrating the almost vanishing dynamic spin polarization.

However, a large ξ_2 parameter for one of the fine structure components of the ^2D line has been found. This has been explained by Lohmann *et al.* (2003b) inspecting two features of the dynamic spin polarization. First, the spin polarization parameter ξ_2 is a function of the sine of the phase shift difference Δ of the emitted partial waves, $\xi_2 \sim f(\sin \Delta)$. I.e., a small relative phase shift between the partial waves automatically results in an almost vanishing dynamic spin polarization. The calculated relative phases of the emitted partial waves are given in Table 4.32. Their analysis yields comparatively large phase shifts Δ between the emitted εf and εp partial waves whereas the phase shift between partial waves having the same orbital angular momentum, e.g. $\Delta_{\varepsilon p_{3/2}-\varepsilon p_{1/2}}$, has been found almost zero. This is because the latter can be generated via relativistic effects, only.

Second, a large ξ_2 parameter requires the Auger decay to proceed via at least two partial waves. Otherwise, we have $\xi_2 = 0$ and thus zero dynamic spin polarization. This requires information about the partial decay widths of the fine structure states of the Auger lines, which are shown in Table 4.33. Almost equal intensities for the εf and εp partial waves for the $^2D_{5/2}$ fine structure component have been found. This feature, together with the comparatively large phase shift eventually yields the calculated large dynamic spin polarization parameter ξ_2 for the fine structure component. On the other hand, the $^2D_{3/2}$ fine structure component decays almost solely via the $\varepsilon p_{1/2}$, $\varepsilon p_{3/2}$ partial waves, having a small relativistic phase shift, only. This results in an almost vanishing DSP for the $^2D_{3/2}$ fine structure state. Unfortunately, the fine structure of the ^2D line is not fully resolved. The data of Lohmann *et al.*

Table 4.33. The calculated partial widths of the $Ar^*(2p_{3/2} \rightarrow 4s_{1/2}) L_3M_{2,3}M_{2,3}$ Auger transitions; after Lohmann $et\ al.$ (2003b)

Final state	Relative partial widths (%)			
	$\varepsilon p_{1/2}$	$\varepsilon p_{3/2}$	$\varepsilon f_{5/2}$	$\varepsilon f_{7/2}$
$\lvert(^3P)4s\,^4P_{5/2}\rangle$	—	99.93	0.01	0.06
$\lvert(^3P)4s\,^4P_{3/2}\rangle$	44.75	55.25	0.00	—
$\lvert(^3P)4s\,^4P_{1/2}\rangle$	68.66	31.34	—	—
$\lvert(^3P)4s\,^2P_{3/2}\rangle$	10.40	89.59	0.00	—
$\lvert(^3P)4s\,^2P_{1/2}\rangle$	84.30	15.70	—	—
$\lvert(^1D)4s\,^2D_{5/2}\rangle$	—	50.62	3.95	45.43
$\lvert(^1D)4s\,^2D_{3/2}\rangle$	60.13	39.18	0.69	—
$\lvert(^1S)4s\,^2S_{1/2}\rangle$	0.00	100.00	—	—

(2003b), however, yield experimental evidence for a non-zero DSP which seemed to stem from the Auger decay to the $(2p^{-2}3d)\,^2D$ final state. Such excited states have not been included in their numerical approach; i.e. the 8 CSF-CI. Therefore, using larger basis sets seems to be a good possibility to numerically reproduce the experimental values for the DSP.

This has been performed utilizing a 36 CSF basis set (Lohmann $et\ al.$ 2005). The numerical results of the extended 36 CSF-CI, calculation (b), for the TSP and DSP are shown in Fig. 4.9. A non-vanishing DSP has been found for most of the non-diagram Auger lines stemming from the transitions to the $Ar^{+*}(3p_{1/2,3/2}^{-2}3d_{3/2,5/2})_{J_f}$ final states not present in the 8 CSF-CI. This non-vanishing DSP gives rise to a new effect which will be discussed in the following section.

4.7.6 Configuration-Interaction Induced Dynamic Spin Polarization

Using equations (4.30) and (4.31) we have been able to plot the spin-up and spin-down partial intensities of the spectrum.

Our results are shown in Figs. 4.10a and 4.10b for the spin resolved spectrum for the TSP and in Figs. 4.10c and 4.10d for the spin resolved DSP spectrum of the $L_{2,3}M_{2,3}M_{2,3}$ Auger decay for the excited intermediate argon $(2p_{1/2}^{-1}4s_{1/2})_{J=1}$ and $(2p_{3/2}^{-1}4s_{1/2})_{J=1}$ states, respectively, along with the results of the extended 36 CSF-CI calculation.

The spectrum has been generated assuming Lorentz profiles with a FWHM = 0.1 eV for the Auger lines folded with the appropriate Gaussian line shape. The partial intensities of the 36 CSF-CI represent the calculated spin polarizations, as shown in Table 4.34, but normalized to the experimental total intensities and shifted by an energy offset. A bar diagram underneath of the partial intensities shows the integral values of the spin polarization for the most prominent Auger lines. The filled

$Ar^*(2p_{3/2}^{-1}4s_{1/2})_{J=1}L_3M_{2,3}M_{2,3}$

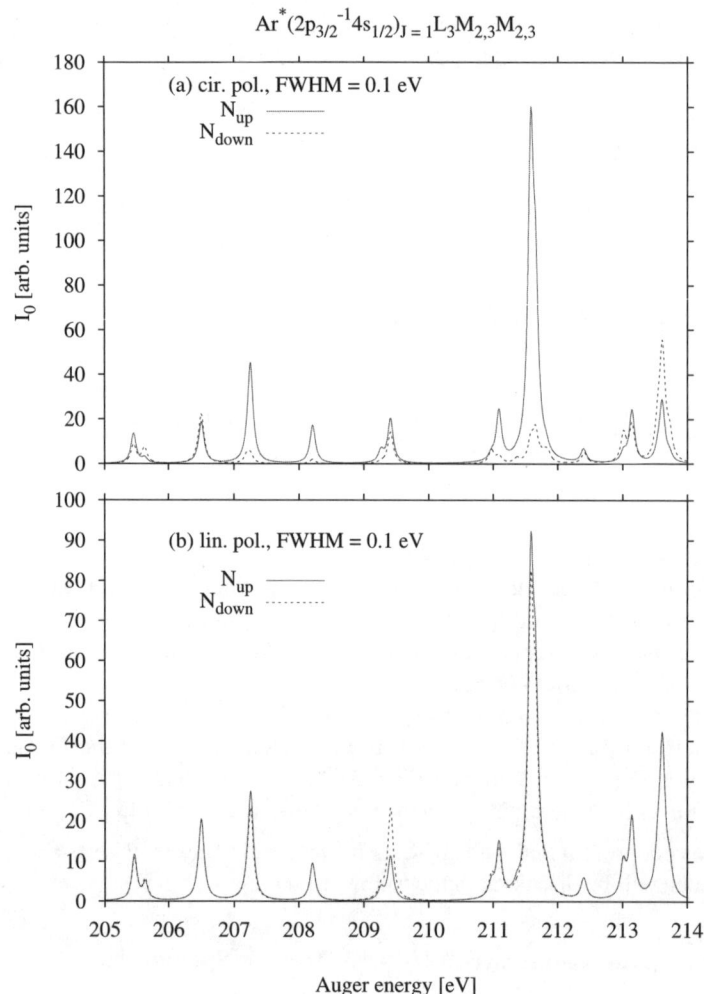

Fig. 4.9. The calculated 36 CSF-CI spin resolved intensities for excitation with (**a**) circularly and (**b**) linearly polarized light showing high degree of transferred spin polarization, and large dynamic spin polarization for most of the non-diagram Auger lines. The data are calculated for the experimental geometry, $\theta_{exp.} = 90^o$, $\phi_{exp.} = 135^o$ (see Fig. 4.7)

bars represent the experimental values, whereas the two open bars refer to our results for the 8 CSF and the extended 36 CSF calculations, respectively. Besides the generation of new lines at lower kinetic energies (see Fig. 4.9), of which only line 3 is shown in Fig. 4.10 and Table 4.34, the comparison reveals an unexpected difference between the two types of spin polarization, transferred and dynamic; whereas for the TSP there is little improvement between the two calculations, only – the 8 CSF-CI reproduces the experimental data already very well, particularly for the $3p_{3/2}$ exci-

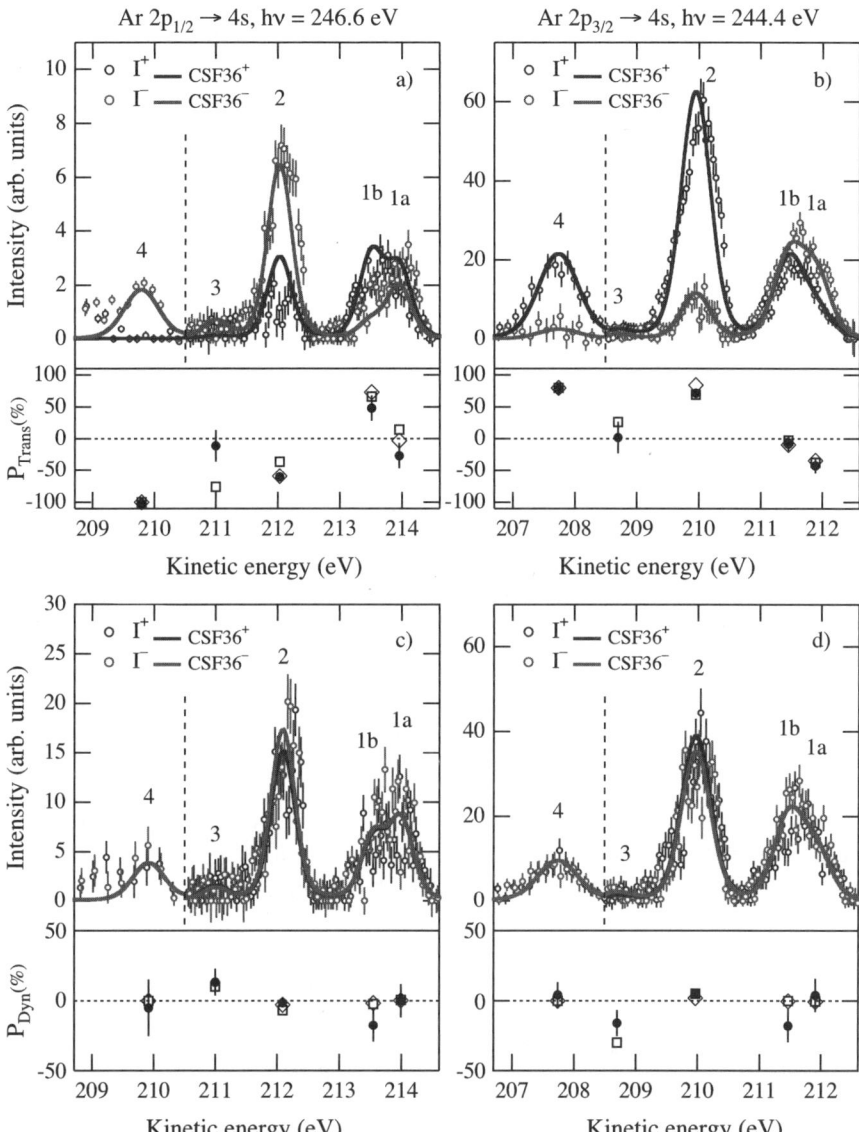

Fig. 4.10. Spin resolved spectrum for the TSP (**a, b**) and DSP (**c, d**) for excitation with circularly and linearly polarized light, respectively. *Full curves*: 36 CSF-CI. *Circles with errorbars*: exp. data. *Blue*: partial intensities for spin-up (I^+). *Red*: partial intensities for spin-down (I^-). The degree of DSP and TSP for the unresolved group of lines is shown in the chart underneath. *Open diamonds*: 8 CSF-CI. *Open squares*: 36 CSF-CI. *Filled circles*: exp. data. The dashed lines in all figures separate spectra taken with different resolution and hence retardation voltage (Lohmann *et al.* 2005)

Table 4.34. The calculated degree of dynamic, $p_{dyn}(\theta_{exp}, \phi_{exp})$, and transferred, $p_{trans}(\theta_{exp})$, spin polarization is given. Note, that transitions to fine structure terms with $J \geq 7/2$ are suppressed due to J-dependent selection rules. (a): Our tentative assignment of the observed and calculated peaks (36 CSF-CI) of the unresolved LSJ fine structure terms. (b): Peak numbers as assigned in the experimental spectrum

Ar* $L_{2,3}M_{2,3}M_{2,3}$		$2p_{3/2} \to 4s$		$2p_{1/2} \to 4s$	
Final states	No.	Spin polarization (%)			
(a)	(b)	p_{dyn}	p_{trans}	p_{dyn}	p_{trans}
$\left.\begin{array}{l}\|(3p^2[^3P]\,4s)\,^4P_{1/2,...,5/2}\rangle \\ \|(3p^2[^3P]\,3d)\,^4D_{1/2,...,7/2}\rangle\end{array}\right\}$	1a	-0.30	-42.51	0.49	8.53
$\|(3p^2[^3P]\,4s)\,^2P_{1/2,3/2}\rangle$	1b	-0.06	-2.57	-2.38	66.31
$\left.\begin{array}{l}\|(3p^2[^3P]\,3d)\,^2P_{1/2,3/2}\rangle \\ \|(3p^2[^3P]\,3d)\,^4P_{1/2,...,5/2}\rangle \\ \|(3p^2[^1D]\,4s)\,^2D_{3/2,5/2}\rangle \\ \|(3p^2[^3P]\,3d)\,^2F_{5/2,7/2}\rangle \\ \|(3p^2[^3P]\,3d)\,^2D_{3/2,5/2}\rangle\end{array}\right\}$	2	4.42	69.50	-4.07	-41.36
$\left.\begin{array}{l}\|(3p^2[^1D]\,3d)\,^2F_{5/2,7/2}\rangle \\ \|(3p^2[^1D]\,3d)\,^2S_{1/2}\rangle\end{array}\right\}$	3	-30.29	25.62	10.14	-75.88
$\|(3p^2[^1S]\,4s)\,^2S_{1/2}\rangle$	4	0.02	79.56	-0.02	-99.98

tation – the corresponding results for the DSP differ qualitatively. In this case, the 8 CSF-CI yields a vanishing spin polarization over the whole spectrum. Reasonable agreement with the experiment, showing non-vanishing spin polarization for some lines (2, 3), can be achieved only by employing the extended 36 CSF-CI. Inspecting more closely the different transitions contributing to the spin polarized part of the spectrum reveals that the CI with 36 CSF, in contrast to the 8 CSF-CI, generates unresolved fine structure components with low and high total angular momentum as shown in Table 4.34. The 36 CSF-CI produces J-components which correspond in a tentative LSJ coupling scheme to $^2L_{J,J'}$ states with $L \geq 3$ and $J, J' \geq 5/2, 7/2$. These virtual fine structure components however, require non-diagram lines in form of Auger satellites in order to populate the high J-part. Stressing the picture of Auger decay as two-electron transition, the non-diagram lines are normally an order of magnitude lower than the corresponding diagram lines, similar to the situation in photoionization concerning main and satellite lines. Consequently, such situation of fine structure multiplets with mixed diagram/non-diagram character gives rise to the possibility of asymmetric cases where the high J-part of certain multiplets is suppressed due to J-dependent selection rules. Therefore, only the low J-components survive with no partner for polarization cancellation. However, both, the statistics of the experimental data as well as the number of configurations included in the CI calculations requires further improvements in order to quantify the observed effect in more detail.

4.7.7 Final Comments

We discussed the photoexcited $Ar^*(2p_{1/2,3/2}^{-1}, 4s_{1/2})_{J=1} L_{2,3}M_{2,3}M_{2,3}$ Auger decay, which has been investigated, both, experimentally and numerically. Emphasis has been laid on the analysis of the spin polarization of the emitted Auger electrons, where a large transferred spin polarization has been numerically predicted for most lines of the spectrum. This has been well confirmed in the experiment. Only small DSP has been calculated in the 8 CSF-CI, which has been found in accordance with derived propensity rules, while experimental data and the 36 CSF-CI point towards large DSP for some lines of the spectrum. The spin resolved, low resolution measurements of the resonant Auger electrons from 2p excited Ar by linearly polarized light have revealed a measurable DSP effect, which is in contrast to the hitherto understanding that a fully resolved fine structure must be a prerequisite in a non-relativistic approximation. It has been demonstrated that under certain conditions this unexpected DSP can be explained as CI-induced effect in the final ionic state. This effect is due to the quenching of the cancellation between different multiplet components as a result of their asymmetric population by internal selection rules for diagram lines.

4.8 Open Shell Atoms – The KLL Auger Spectra of Alkali Atoms

As has been seen from the discussion in the previous sections, most studies investigated closed shell systems. In the last two sections we will consider Auger transitions from open shell atoms. For open shell atoms, the theoretical description of angle and spin resolved Auger emission remains principally the same, though, its numerical description leaves us with the task to deal with generally more than two open shells for the final or even the initial state of the Auger transition. This usually requires non-trivial approaches. On the other hand, diagram open shell Auger transitions can be seen as the ionic counterpart of resonant closed shell Auger emission from neutral but excited atomic states.

Particularly, we will focus on KLL open shell Auger transitions, which, due to the presence of additional open shell angular momenta, can exhibit an anisotropic angular distribution, an effect which cannot be observed for diagram transitions of closed shell atoms. A similar effect has been investigated by Dill *et al.* (1975) for the angular anisotropy of photoelectrons emitted from the K shell of open shell atoms. They pointed out that even in extreme LS coupling the electron-ion interaction is generally anisotropic for open shell atoms.

In this section we are concentrating on the KLL Auger transitions of the open shell alkali atoms and their angular distribution; e.g. see Lohmann and Fritzsche (1994), Lohmann *et al.* (1996). While in closed shell atoms the KLL Auger spectra has to be isotropic since no alignment can be generated this is no longer valid for open shell systems like the alkalis. For the alkali elements, the electrons of the outermost open shell can couple their angular momentum with the angular momentum of the inner shell hole which can result in a total angular momentum $J > 1/2$. Thereby an alignment unequal to zero and thus, an observable anisotropic angular distribution is possible.

Applying the results of Chaps. 2 and 3 we are investigating the KLL Auger spectra of the alkali atoms Na, K, Rb, and Cs within the two-step model. Ab-initio MCDF calculations have been performed to obtain the Auger energies, the absolute and relative Auger rates, and the angular anisotropy coefficients of the Auger decay, where attention has been drawn to the influence of different potentials for the wavefunction of the emitted electron on the Auger line intensities and angular anisotropy parameters (see Sect. 3.2). The obtained data allow for recommending whether an angular distribution measurement of the Auger electrons seems to be reasonable or not. This depends strongly on the energy resolution and the intensity of the Auger multiplets.

4.8.1 General Considerations

We are describing the Auger emission process in the well observed two-step model which is illustrated here for the KLL Auger decay of sodium

$$\gamma_{Syn} + \text{Na}\,^2\text{S}_{1/2} \longrightarrow e_e + \text{Na}^+\left(1s^{-1}3s\right)^{1,3}\text{S}_{0,1} \tag{4.33a}$$

$$\longrightarrow \text{Na}^{++} + e_{Auger}. \tag{4.33b}$$

Here, the sodium atom is ionized in the inner K shell in a first step, for instance by a synchrotron photon beam, while the primary emitted electron e_e is not observed. In the second step, the Auger emission takes place. The resonant intermediate state Na^+ is usually referred to as the initial state of the Auger decay and assumed to be independent of the specific ionization process, i.e. (4.33a) and (4.33b) can be treated as independent processes. In the following we will focus on the Auger process, only.

The computational method, applying the RATR program, has been extensively described in Sect. 3.2; also see e.g. Lohmann and Fritzsche (1994). Therefore, we will only give a brief outline here. We obtained the continuum spinors for the construction of the final scattering states by solving the Dirac–Fock equations for the emitted electron in the stationary field of the doubly ionized ion where terms for the exchange interaction between the continuum electron and the remaining bound state electrons are included. Orthogonality is automatically ensured in our approach (see Sect. 3.2). To obtain the normalization and the phase of the continuum spinors we adapted a WKB method of Ong and Russek (1978).

The resonant $[1s^{-1}ns](J = 0, 1)$ initial states of the alkali KLL spectra have been calculated in a single-configuration approximation. Here, $n = 3, \ldots, 6$ correspond to the open s-shell in dependence on the alkali atom. Virtual excitations of the ns electron into CSFs of the same symmetry $(J = 0, 1)$ and even parity were not included. For the $[2s^0ns]$, $[2s2p^5ns]$, and $[2p^4ns]$ final states all CSF from the

KLL and KLX ($X = M_1, N_1, O_1$, or P_1 for Na, K, Rb, and Cs, respectively) have
been taken into account. The coupling of the two L-holes with the outer s-electron
results in 16 CSF. Thus, together with 3 CSFs from the KLX spectrum, a total of 19
CSF has been considered. We obtained the mixing coefficients between CSF of the
same symmetry in the average-level scheme (Dyall *et al.* 1989). For the evaluation
of the Auger amplitudes we used the bound orbitals of the initial resonant state and
orthogonalized the continuum spinors with respect to the bound ones. Note, that we
have to orthogonalize in any case since we are using different bound state orbitals
for the description of the initial and final states.

Four different calculations have been performed. In a first calculation the continuum wavefunction is calculated by using only the pure Coulomb potential, and
exchange interaction is neglected, that is $\xi_a^{(P)}(r) = \xi_a^{(Q)}(r) = 0$; see (3.73) and
(3.80) and the discussion in Sect. 3.2 (Note that spin-orbit terms etc. are automatically included since we solve the Dirac equation in full.). Secondly, we can include
the Breit interaction $b(i, j)$ of (3.4) in the calculation. As a third possibility the
Dirac equation is calculated including again only the Coulomb potential but also
the exchange interaction in the generation of the continuum orbitals in (3.73a) and
(3.73b). Finally, which is from our point of view the best approach, we include all
interactions in the Dirac equation that is Coulomb, Breit, and exchange interaction
in the continuum.

By obtaining the continuum and the bound states wavefunctions the Auger transition matrix elements can be evaluated. The calculation of the Auger rates have
been performed with all four different types of calculation. The relative intensities
are easily obtained of them. They have been normalized to the total rates. The calculation of the anisotropy parameter α_2 have been performed including only the
Coulomb interaction, and with the full approach, too. The calculation of α_2 requires
an explicit knowledge of the scattering phase.

4.8.2 Alkali KLL Spectra and Designation of States

Auger spectra of open shell systems are often complex and cause a great deal of
difficulties in their interpretation compared to the well known closed shell atoms;
mainly the rare gases. Even for atoms with a simple shell structure the number of
possible decay lines increases enormously often showing a rather small fine structure splitting. This also complicates the experimental identification of the individual
Auger lines. Therefore, the fine structure splitting was not properly resolved in most
previous experiments. So far, most data to compare with are available for the Na
KLL Auger transitions. There are also investigations on Cs KLL, because these
Auger transitions have importance in several applications in nuclear physics (Erman and Sujkowski 1961). For K and Rb there are only a few data known for the
KLL Auger transitions. Our work will therefore provide a serious of consistent data
which might initiate further investigations in this field. Even by the fact that nowadays, the application of high resolution experimental techniques allows to resolve
the fine structure in some of the considered spectra.

Although all our calculations have been performed in an intermediate coupling scheme we will designate the individual Auger transitions in the LSJ notation. This is because most of the available data from the literature are classified in this scheme. To identify the various final states in our calculation beyond their total angular momentum and parity we used their main contribution in the CSF expansion of the MCDF wavefunctions and their sequence of total energies. In our discussion we consider only the KLL Auger lines following a resonant excitation of the $^3S_1$ initial state. Carrying out the calculations with the initial $^1S_0$ state results in an isotropic decay for all related KLL transitions. Within the chosen classification scheme we denote our possible final states as KL_1L_1 $(^1S)ns$ $^2S_{1/2}$, and $KL_1L_{2,3}$ $(^1P)ns$ $^2P^o_{1/2,3/2}$, $(^3P)ns$ $^2P^o_{1/2,3/2}$, and $(^3P)ns$ $^4P^o_{1/2,3/2,5/2}$ transitions where the superscript "o" characterizes the odd parity of the final states. The ^2P states occur twice due to the possible coupling. The $KL_{2,3}L_{2,3}$ Auger series lead to $^2S_{1/2}$, $^2D_{1/2,3/2}$, $^2P_{1/2,3/2}$, and $^4P_{1/2,3/2,5/2}$ final states where the ^2P and ^4P states are forbidden in extreme LS coupling, because the selection rule $\Delta L = 0$ cannot be fulfilled without parity violation (e.g. see the book by Chattarji 1976). Since the KLM_1 Auger transitions in sodium are energetically close to the KLL lines, we have also included these transitions in our calculations, i.e. the KL_1M_1 $^2S_{1/2}$, KL_2M_1 $^2P^o_{1/2}$, and KL_3M_1 $^2P^o_{3/2}$ final states. For the heavier alkali elements M must be replaced by N, O, and P, respectively. However, the comparison with experimental data is not without problems since most experiments have not fully resolved the fine structure of the final states.

4.8.3 Auger Transition Energies

The initial and final ionic bound states are generated by using the GRASP atomic structure package. Beyond the Coulomb interaction in the self-consistent-field these calculations include the transverse Breit interaction as well as estimates of the main QED corrections as perturbation. From the independent computations for the initial and final state configurations the Auger energies are obtained as

$$E_{Auger} = E_f - E_i. \tag{4.34}$$

The Auger transitions energies are given in atomic units (a.u.) and are included in Tables 4.36–4.38 for all calculated transitions. Unfortunately, the experimental data to compare with theory are rare. Most energies are available for the sodium spectra (Aksela *et al.* 1984a; Hillig *et al.* 1974; Fahlman *et al.* 1966) where the latter ones by Fahlman *et al.* were obtained for sodium in solid Na_2O, and thus show a chemical shift with respect to the free atom case. Generally good agreement with the experiment has been found. The deviations do not exceed 2 eV compared with recently measured Auger energies. The good agreement with the experimental data by Aksela *et al.* (1984a) should be noted. The stronger deviations from earlier experimental data may be caused by an insufficient energy calibration which has been pointed out by Aksela *et al.* (1984a). There is only one experiment carried out

Table 4.35. Comparison of theoretical and experimental KLL Auger energies for selected transitions in sodium and cesium. For the listed LS terms the energies are averaged over the various $^{2S+1}L_J$ states ($ns = 3s$ or $6s$, respectively). (a): Lohmann and Fritzsche (1994); (b): Aksela et al. (1984a), T and E denote theoretical and experimental data, respectively; (c): Hillig et al. (1974); (d): Larkins (1977) using the experimental Na solid-state shift of 14.7 eV of Wagner (1975); (e): Erman and Sujkowski (1961); (f): Asaad and Burhop (1958)

| Transition | KLL Auger energies [eV] | | | | | | |
| | Na | | | | Cs | | |
	(a)	(b)	(c)	(d)	(a)	(e)	(f)
KL_1L_1 (^{1}S)$ns\,^2$S	906.74		909.1(3)	911.4	24379.9	24400(20)	24344
$KL_1L_{2,3}$ (^{1}P)$ns\,^2$P	937.59		937.6(3)	939.6	24744.1	24740(20)	24710
(^{3}P)$ns\,^2$P	950.29		948.1(3)	952.3	24976.5		
(^{3}P)$ns\,^4$P	951.98		949.9(3)	952.3	25121.7	25070(30)	25040
$KL_{2,3}L_{2,3}$ (^{1}S)$ns\,^2$S	975.14		972.2(3)	974.5	25071.4		
(^{1}D)$ns\,^2$D	979.07 {977.2 T / 979.12 E}		976.7(3)	977.6	25432.0	25430(20)	25387
(^{3}P)$ns\,^2$P	982.39			981.0	25448.5		
(^{3}P)$ns\,^4$P	983.28			981.0	25788.8	25770(20)	25747
$KL_1M_1\,^2$S	990.46		993.8(3)				
$KL_{2,3}M_1\,^2$P	1026.30		1027.7(6)				

for cesium (Erman and Sujkowski 1961). Again our values have been found within the error bars.

There are some theoretical calculations to compare with (Aksela et al. 1984a; Asaad and Burhop 1958; Hörnfeldt et al. 1962; Hörnfeldt 1962; Larkins 1977; Shirley 1973). However, most of them have used less sophisticated methods and thus less agreement have been found. The most reliable calculations from our point of view are those of Aksela et al. (1984a) and Larkins (1977). Unfortunately, the first group calculated only the ^{2}D line of sodium, where the latter calculated the energies with respect to the Fermi level of a solid state. By using the experimentally obtained solid-state shift for Na of 14.7 eV of Wagner (1975) we are able to compare the Na energies. Good agreement have been found with both calculations. A comparison of our energies with some of the cited data is given in Table 4.35.

4.8.4 Auger Rates and Relative Intensities

General Considerations

The absolute Auger rates and relative intensities have been calculated in four different approximations as has been discussed in Sect. 3.2. The relative intensities of the

Auger lines for the alkali elements are shown in Tables 4.36–4.37. The total Auger rates which correspond to 100% intensity are also given.

In calculation (a) we have only taken the static Coulomb repulsion in the evaluation of the Auger matrix elements into account. The analogous computation (b) includes additionally the (transverse) Breit interaction in the transition operator. Such relativistic corrections are particularly important for weak lines and/or heavier elements and may either increase or reduce the line intensities. By contrast, the exchange interaction between the emitted electron and the final ionic core in the solution of the continuum wavefunctions reduces the absolute Auger rates in most cases. Calculation (c) shows the influence of the exchange interaction with respect to approach (a) and calculation (d) with respect to calculation (b). Since the effects of exchange depend on the individual transition one should note, that this may also result in a slightly higher intensity for some of the Auger lines.

Sodium

The sodium results are presented in Table 4.36. By comparing calculations with and without exchange interaction in the solution of the continuum spinors, i.e. the calculations (c) and (a), it is shown that the effects of exchange ranges from 2% up to 15%. Although the absolute difference is less than 10^{-4} a.u. some of the weaker Auger lines can change their intensity by a factor of two. Due to the exchange most of the absolute Auger rates of the Na KLL spectra are clearly reduced. The influence of the Breit interaction does hardly reach 10^{-5} a.u. and therefore can be neglected. A similar result has been obtained for the KLL spectra of neon (Fritzsche *et al.* 1991) and can be generally expected for light atoms. The KLL Auger spectrum after resonant excitation of the intermediate $^3S_1$ state and subsequent photoionization of the 1s shell has been plotted in Fig. 4.11.

The line profiles have been obtained assuming Lorentz profiles for the relative Auger line intensities using (4.14), and normalizing the total spectrum to 100. Out of this, the absolute line profiles can be obtained (Mayer-Kuckuk 1977) as

$$I_0^{abs}(E) = \frac{\pi \Gamma}{2} \frac{\Gamma_{tot}}{100} I_0^{rel}(E) = \frac{\Gamma_{tot}}{100} \frac{\Gamma^2/4}{(E - E_0)^2 + \Gamma^2/4}, \qquad (4.35)$$

where E_0 denotes the Auger energy, Γ is the decay width, and Γ_{tot} is the total Auger decay rate. A numerical FWHM$_{Na}$ of $\Gamma = 0.45$ eV has been assumed for generating the Lorentz profiles of the Auger lines. The LS coupling character of the spectrum is clearly seen, showing unresolved groups of fine structure multiplets, only.

Cesium

The cesium data are given in Table 4.37. For $Z = 55$ the Breit interaction may change the decay rates up to 35% for some transitions. The influence of the exchange interaction is slightly reduced but still of the same order as in the sodium case. This can be seen as a constant behaviour which we have found of the same

Table 4.36. Relative intensities and total Auger rates for the KLL Auger transitions of sodium. The Auger energies and rates are given in atomic units. ~ 0 denotes an absolute rate $< 10^{-6}$. Note, that the absolute rates for columns (a)–(d) must be multiplied by 10^{-4}. An odd parity of the states is indicated by an ^{o} superscript. Occupied orbitals are shown only if they are necessary to classify the state. The columns are denoted as follows: (a): MCDF calculation which only includes the Coulomb interaction in the (reduced) Auger matrix elements; see equation (2.132). The exchange interaction between the emitted electron and the bound ionic core is neglected in the generation of the continuum spinors. (b): Same calculation as (a) but including both the Coulomb and transverse Breit interaction in the Auger matrix elements. (c): Same calculation as (a) but including the exchange interaction in the generation of the continuum orbitals in (3.73). (d): Same as (b) including the exchange interaction

	Na KLL Auger transitions				
	Auger	Relative intensities			
Final states	energies	(a)	(b)	(c)	(d)
$\left\vert 2p^6(^1S)3s\,^2S_{1/2}\right\rangle$	33.323	7.80	7.89	8.48	8.58
$\left\vert 2s2p^5(^1P)3s\,^2P^o_{1/2}\right\rangle$	34.457	7.12	7.06	8.03	7.98
$\left\vert 2s2p^5(^1P)3s\,^2P^o_{3/2}\right\rangle$	34.457	14.35	14.22	16.32	16.20
$\left\vert 2s2p^5(^3P)3s\,^2P^o_{1/2}\right\rangle$	34.921	0.50	0.53	0.22	0.24
$\left\vert 2s2p^5(^3P)3s\,^2P^o_{3/2}\right\rangle$	34.925	0.94	0.95	0.46	0.46
$\left\vert 2s2p^5(^3P)3s\,^4P^o_{1/2}\right\rangle$	34.981	1.48	1.56	0.68	0.73
$\left\vert 2s2p^5(^3P)3s\,^4P^o_{3/2}\right\rangle$	34.984	2.93	3.01	1.21	1.27
$\left\vert 2s2p^5(^3P)3s\,^4P^o_{5/2}\right\rangle$	34.988	4.24	4.29	2.79	2.83
$\left\vert 2s^22p^4(^1S)3s\,^2S_{1/2}\right\rangle$	35.837	6.67	6.69	5.78	5.81
$\left\vert 2s^22p^4(^1D)3s\,^2D_{3/2}\right\rangle$	35.981	21.22	21.18	22.23	22.20
$\left\vert 2s^22p^4(^1D)3s\,^2D_{5/2}\right\rangle$	35.981	31.87	31.74	32.77	32.65
$\left\vert 2s^22p^4(^3P)3s\,^2P_{1/2}\right\rangle$	36.100	~ 0	~ 0	~ 0	~ 0
$\left\vert 2s^22p^4(^3P)3s\,^2P_{3/2}\right\rangle$	36.105	0.01	0.01	0.01	0.01
$\left\vert 2s^22p^4(^3P)3s\,^4P_{1/2}\right\rangle$	36.132	~ 0	~ 0	~ 0	~ 0
$\left\vert 2s^22p^4(^3P)3s\,^4P_{3/2}\right\rangle$	36.134	~ 0	~ 0	~ 0	~ 0
$\left\vert 2s^22p^4(^3P)3s\,^4P_{5/2}\right\rangle$	36.138	0.01	0.01	0.01	0.01
$\left\vert 2s2p^6\,^2S_{1/2}\right\rangle$	36.400	0.79	0.80	0.78	0.79
$\left\vert 2s^22p^5\,^2P^o_{1/2}\right\rangle$	37.713	0.01	0.01	0.05	0.05
$\left\vert 2s^22p^5\,^2P^o_{3/2}\right\rangle$	37.719	0.02	0.02	0.10	0.09
Σ		100	100	100	100
Abs. tot. rates $\times[10^{-4}]$		119.98	120.10	104.47	104.55

Na KLL

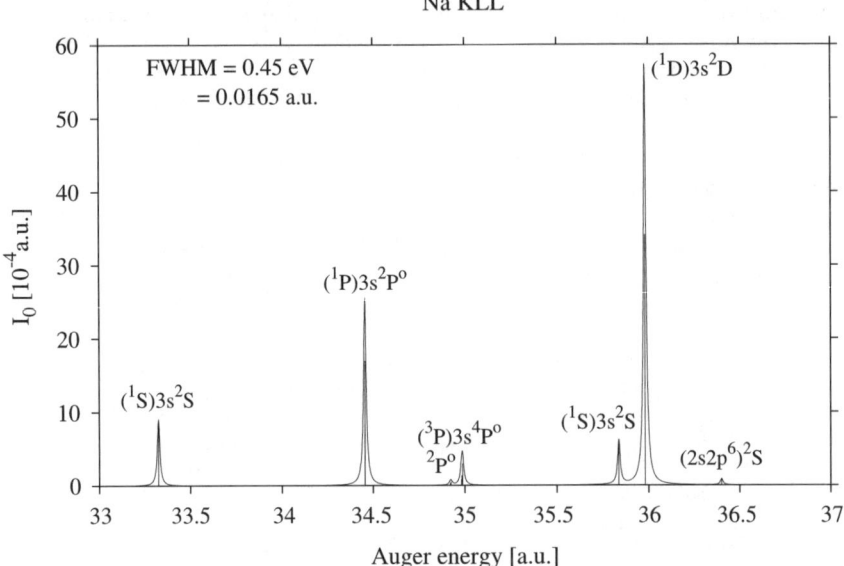

Fig. 4.11. The KLL Auger spectrum of sodium after ionization of the 1s shell. The Auger decay propagates from the resonantly excited $^3S_1$ state

order for all alkali elements. Thus, the influence of the exchange interaction on the continuum spinors should not be neglected in an accurate calculation.

Our calculation including both, Breit and exchange interaction (calc. d), shows a non-vanishing transition probability for a $KL_{2,3}L_{2,3}(^3P)6s\ ^2P_{1/2,3/2}$ doublet which we would interpret to stem from the non-relativistically forbidden $^3P_1$ Auger line. In closed shell systems the, in principle possible, $KL_{2,3}L_{2,3}\ ^3P_1$ Auger line is forbidden in any pure coupling scheme due to its generally single channel character. This is no longer valid in the studied case for two reasons. First, the coupling of the outer s electron leads to the final $^2P_{1/2,3/2}$ doublet states, which will couple with the $^4P_{1/2,3/2}$ and $^2D_{3/2}$ states, respectively. Further, the same decay mode is caused by the relativistic Breit interaction. Although it would be difficult to separate the two doublet lines energetically, we would predict from our results a further Auger line, even for an unresolved fine structure of the cesium open shell KLL Auger spectrum which, considering the closed shell atom case, is usually interpreted on the basis of a 9-line spectrum[7] in the intermediate coupling region.

In Fig. 4.12, the KLL Auger spectrum of cesium is shown, assuming a decay width of $FWHM_{Cs} = 12.716\,eV$. The strong fine structure splitting for the $(^3P)6s\ ^{2,4}P_J$ multiplets can be clearly seen. While the decay channel into the $(^3P)6s\ ^4P_J$ fine structure states has fully opened, the decay into the non-relativistically

---

[7] For the general interpretation of the KLL Auger spectra we like to refer the reader to the book by Chattarji (1976, Chap. 4).

Table 4.37. Relative intensities and total Auger rates for the KLL Auger transitions of cesium. For explanations see Table 4.36

		Cs KLL Auger transitions			
	Auger	Relative intensities			
Final states	energies	(a)	(b)	(c)	(d)
$\lvert 2p^6(^1S)6s\,^2S_{1/2}\rangle$	895.948	9.03	11.19	9.16	11.36
$\lvert 2s2p^5(^1P)6s\,^2P^o_{1/2}\rangle$	909.330	2.84	1.76	3.02	1.86
$\lvert 2s2p^5(^1P)6s\,^2P^o_{3/2}\rangle$	909.332	11.68	10.65	11.84	10.73
$\lvert 2s2p^5(^3P)6s\,^2P^o_{1/2}\rangle$	910.523	1.30	4.31	1.16	4.04
$\lvert 2s^22p^4(^1S)6s\,^2S_{1/2}\rangle$	921.360	2.66	2.67	2.53	2.56
$\lvert 2s2p^5(^3P)6s\,^2P^o_{3/2}\rangle$	921.549	6.40	6.20	6.63	6.42
$\lvert 2s2p^5(^3P)6s\,^4P^o_{1/2}\rangle$	921.549	4.42	4.79	4.50	4.92
$\lvert 2s2p^5(^3P)6s\,^4P^o_{3/2}\rangle$	923.540	0.64	0.96	0.54	0.83
$\lvert 2s2p^5(^3P)6s\,^4P^o_{5/2}\rangle$	923.543	2.53	3.75	2.39	3.58
$\lvert 2s^22p^4(^1D)6s\,^2D_{3/2}\rangle$	934.612	15.99	14.62	16.00	14.69
$\lvert 2s^22p^4(^1D)6s\,^2D_{5/2}\rangle$	934.612	25.25	22.21	25.23	22.28
$\lvert 2s^22p^4(^3P)6s\,^2P_{1/2}\rangle$	935.219	0.01	0.17	0.01	0.17
$\lvert 2s^22p^4(^3P)6s\,^2P_{3/2}\rangle$	935.220	0.02	0.36	0.03	0.37
$\lvert 2s^22p^4(^3P)6s\,^4P_{1/2}\rangle$	946.579	2.77	3.05	2.66	2.95
$\lvert 2s^22p^4(^3P)6s\,^4P_{3/2}\rangle$	947.952	5.88	5.90	5.83	5.88
$\lvert 2s^22p^4(^3P)6s\,^4P_{5/2}\rangle$	947.954	8.55	7.38	8.45	7.35
$\lvert 2s2p^6\,^2S_{1/2}\rangle$	1110.64	0.01	0.01	0.01	0.01
$\lvert 2s^22p^5\,^2P^o_{1/2}\rangle$	1124.04	~ 0	~ 0	~ 0	~ 0
$\lvert 2s^22p^5\,^2P^o_{3/2}\rangle$	1136.77	~ 0	~ 0	~ 0	~ 0
Σ		100	100	100	100
Abs. tot. rates $\times[10^{-4}]$		314.59	328.77	300.21	313.73

forbidden $(^3P)6s\,^2P$ state (see above) can be seen as a further broadening of the right shoulder of the dominating $(^1D)6s\,^2D$ line.

Potassium and Rubidium

For potassium and rubidium the effects of relativity and exchange interaction are expected to link the different behaviour of sodium and cesium in the previous sections. This is confirmed by our results shown in Tables 4.38 and 4.39. Potassium has been found similar to the sodium case. The Auger rates are more influenced by exchange effects than by the Breit interaction. In rubidium the corrections to the

Cs KLL

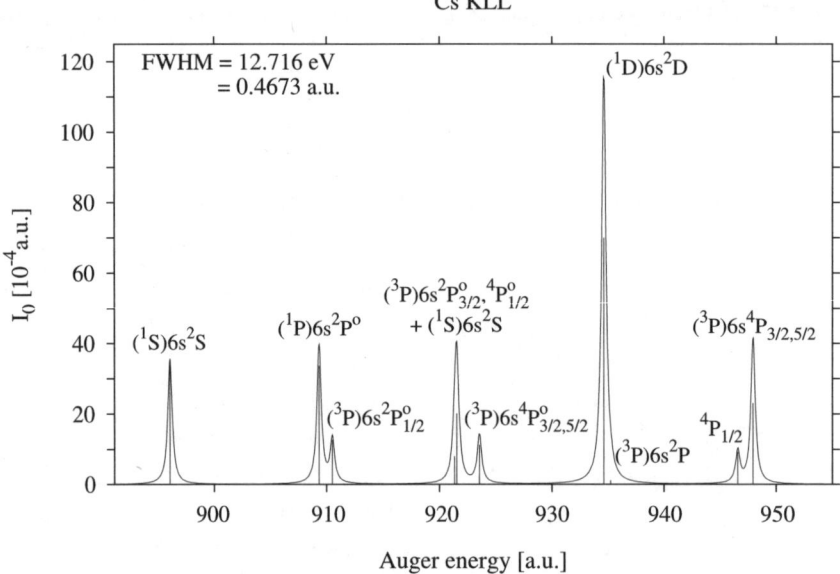

Fig. 4.12. The KLL Auger spectrum of cesium after ionization of the 1s shell. The Auger decay propagates from the resonantly excited $^3S_1$ state

K KLL

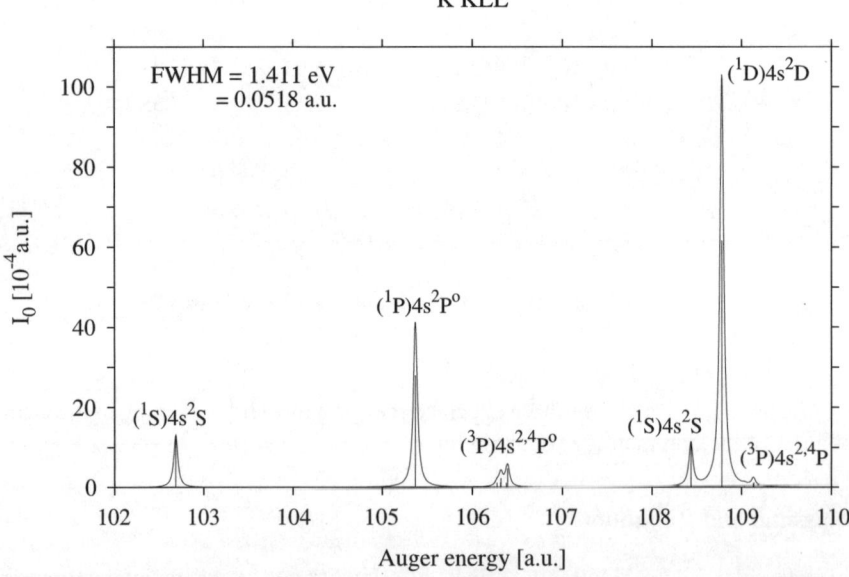

Fig. 4.13. The KLL Auger spectrum of potassium after ionization of the 1s shell. The Auger decay propagates from the resonantly excited $^3S_1$ state

Table 4.38. Relative intensities and total Auger rates for the KLL Auger transitions of potassium. For explanations see Table 4.36

Final states	Auger energies	Relative intensities (a)	(b)	(c)	(d)
$\|2p^6(^1S)4s\,^2S_{1/2}\rangle$	102.683	6.67	6.91	6.99	7.26
$\|2s2p^5(^1P)4s\,^2P^o_{1/2}\rangle$	105.362	6.86	6.69	7.62	7.47
$\|2s2p^5(^1P)4s\,^2P^o_{3/2}\rangle$	105.362	14.25	13.89	15.75	15.40
$\|2s2p^5(^3P)4s\,^2P^o_{1/2}\rangle$	106.271	0.68	0.86	0.45	0.60
$\|2s2p^5(^3P)4s\,^4P^o_{1/2}\rangle$	106.309	0.82	0.95	0.53	0.63
$\|2s2p^5(^3P)4s\,^2P^o_{3/2}\rangle$	106.315	1.63	1.83	1.07	1.24
$\|2s2p^5(^3P)4s\,^4P^o_{3/2}\rangle$	106.374	0.95	1.01	0.55	0.59
$\|2s2p^5(^3P)4s\,^4P^o_{5/2}\rangle$	106.388	3.00	3.14	2.44	2.57
$\|2s^22p^4(^1S)4s\,^2S_{1/2}\rangle$	108.434	6.63	6.69	5.97	6.03
$\|2s^22p^4(^1D)4s\,^2D_{3/2}\rangle$	108.777	22.80	22.70	22.96	22.89
$\|2s^22p^4(^1D)4s\,^2D_{5/2}\rangle$	108.777	34.36	33.98	34.32	33.98
$\|2s^22p^4(^3P)4s\,^2P_{1/2}\rangle$	109.031	0.16	0.17	0.15	0.15
$\|2s^22p^4(^3P)4s\,^4P_{1/2}\rangle$	109.049	0.01	0.01	0.01	0.01
$\|2s^22p^4(^3P)4s\,^2P_{3/2}\rangle$	109.053	~ 0	0.01	~ 0	0.01
$\|2s^22p^4(^3P)4s\,^4P_{3/2}\rangle$	109.129	0.47	0.48	0.47	0.48
$\|2s^22p^4(^3P)4s\,^4P_{5/2}\rangle$	109.137	0.65	0.63	0.65	0.63
$\|2s2p^6\,^2S_{1/2}\rangle$	117.952	0.05	0.05	0.05	0.05
$\|2s^22p^5\,^2P^o_{1/2}\rangle$	121.045	~ 0	~ 0	0.01	0.01
$\|2s^22p^5\,^2P^o_{3/2}\rangle$	121.146	0.01	0.01	0.01	0.01
Σ		100	100	100	100
Abs. tot. rates $\times[10^{-4}]$		202.04	202.30	181.04	181.26

Auger transitions are dominated by the Breit interaction rather than by exchange effects. However, the effects are smaller than in cesium.

The KLL Auger spectra of potassium and rubidium have been plotted in Figs. 4.13 and 4.14, applying decay widths of $FWHM_K = 1.411$ eV and $FWHM_{Rb} = 5.722$ eV, respectively. For potassium, the $(^3P)4s\,^{2,4}P$ channel, which is forbidden in the almost pure LS coupling sodium case, begins slightly to open. However, all lines still reveal the unresolved fine structure of the multiplets demonstrating that even potassium is more close to the LS than to the intermediate coupling case. For rubidium, the $(^3P)5s\,^4P$ channel has opened and an increasing fine structure split-

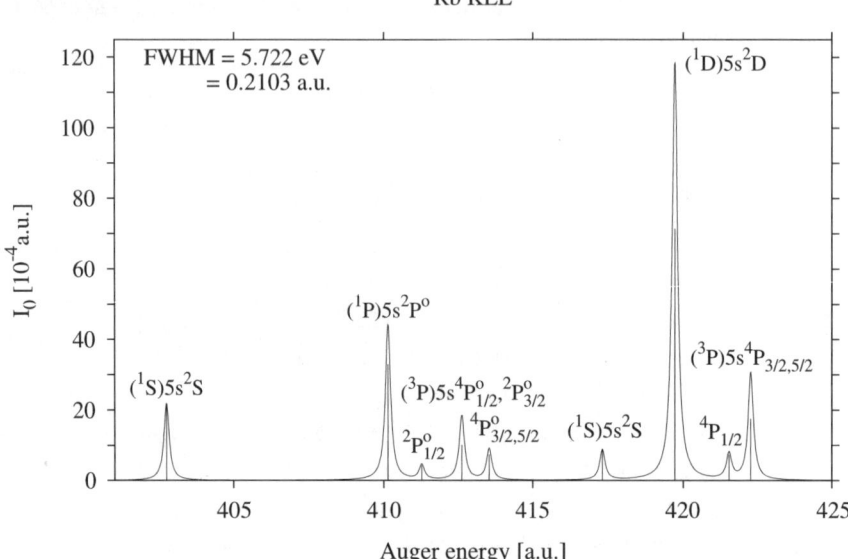

Fig. 4.14. The KLL Auger spectrum of rubidium after ionization of the 1s shell. The Auger decay propagates from the resonantly excited $^3S_1$ state

ting, particularly for the lines stemming from the $(^3P)5s\,^{2,4}P_J$ multiplets begins to exhibit.

4.8.5 Comparison with Other Data

Concerning the validity of our results it is worth to consider the special case of extreme LS coupling which results should be close to mainly the sodium calculations including only the Coulomb repulsion in the Auger matrix (calc. a). The $KL_{2,3}L_{2,3}(^3P)ns\ ^2P_{1/2,3/2}$ and $^4P_{1/2,3/2,5/2}$ Auger transitions are strongly forbidden in LS coupling.[8] In sodium we therefore expect an Auger rate which is almost zero. This is confirmed by our results. For the KLL Auger transitions of neon, which is close to sodium with respect of its coupling scheme, Siegbahn *et al.* (1967) pointed out that there is evidence for a non-vanishing transition probability of the $KL_{2,3}L_{2,3}\ ^3P_2$ Auger line. This is confirmed by our calculation for sodium where the $KL_{2,3}L_{2,3}(^3P)3s\ ^4P_{5/2}$ Auger line, which can be seen to stem from the $^3P_2$ line, shows an Auger rate slightly different from zero. This decay channel becomes stronger in the potassium case where we can predict non-zero Auger rates for the two $^4P_{3/2,5/2}$ lines. However, high resolution techniques would be necessary to measure these intensities. With increasing atomic number Z the possible $KL_{2,3}L_{2,3}\ ^3P$ channels open further which results in the well known fact of an observable $KL_{2,3}L_{2,3}(^3P)ns\ ^4P_J$ quartet in the rubidium and cesium case.

---

[8] For further explanation we refer again to Chattarji (1976).

Table 4.39. Relative intensities and total Auger rates for the KLL Auger transitions of rubidium. For explanations see Table 4.36

	Rb KLL Auger transitions				
	Auger	Relative intensities			
Final states	energies	(a)	(b)	(c)	(d)
$\lvert 2p^6(^1S)5s\,^2S_{1/2}\rangle$	402.740	7.13	8.07	7.34	8.31
$\lvert 2s2p^5(^1P)5s\,^2P^o_{1/2}\rangle$	410.141	4.69	3.94	5.06	4.29
$\lvert 2s2p^5(^1P)5s\,^2P^o_{3/2}\rangle$	410.144	13.07	11.96	13.61	12.49
$\lvert 2s2p^5(^3P)5s\,^2P^o_{1/2}\rangle$	411.280	0.87	1.82	0.72	1.61
$\lvert 2s2p^5(^3P)5s\,^4P^o_{1/2}\rangle$	412.621	2.67	3.14	2.63	3.11
$\lvert 2s2p^5(^3P)5s\,^2P^o_{3/2}\rangle$	412.622	3.34	3.91	3.33	3.87
$\lvert 2s2p^5(^3P)5s\,^4P^o_{3/2}\rangle$	413.533	0.64	0.78	0.49	0.61
$\lvert 2s2p^5(^3P)5s\,^4P^o_{5/2}\rangle$	413.539	2.50	3.03	2.28	2.79
$\lvert 2s^22p^4(^1S)5s\,^2S_{1/2}\rangle$	417.316	3.42	3.48	3.24	3.31
$\lvert 2s^22p^4(^1D)5s\,^2D_{3/2}\rangle$	419.719	18.57	18.09	18.50	18.06
$\lvert 2s^22p^4(^1D)5s\,^2D_{5/2}\rangle$	419.720	28.41	27.21	28.24	27.09
$\lvert 2s^22p^4(^3P)5s\,^2P_{1/2}\rangle$	420.178	~ 0	0.04	~ 0	0.04
$\lvert 2s^22p^4(^3P)5s\,^2P_{3/2}\rangle$	420.181	0.01	0.08	0.01	0.08
$\lvert 2s^22p^4(^3P)5s\,^4P_{1/2}\rangle$	421.540	2.79	2.92	2.62	2.74
$\lvert 2s^22p^4(^3P)5s\,^4P_{3/2}\rangle$	422.267	4.83	4.89	4.84	4.93
$\lvert 2s^22p^4(^3P)5s\,^4P_{5/2}\rangle$	422.270	7.09	6.65	7.08	6.66
$\lvert 2s2p^6\,^2S_{1/2}\rangle$	481.297	0.01	0.01	0.01	0.01
$\lvert 2s^22p^5\,^2P^o_{1/2}\rangle$	488.968	~ 0	~ 0	~ 0	~ 0
$\lvert 2s^22p^5\,^2P^o_{3/2}\rangle$	491.149	~ 0	~ 0	~ 0	~ 0
Σ		100	100	100	100
Abs. tot. rates $\times[10^{-4}]$		279.58	281.77	261.36	263.58

A comparison of our present data with results by other groups is given in Table 4.40. To our knowledge, there are two experimental data sets for sodium (Hillig *et al.* 1974; Fahlman *et al.* 1966), where the latter have been obtained for Na in Na_2O which results in a solid state and/or chemical shift and thus agrees less with both, our results, and the experiment of Hillig *et al.* (1974), too. However, there is still a strong discrepancy between our calculation and the more reliable results of Hillig *et al.* We have calculated approximately only half the measured intensity of the most intense ^2D line. There are several explanations for this behaviour. First, with respect to our calculation, it is not a good choice to normalize the spectra to the KL_1L_1 Auger line since it is nearest to the ionization threshold and thus can be

Table 4.40. Relative intensities (calc. d) of some Auger transitions in comparison with other data. Our data are renormalized to the KL_1L_1 Auger transition. For the listed LS terms the data are averaged over the various $^{2S+1}L_J$ multiplets as indicated. See Sect. 4.8.5 for further explanations. (a): Lohmann and Fritzsche (1994); (b): Hillig *et al.* (1974); (c): Fahlman *et al.* (1966), data obtained for Na in solid Na_2O; (d): Chen and Crasemann (1973); (e): Erman and Sujkowski (1961); (f): Asaad and Burhop (1958)

Transition	KLL Auger intensities			
Na	(a)	(b)	(c)	(d)
KL_1L_1 (^1S)$3s\,^2S$	1	1.0 ± 0.1	1	1.0
$KL_1L_{2,3}$ (^1P)$3s\,^2P^o$	2.82	3.4 ± 0.2	4.3 ± 0.5	3.25
(^3P)$3s\,^2P^o$	0.08	0.39 ± 0.15	1.5 ± 0.4	0.98
(^3P)$3s\,^4P^o$	0.56	0.93 ± 0.15		
$KL_{2,3}L_{2,3}$ (^1S)$3s\,^2S$	0.68	1.65 ± 0.2	1.7 ± 0.5	1.08
(^1D)$3s\,^2D$	6.39	12.3 ± 0.4	16.6 ± 1.7	9.40
(^3P)$3s\,^{2,4}P$	0.001	<0.15		0.007
$KL_1M_1\,^2S$	0.09	0.14 ± 0.04		
$KL_{2,3}M_1\,^2P^o$	0.02	0.20 ± 0.05		
KL_1L_1	8.67	5.1 ± 0.5	4.0 ± 1.0	6.36
$KL_1L_{2,3}$	30.02	24.0 ± 1.6	23.1 ± 3.0	26.9
$KL_{2,3}L_{2,3}$	61.31	70.9 ± 3.1	72.9 ± 9.0	66.7
Cs	(a)	(e)	(f)	
KL_1L_1 (^1S)$6s\,^2S_{1/2}$	1	1	1	
$KL_1L_{2,3}$				
(^1P)$6s\,^2P^o_{1/2,3/2}$, (^3P)$6s\,^2P^o_{1/2}$	1.464	1.41 ± 0.10	1.56	
(^3P)$6s\,^2P^o_{3/2}$, $^4P^o_{1/2}$	0.998	†	1.67	
(^3P)$6s\,^4P^o_{3/2,5/2}$	0.388	†	0.23	
†	1.611	1.73 ± 0.15	2.11	
$KL_{2,3}L_{2,3}$				
(^1S)$6s\,^2S_{1/2}$	0.225	†	0.21	
(^1D)$6s\,^2D_{3/2,5/2}$, (^3P)$6s\,^2P_{1/2,3/2}$	3.302	3.39 ± 0.25	4.55	
(^3P)$6s\,^4P_{1/2,3/2,5/2}$	1.424	1.44 ± 0.10	2.10	

†: Unresolved experimental value summed over the marked states.

more influenced by correlation effects than the others. In fact, by renormalizing the results of Hillig *et al.* to 100% for the full spectra, we obtain 61.47% which is closer to our value of 54.85%. However, there still remains a discrepancy. Secondly, since we have only considered the initial $^3S_1$ state in detail there might be some overlap with lines which stem from the initial $^1S_0$ states. Finally, as pointed out by Hillig *et al.*, too there might be some superposition with satellite lines.

There are some theoretical results by other groups (Chen and Crasemann 1973; McGuire 1969, 1970; Walters and Bhalla 1971) all using Hartree–Fock–Slater functions and thus their results are close to each other. However, they all show discrepancies with the experiment as well as with our data.

Considering the group intensities our value obtained for the $KL_{2,3}L_{2,3}$ group is much lower than the other experimental and theoretical data whereas we have obtained larger intensities for the KL_1L_1 and $KL_1L_{2,3}$ groups.

There is, to our knowledge, only one experiment on cesium by Erman and Sujkowski (1961) and our relative intensities are in good agreement with their results whereas the theoretical results of Asaad and Burhop (1958) show some deviations.

4.8.6 Angular Distribution – Anisotropy Coefficients α_2

Generally, we do point out that in open shell KLL Auger transitions a non-vanishing anisotropy parameter α_2 can only occur due to the coupling of the angular momenta of the inner shell hole and the outer open shell electrons. In alkali elements we have one s electron in the outer shell which couples with the inner shell 1s hole in two different ways. The coupling leading to an initial total angular momentum $J = 0$ is of no further interest because no alignment can be produced during the ionization process and thus, no anisotropic angular distribution can occur. As can be seen from (2.139), by inserting $K = 2$, the anisotropy parameter α_2 vanishes in this case. If the resonant initial state has a total angular momentum $J = 1$, both, a non-zero alignment $\mathcal{A}_{20}$ and an anisotropy coefficient $\alpha_2 \neq 0$ are allowed. Using the LSJ notation for the designation of the states, e.g. see Sect. 4.8.2, this initial state has a $^3S_1$ symmetry. For a primary photoionization and, neglecting the spin-orbit interaction in the generation of the photoelectron spectrum, Berezhko and Kabachnik (1977) showed $\mathcal{A}_{20} = 0$. However, in a full-relativistic treatment a non-zero alignment even for KLL Auger transitions is possible.

Angular Anisotropy of KLL Open Shell Auger Lines

For the interpretation of our data it is useful to investigate the α_2 parameters in the LS coupling limit, too. The angular distribution anisotropy parameter α_2 depends on the total angular momentum J of the initial state. According to (2.139) an anisotropy coefficient $\alpha_2 \neq 0$ is only possible for a total angular momentum $J > 1/2$. In closed shell atoms, e.g. the rare gases, the initial total angular momentum is $J = 1/2$ and thus we expect $\alpha_2 = 0$ for all KLL Auger transitions.

On the other hand, Auger transitions in open shell systems can allow a non-zero anisotropy even for KLL transitions. This is because the open shell electrons can provide an additional angular momentum which can result in a total initial angular momentum $J > 1/2$. For the alkali KLL Auger transitions we have a single s electron in the outer open shell which can couple with the K-hole in two ways, which results in either a $^1S_0$ or $^3S_1$ initial state. For the $^1S_0$ state no anisotropic angular distribution can be obtained ($J = 0$). However, the $^3S_1$ initial state is of particular interest since the angular momentum coupling rules allow a non-zero angular distribution in this case. Therefore, we are investigating the KLL Auger transitions from the $^3S_1$ initial state in the following in more detail, mainly under the aspect of the anisotropy parameter α_2.

For atoms having only one electron in the outermost shell it is a first (however rather crude) attempt to use the so called *resonant spectator model* (Lohmann 1991). It is investigated in detail by several authors for the case of resonantly excited noble-gas spectra (e.g. Schmidt 1992; Lohmann 1991; Hergenhahn *et al.* 1991). Even though the spectator model might not be sufficient for a description of open shell transitions it allows some interesting predictions. Applying equation (24) of Lohmann (1991) to Auger transitions with a total angular momentum of the closed shells $J_c = 0$, i.e. considering the final state without the outer electron, (e.g. the KL_1L_1 alkali Auger transitions) we remain with $\alpha_2^{res} = 0$ since the 6j-symbol vanishes due to selection rules. Further, summing over the fine structure of the final states (i.e. applying equation (22) of Hergenhahn *et al.* 1991) we obtain $\alpha_2^{res} = 0$ by the same reason.

Now, consider the case of the KLL open shell transitions in the LS coupling scheme. General expressions for the parameter α_2^{LS} may be found in Klar (1980) or Kabachnik *et al.* (1988). In both articles, however, the discussion has been restricted to the case of the rare gases. Using the same formalism as in (2.139) we obtain

$$
\alpha_2^{(LS)} = -\left[\sum_{\ell j}(2j+1) \begin{Bmatrix} J & j & J_f \\ L & \ell & L_f \\ S & 1/2 & S_f \end{Bmatrix}^2 |\langle (L_f\ell)L\|V\|L\rangle|^2 \right]^{-1}
$$

$$
\times \sqrt{5(2J+1)} \sum_{\ell\ell'jj'} (-1)^{J_f+J+1/2+j+j'+\ell'} i^{\ell+\ell'}
$$

$$
\times \cos(\sigma_{\ell'} - \sigma_\ell) \langle (L_f\ell)L\|V\|L\rangle \langle (L_f\ell')L\|V\|L\rangle
$$

$$
\times \sqrt{(2j+1)(2j'+1)} \begin{pmatrix} j' & j & 2 \\ 1/2 & -1/2 & 0 \end{pmatrix} \begin{Bmatrix} j' & j & 2 \\ J & J & J_f \end{Bmatrix}
$$

$$
\times \sqrt{(2j+1)(2j'+1)} \begin{Bmatrix} J & j & J_f \\ L & \ell & L_f \\ S & 1/2 & S_f \end{Bmatrix} \begin{Bmatrix} J & j' & J_f \\ L & \ell' & L_f \\ S & 1/2 & S_f \end{Bmatrix}, \quad (4.36)
$$

where L, S and L_f, S_f denote orbital and spin angular momenta of the initial and final state, respectively. Note that the scattering phase depends only on the angular momentum ℓ in the above equation and not on the total angular momentum j as in (2.139).

Inserting $S = J = 1$ and $L = 0$ for the $^3S_1$ state the 9j-symbols reduce to 6j-symbols, and the triangular relations of the 9j-symbols yield $\ell = \ell' = L_f$. Thus, the sum over ℓ and ℓ' can be omitted and the anisotropy coefficient becomes independent of the Auger transition matrix elements which yields

$$
\alpha_2^{(LS)}(^3S_1) = \left[\sum_j (-1)^{2j}(2j+1) \begin{Bmatrix} 1/2 & S_f & 1 \\ J_f & j & L_f \end{Bmatrix}^2 \right]^{-1}
$$

$$
\times \sqrt{15} \sum_{jj'} (-1)^{J_f+1/2+2j+2j'} \sqrt{(2j+1)(2j'+1)}
$$

$$
\times \begin{pmatrix} j' & j & 2 \\ 1/2 & -1/2 & 0 \end{pmatrix} \begin{Bmatrix} j' & j & 2 \\ 1 & 1 & J_f \end{Bmatrix}
$$

$$
\times \begin{Bmatrix} 1/2 & S_f & 1 \\ J_f & j & L_f \end{Bmatrix} \begin{Bmatrix} 1/2 & S_f & 1 \\ J_f & j' & L_f \end{Bmatrix}. \tag{4.37}
$$

Applying twice the sum rules of the $6j$-symbols (e.g. Brink and Satchler 1962) the summations over j and j' can be carried out. The denominator in (4.37) equals one and we remain with the numerator after some algebra as

$$
\alpha_2^{(LS)}(^3S_1 \longrightarrow {}^{2S_f+1}L_{f\,J_f}) = (-1)^{J_f+1/2}\sqrt{15}\,(2S_f+1)\,(2L_f+1)
$$

$$
\times \begin{pmatrix} L_f & L_f & 2 \\ 0 & 0 & 0 \end{pmatrix} \begin{Bmatrix} L_f & L_f & 2 \\ S_f & S_f & J_f \end{Bmatrix} \begin{Bmatrix} S_f & S_f & 2 \\ 1 & 1 & 1/2 \end{Bmatrix}. \tag{4.38}
$$

Thus, as a general result, the anisotropy parameter $\alpha_2^{(LS)}(^3S_1)$ is independent of the Auger transition matrix elements in the LS coupling approximation for all KLL Auger transitions. The simple structure of (4.38) allows further predictions for the anisotropy coefficient in the LS coupling scheme. Considering the possible final states (as discussed in Sect. 4.8.2), and using the coupling rules of the $3j$- and $6j$-symbols, (4.38) yields

$$
\alpha_2^{(LS)}(^3S_1 \longrightarrow {}^2L_{f\,J_f}) = 0 \tag{4.39}
$$

for all Auger transitions to a final doublet state. Only the possible quartet states $^4P_{1/2,3/2,5/2}$ can have a non-vanishing angular distribution if extreme LS coupling is considered. Their numbers are listed in Table 4.41. Finally, if we sum over the fine structure of the final state (4.38) yields

$$
\alpha_2^{\Sigma(LS)} = \sum_{J_f} \alpha_2^{(LS)}(^3S_1 \longrightarrow {}^{2S_f+1}L_{f\,J_f}) = 0. \tag{4.40}
$$

Thus, in extreme LS coupling, a non-zero angular anisotropy can only be expected in high resolution experiments which allow to resolve the fine structure. Further, Berezhko et al. (1978b) showed that no alignment can occur for S states in this approximation.

However, the most general equation (2.139) allows an anisotropy coefficient $\alpha_2 \neq 0$ for KLL Auger transitions from open shell atoms. Further, following the

Table 4.41. The anisotropy coefficients $\alpha_2^{(LS)}$ for the $^3S_1 \longrightarrow {}^{2S_f+1}L_{f\,J_f}$ KLL Auger transitions of alkali atoms in extreme LS coupling. The data are independent of the Auger energy and of the nuclear charge of the atom

KLL Auger transitions	
Final states	$\alpha_2^{(LS)}$
$^2S_{1/2}$	0
$^2P_{1/2,3/2}$	0
$^2D_{3/2,5/2}$	0
$^4P_{1/2}$	$\frac{-1}{\sqrt{2}} \sim -0.7071$
$^4P_{3/2}$	$\sqrt{\frac{8}{25}} \sim 0.5657$
$^4P_{5/2}$	$-\sqrt{\frac{1}{50}} \sim -0.1414$
$\sum_{J_f} {}^4P_{J_f}$	0

more general considerations of Kabachnik and Sazhina (1976), and Bussert and Klar (1983) a non-zero alignment is possible, too. This is due to the fact, that the resonant spectator model (see Sect. 4.5) as well as the LS coupling calculation *are not able to handle correlation effects properly*. Thus an anisotropy parameter $\alpha_2 \neq 0$ gives a direct measure for the strength of such effects. This is an important topic of the following discussion.

In pure LS coupling the KLL Auger decay proceeds via one single channel and is thus independent of the Auger matrix elements, though its dependence on the angular momentum coupling coefficients is different of that for single-channel transitions in closed shell atoms.[9] Generally, we have found all Auger transitions resulting in a final doublet state with a vanishing anisotropy parameter $\alpha_2 = 0$. Only the Auger transitions to a final quartet state can have an $\alpha_2 \neq 0$.

Thus, even by means of our full-relativistic intermediate coupled MCDF approach, we would expect a behaviour more similar to the LS coupling case for the relatively light sodium and potassium atoms. This provides us with a simple test for our calculations. For the angular anisotropy parameters we have performed the two calculations (b) and (d). Both include the Coulomb and Breit interaction in the Auger matrix. Calculation (b) neglects the exchange interaction in the solution of the continuum waves and calculation (d) includes it.

Since there are no experimental data available so far, the following discussion should be understood as a guide for possible experiments. We have found almost vanishing angular anisotropy for the KL_1L_1, $KL_{2,3}L_{2,3}$, KL_1X_1 $^2S_{1/2}$, (X = M, N, O, P) and $KL_{2,3}L_{2,3}$ $^2D_{3/2,5/2}$ Auger transitions for all alkali elements. This is not changed by the inclusion of the exchange interaction. Therefore, the electron-electron correlation between the inner 1s hole and the outer s electron can be as-

---

[9] Compare for instance (4.38) with equation (25) of Lohmann (1990).

sumed as small for these transitions. Although the $KL_{2,3}X_1$ $^2P_{1/2,3/2}$ Auger lines show a remarkable angular anisotropy parameter mainly for the heavy atoms their transition rate is almost zero for the light atoms and becomes negligible for Rb and Cs.

Sodium

The anisotropy parameters for the KLL Auger transitions in sodium are shown in Table 4.42. Considering that sodium is a light atom we would expect our calculated data close to the LS coupling values of Table 4.41. The largest anisotropy coefficients have been found for the $KL_1L_{2,3}$ $^4P_{1/2,3/2,5/2}$ states. Comparing the values with the LS coupling data, we find the LS coupling scheme well represented. There

Table 4.42. The anisotropy coefficients α_2 for the KLL Auger transitions of sodium. The columns are denoted as in Table 4.36. †: The Auger transitions from which an observable angular anisotropy can be expected (Tables 4.42–4.45); for further information see text

Na KLL Auger transitions		
Final states	**α_2 parameters**	
	(b)	(d)
$\left\vert 2p^6(^1S)3s\,^2S_{1/2}\right\rangle$	~ 0	~ 0
$\left\vert 2s2p^5(^1P)3s\,^2P^o_{1/2}\right\rangle$	0.006	0.002
$\left\vert 2s2p^5(^1P)3s\,^2P^o_{3/2}\right\rangle$	-0.003	-0.003
$\left\vert 2s2p^5(^3P)3s\,^2P^o_{1/2}\right\rangle$	0.099	0.203
$\left\vert 2s2p^5(^3P)3s\,^2P^o_{3/2}\right\rangle$	-0.044	-0.087
$\left\vert 2s2p^5(^3P)3s\,^4P^o_{1/2}\right\rangle$ †	-0.734	-0.624
$\left\vert 2s2p^5(^3P)3s\,^4P^o_{3/2}\right\rangle$ †	0.581	0.512
$\left\vert 2s2p^5(^3P)3s\,^4P^o_{5/2}\right\rangle$ †	-0.141	-0.141
$\left\vert 2s^22p^4(^1S)3s\,^2S_{1/2}\right\rangle$	~ 0	~ 0
$\left\vert 2s^22p^4(^1D)3s\,^2D_{3/2}\right\rangle$	-0.001	-0.024
$\left\vert 2s^22p^4(^1D)3s\,^2D_{5/2}\right\rangle$	0.001	0.015
$\left\vert 2s^22p^4(^3P)3s\,^2P_{1/2}\right\rangle$	-0.062	0.019
$\left\vert 2s^22p^4(^3P)3s\,^2P_{3/2}\right\rangle$	-0.027	-0.051
$\left\vert 2s^22p^4(^3P)3s\,^4P_{1/2}\right\rangle$	-0.034	-0.126
$\left\vert 2s^22p^4(^3P)3s\,^4P_{3/2}\right\rangle$	-0.456	-0.487
$\left\vert 2s^22p^4(^3P)3s\,^4P_{5/2}\right\rangle$	0.112	0.136
$\left\vert 2s2p^6\,^2S_{1/2}\right\rangle$	~ 0	0.0002
$\left\vert 2s^22p^5\,^2P^o_{1/2}\right\rangle$	-0.031	-0.112
$\left\vert 2s^22p^5\,^2P^o_{3/2}\right\rangle$	0.012	0.058

are small deviations if relativity and exchange have been taken into account. All other anisotropy coefficients should be zero in extreme LS coupling which is confirmed by our calculations for most of the Na transitions where the α_2 parameters have been found at least a magnitude smaller than for the quartet states. By including exchange effects the anisotropy parameters may be lowered or increased. The relatively large $KL_1L_{2,3}$ $^4P_{J_f}$ anisotropy parameters are affected up to 15%. The smaller anisotropy coefficients have been found by a factor of two or three larger for some transitions. This is in accordance with the behaviour of the relative intensities where considerable effects, caused by exchange interaction, have been found for the sodium transitions, too. However, the inclusion of exchange shows a larger effect to the anisotropy coefficients. Although the $KL_1L_{2,3}$ $^4P_{J_f}$ states provide a non-zero angular anisotropy its experimental determination might become somewhat tedious because the $^4P_{J_f}$ fine structure states must be energetically resolved, which requires an energy resolution of $\sim 10^{-3}$ a.u. If the fine structure is not resolved, this results in an anisotropy coefficient $\alpha_2 = -0.0421$[10] which is even hard to determine in an experiment. The result is close to the LS coupling prediction $\alpha_2^{\Sigma(LS)} = 0$ of (4.40). Its non-vanishing value indicates the influence of an intermediate coupling scheme and exchange effects.

Finally, we point out, that the discussion of the relatively large anisotropy parameters which have been obtained for the $KL_{2,3}L_{2,3}$ $^2P_{1/2,3/2}$ and $^4P_{1/2,3/2,5/2}$ Auger transitions is useless because our calculation of the Auger rates show them as strongly forbidden (see the discussion in Sect. 4.8.4, and Table 4.36 or Fig. 4.11) even by the inclusion of the Breit and exchange interaction.[11]

Potassium

The potassium data, see Table 4.43, are similar to the previous case and the general discussion holds here, too. The $KL_1L_{2,3}$ $^4P_{1/2,3/2,5/2}$ states show again a non-vanishing anisotropy. In particular, the large value of the $KL_1L_{2,3}$ $^4P_{1/2}$ Auger transitions should be noted. In our calculation we have found this Auger line energetically separated from its next neighbours by ~ 0.04 and 0.006 a.u., respectively. A non-zero anisotropy $\alpha_2 \sim -0.6$ should however remain even if the Auger line can not be resolved from its neighbouring lines. This is due to the fact that these lines belong to another multiplet, and thus overlap incoherently. It might be also possible to resolve the $KL_1L_{2,3}(^3P)4s$ $^2P_{1/2}$ Auger line which shows a measurable anisotropy in our calculation. LS coupling seems to be no longer valid for potassium because our calculation predicts a non-zero angular anisotropy even for the $KL_1L_{2,3}(^3P)4s$ $^2P_{1/2}$ Auger transitions which should be zero otherwise. Further, the deviations from the LS coupling values for the $KL_1L_{2,3}$ $^4P_{J_f}$ anisotropy coefficients become stronger. This behaviour is confirmed by changes in the energetic order of the KLL final states.

---

[10] The value can be easily obtained by using Tables 4.36 and 4.42.

[11] For Na the LS coupling represents a good approximation. Therefore the selection rule $\Delta L = 0$ applies.

Table 4.43. The anisotropy coefficients α_2 for the KLL Auger transitions of potassium. The columns are denoted as in Table 4.36

	K KLL Auger transitions	
Final states	α_2 parameters	
	(b)	(d)
$\left\|2p^6(^1S)4s\ ^2S_{1/2}\right\rangle$	~ 0	~ 0
$\left\|2s2p^5(^1P)4s\ ^2P^o_{1/2}\right\rangle$	0.033	0.020
$\left\|2s2p^5(^1P)4s\ ^2P^o_{3/2}\right\rangle$	-0.016	-0.012
$\left\|2s2p^5(^3P)4s\ ^2P^o_{1/2}\right\rangle$ †	0.358	0.356
$\left\|2s2p^5(^3P)4s\ ^4P^o_{1/2}\right\rangle$ †	-1.403	-1.405
$\left\|2s2p^5(^3P)4s\ ^2P^o_{3/2}\right\rangle$ †	0.456	0.457
$\left\|2s2p^5(^3P)4s\ ^4P^o_{3/2}\right\rangle$ †	0.630	0.632
$\left\|2s2p^5(^3P)4s\ ^4P^o_{5/2}\right\rangle$ †	-0.140	-0.139
$\left\|2s^22p^4(^1S)4s\ ^2S_{1/2}\right\rangle$	~ 0	~ 0
$\left\|2s^22p^4(^1D)4s\ ^2D_{3/2}\right\rangle$	-0.003	-0.019
$\left\|2s^22p^4(^1D)4s\ ^2D_{5/2}\right\rangle$	0.002	0.012
$\left\|2s^22p^4(^3P)4s\ ^2P_{1/2}\right\rangle$	-0.001	0.007
$\left\|2s^22p^4(^3P)4s\ ^4P_{1/2}\right\rangle$	-0.104	-0.215
$\left\|2s^22p^4(^3P)4s\ ^2P_{3/2}\right\rangle$	0.453	0.441
$\left\|2s^22p^4(^3P)4s\ ^4P_{3/2}\right\rangle$	-0.070	-0.088
$\left\|2s^22p^4(^3P)4s\ ^4P_{5/2}\right\rangle$	0.048	0.061
$\left\|2s2p^6\ ^2S_{1/2}\right\rangle$	0.0001	0.0007
$\left\|2s^22p^5\ ^2P^o_{1/2}\right\rangle$	-0.067	-0.100
$\left\|2s^22p^5\ ^2P^o_{3/2}\right\rangle$	0.023	0.046

Rubidium and Cesium

The heavier elements rubidium and cesium, see Tables 4.44 and 4.45, have to be described in an intermediate coupling scheme. In both elements the corrections due to the Breit interaction become more important than those due to exchange. The transitions to the $KL_{2,3}L_{2,3}\ ^2P_{J_f}$ and particularly to the $^4P_{J_f}$ final states are now comparable with the stronger Auger lines. Even though most of these Auger lines and also the $KL_1L_{2,3}$ transitions show non-zero anisotropy coefficients it becomes less likely to measure their anisotropy. This follows from the small fine structure splitting of the multiplets which can be explained by the overlap of the different s orbitals. In rubidium, for instance, the overlap between the 1s and 5s electrons is much smaller than between the corresponding 1s and 3s in sodium. The behaviour of the Auger spectra is assumed to be similar to the KLL Auger transitions of the neighbouring rare gases Kr or Xe which finally results in an isotropic angular distribution.

Table 4.44. The anisotropy coefficients α_2 for the KLL Auger transitions of rubidium. The columns are denoted as in Table 4.36

Final states	α_2 parameters	
	(b)	(d)
$\left\vert 2p^6(^1S)5s\ ^2S_{1/2}\right\rangle$	~ 0	~ 0
$\left\vert 2s2p^5(^1P)5s\ ^2P^o_{1/2}\right\rangle$	0.358	0.316
$\left\vert 2s2p^5(^1P)5s\ ^2P^o_{3/2}\right\rangle$	-0.119	-0.112
$\left\vert 2s2p^5(^3P)5s\ ^2P^o_{1/2}\right\rangle$	0.010	0.011
$\left\vert 2s2p^5(^3P)5s\ ^4P^o_{1/2}\right\rangle$ †	-0.741	-0.694
$\left\vert 2s2p^5(^3P)5s\ ^2P^o_{3/2}\right\rangle$ †	0.591	0.557
$\left\vert 2s2p^5(^3P)5s\ ^4P^o_{3/2}\right\rangle$	0.535	0.531
$\left\vert 2s2p^5(^3P)5s\ ^4P^o_{5/2}\right\rangle$	-0.134	-0.134
$\left\vert 2s^22p^4(^1S)5s\ ^2S_{1/2}\right\rangle$	~ 0	~ 0
$\left\vert 2s^22p^4(^1D)5s\ ^2D_{3/2}\right\rangle$	0.001	-0.008
$\left\vert 2s^22p^4(^1D)5s\ ^2D_{5/2}\right\rangle$	-0.001	0.005
$\left\vert 2s^22p^4(^3P)5s\ ^2P_{1/2}\right\rangle$	-0.624	-0.570
$\left\vert 2s^22p^4(^3P)5s\ ^2P_{3/2}\right\rangle$	0.310	0.292
$\left\vert 2s^22p^4(^3P)5s\ ^4P_{1/2}\right\rangle$	~ 0	~ 0
$\left\vert 2s^22p^4(^3P)5s\ ^4P_{3/2}\right\rangle$	-0.082	-0.092
$\left\vert 2s^22p^4(^3P)5s\ ^4P_{5/2}\right\rangle$	0.060	0.068
$\left\vert 2s2p^6\ ^2S_{1/2}\right\rangle$	~ 0	0.001
$\left\vert 2s^22p^5\ ^2P^o_{1/2}\right\rangle$	-0.381	-0.325
$\left\vert 2s^22p^5\ ^2P^o_{3/2}\right\rangle$	0.079	0.087

Rb KLL Auger transitions

However, if we trust our classification of states, the $KL_1L_{2,3}$ $(^3P)ns\ ^4P_{1/2}$ and $(^3P)ns\ ^2P_{3/2}$ Auger lines, with $n = 5$ and 6 for Rb and Cs, respectively, which have almost the same energy, overlap incoherently and therefore a small negative non-zero angular anisotropy $\alpha_2 \sim -0.1$ should be observed for the unresolved lines.

4.9 Open Shell Atoms – The KLL Auger Spectrum of Atomic Oxygen

Eventually, we consider Auger transitions from oxygen atoms. As has been discussed in the last section, Lohmann and Fritzsche (1994) predicted large anisotropy parameters for many of the KLL Auger transitions in alkali elements. Though, their

Table 4.45. The anisotropy coefficients α_2 for the KLL Auger transitions of cesium. The columns are denoted as in Table 4.36

Cs KLL Auger transitions		
Final states	α_2 parameters	
	(b)	(d)
$\left|2p^6(^1S)6s\,^2S_{1/2}\right\rangle$	~ 0	~ 0
$\left|2s2p^5(^1P)6s\,^2P^o_{1/2}\right\rangle$	0.694	0.698
$\left|2s2p^5(^1P)6s\,^2P^o_{3/2}\right\rangle$	-0.115	-0.123
$\left|2s2p^5(^3P)6s\,^2P^o_{1/2}\right\rangle$	0.001	0.001
$\left|2s^22p^4(^1S)6s\,^2S_{1/2}\right\rangle$	~ 0	~ 0
$\left|2s2p^5(^3P)6s\,^2P^o_{3/2}\right\rangle$ †	0.418	0.398
$\left|2s2p^5(^3P)6s\,^4P^o_{1/2}\right\rangle$ †	-0.541	-0.519
$\left|2s2p^5(^3P)6s\,^4P^o_{3/2}\right\rangle$	0.503	0.497
$\left|2s2p^5(^3P)6s\,^4P^o_{5/2}\right\rangle$	-0.127	-0.127
$\left|2s^22p^4(^1D)6s\,^2D_{3/2}\right\rangle$	0.010	0.005
$\left|2s^22p^4(^1D)6s\,^2D_{5/2}\right\rangle$	-0.007	-0.003
$\left|2s^22p^4(^3P)6s\,^2P_{1/2}\right\rangle$	-0.721	-0.680
$\left|2s^22p^4(^3P)6s\,^2P_{3/2}\right\rangle$	0.345	0.334
$\left|2s^22p^4(^3P)6s\,^4P_{1/2}\right\rangle$	~ 0	~ 0
$\left|2s^22p^4(^3P)6s\,^4P_{3/2}\right\rangle$	-0.153	-0.160
$\left|2s^22p^4(^3P)6s\,^4P_{5/2}\right\rangle$	0.123	0.129
$\left|2s2p^6\,^2S_{1/2}\right\rangle$	-0.0001	0.0001
$\left|2s^22p^5\,^2P^o_{1/2}\right\rangle$	-1.289	-1.167
$\left|2s^22p^5\,^2P^o_{3/2}\right\rangle$	0.190	0.188

experimental determination might become tedious because of the fact that an alignment can only be generated by relativistic effects. This is however different if the open shell electron(s) is(are) in the valence p shell. We have discussed this for the case of laser excited sodium (Lohmann *et al.* 1996). Due to the excitation of the valence electron to the p shell the alignment can be expected different from zero even in a non-relativistic treatment. This has been confirmed by Dorn *et al.* (1995) who investigated the angle resolved spectrum of the Na $2s^{-1}2p^53s3p$ Auger transitions, where our calculations (Lohmann *et al.* 1996) predicted large anisotropy parameters for the Na $1s^{-1}3p$ KLL Auger transitions.

In singly inner-shell ionized atomic oxygen we have four electrons in the outer p shell which can couple their angular momenta with the inner shell s hole. This can result in a total angular momentum $J \geq 1$ and thus, non-zero alignment and anisotropy parameters are both possible.

The KLL Auger spectra has been theoretically analyzed by Saha (1994) using a multiconfigurational Hartree–Fock method (MCHF). The KLL Auger spectrum of atomic oxygen has been obtained by Caldwell and Krause (1993). Due to improvements in the experimental techniques high resolution experiments have become possible for atomic oxygen (Becker 1994). Thus, it is our aim to provide theoretical data for an angle resolved experiment, i.e. to provide information on the angular anisotropy.

In this section, we are investigating the ground state KLL Auger transitions of atomic oxygen. The Auger energies, absolute and relative intensities, have been calculated. The main focus is on the calculation of the angular anisotropy parameters using an *ab initio* MCDF approach.

We discuss and compare the energies and intensities with theoretical and experimental data. The anisotropy parameters are discussed. Here, attention is also drawn to the problem of an unresolved initial and final state fine structure. Recommendations are given for the observation of anisotropic KLL Auger lines.

The Auger emission process can be described in the well observed two-step model, i.e. in a first step, the inner shell K-hole is created, for instance by a synchrotron beam. In a second step, the Auger electron is emitted:

$$\gamma_{Syn} + O \longrightarrow e_e + O^+\left(1s^{-1}\right) \tag{4.41}$$

$$\phantom{\gamma_{Syn} + O \longrightarrow}\; \hookrightarrow O^{++} + e_{Auger}. \tag{4.42}$$

In the following we will consider the Auger process only.

We described the computational method in Sect. 3.2 as well as in Lohmann and Fritzsche (1994) and a brief outline has been given in Sect. 4.8. Therefore, we concentrate on the specifics of the initial and final states, only.

For the initial state, the eight resonantly excited $[1s^{-1}2p^4](J = 1/2, 3/2, 5/2)$ states of the ground state oxygen KLL spectra have been included in our calculation. Virtual excitations of the 2p electrons into CSF of the same symmetry ($J = 1/2, 3/2, 5/2$) and even parity were not included.

For the $1s^2 2s^0 2p^4$, $1s^2 2s 2p^3$, and $1s^2 2s^2 2p^2$ final state configurations all CSF from the KLL spectra have been taken into account. The coupling of the two L holes with the valence shell 2p electrons results in 10 CSF of even and 10 CSF of odd parity, i.e. a total of 20 CSF has been considered. We obtained the mixing coefficients between CSF of the same symmetry in the average-level scheme (Dyall *et al.* 1989). For the evaluation of the Auger amplitudes we used the bound orbitals of the initial resonant state and orthogonalized the continuum spinors with respect to this orbital set. As pointed out in Sect. 4.8, the calculation of α_2 requires an explicit knowledge of the scattering phase.

4.9.1 The Structure of the Spectrum

Auger spectra of open shell systems are often complex and cause a great deal of difficulties in their interpretation compared to the spectra of the rare gases. This also applies to the oxygen Auger spectrum where the valence $2p^4$ configuration couples to the inner shell K-hole. The fine structure splitting is often very small also resulting in difficulties in the identification of the measured Auger lines. Therefore, the fine structure splitting has not been properly resolved in most of the previous experiments. Most of the experimental data for oxygen are from Caldwell and Krause (1993) who applied high resolution techniques. However, even they have been able to resolve a total of 10 Auger lines, only.

Angle resolved experiments yield refined information on the Auger decay dynamics and may even allow for the classification of further oxygen KLL Auger lines. By providing a set of consistent data for Auger intensities and angular distribution parameters we, therefore, hope to initiate further investigations in this field.

Although all our calculations have been performed in an intermediate coupling scheme we will designate the individual Auger transitions in the LSJ notation. This scheme has been used to classify most of the available data in the literature. Furthermore, oxygen is a light atom, i.e. LS coupling can be assumed as a good approximation. To denote the final states, we used their main contribution in the configuration expansion and their energies apart from their obvious classification due to the total angular momentum and parity. In our discussion we consider the KLL Auger lines following a resonant excitation of the eight possible, even parity $[1s^{-1}2p^4]$ ${}^4P_{5/2,3/2,1/2}$, ${}^2P_{3/2,1/2}$, ${}^2D_{5/2,3/2}$, and ${}^2S_{1/2}$ initial states. We obtain a total of 20 allowed final states with either 10 of them having even or odd parity, respectively. The even parity states belong to the KL_1L_1 and $KL_{2,3}L_{2,3}$ part of the KLL spectrum. Within the chosen classification scheme we denote them as $1s^22s^02p^4$ and $1s^22s^22p^2$ ${}^3P_{0,1,2}$, ${}^1D_2$, and ${}^1S_0$ final states. The remaining 10 odd parity states (superscript ^{o}) belong to the $KL_1L_{2,3}$ Auger transitions and are classified as $1s^22s^12p^3$ ${}^5S_2$, ${}^3D_{1,2,3}$, ${}^3P_{0,1,2}$, ${}^3S_1$, ${}^1D_2$, and ${}^1P_1$ transitions.

Note, that only an Auger decay from the initial ${}^2P$ and ${}^4P$ states has been observed experimentally. This is since ionization of the K shell occurs solely from the ${}^3P$ oxygen ground state (Caldwell and Krause 1993). An observation of Auger lines originating from the ${}^2D$ and ${}^2S$ initial oxygen states might be possible by using laser excited atoms.

4.9.2 Auger Energies of the Oxygen KLL Spectra

The initial and final ionic bound states have been calculated with the GRASP atomic structure package. The energy splitting of the eight possible initial fine structure states, obtained in this approximation, are shown in Table 4.46. Where the different multiplets are energetically well separated by several eV, the fine structure splitting is usually around $\Delta E \sim 10^{-3}$ a.u. which is hard to resolve within an experiment.

The total decay rate is almost constant for the fine structure states of a multiplet. For the different initial multiplet states its value Γ varies between 5×10^{-3} a.u. and 7×10^{-3} a.u. (see also the next section). Thus, it is almost impossible to resolve the fine structure levels of the different multiplet states.

From the independent computations for the initial and final state configurations the Auger transition energies are obtained as differences of the total energies,

Table 4.46. The energy splitting of the eight possible initial fine structure states of the singly ionized oxygen atom relative to the lowest lying $^4P_{5/2}$ state. The data are given in both, au and eV. Data from Lohmann and Fritzsche (1996)

	O$^+$ states	
	Energy splitting	
Initial states	$\times 10^{-3}$ au	$\times 10^{-2}$ eV
$^4P_{5/2}$	0	0
$^4P_{3/2}$	1.5	4.0
$^4P_{1/2}$	1.9	5.3
$^2D_{5/2}$	160.8	437.6
$^2D_{3/2}$	161.2	438.7
$^2P_{3/2}$	171.8	467.5
$^2P_{1/2}$	173.0	470.8
$^2S_{1/2}$	315.9	859.5

$E_{Auger} = E_f - E_i$. For completeness, they are included in Table 4.47 for all calculated transitions. Note, that the transition energies are given with respect to the lowest lying $^4P_{5/2}$ level of the initial state multiplet.

In an experiment, Caldwell and Krause (1993) have been able to energetically resolve 10 Auger lines. There are also recent theoretical calculations by Saha (1994) to compare with. This author further cited numerical data by Armen and Larkins (1994) who performed a Hartree–Fock calculation, and by Chen (1994) who did an MCDF approach, as well as the optical data by Moore (1976). Since we used the same approach as Chen, the energy splitting obtained in this work is similar to his data. Our MCDF calculation simply includes the main configurations and can therefore not compete with the more extensive calculations by Saha (1994) for the representation of the energy splitting.

4.9.3 Auger Rates and Relative Intensities

General Considerations

The absolute Auger rates and relative intensities have been calculated for all transitions between the eight initial and the allowed final states. The relative intensities for the oxygen KLL Auger lines are shown in Table 4.47. The total Auger rates for the initial fine structure states of each multiplet, each state corresponding to 100% intensity, are also given. They are approximately constant for a multiplet confirming the small effect of correlations to the total decay widths.

Γ ranges from $\sim 50 \times 10^{-4}$ a.u. to $\sim 70 \times 10^{-4}$ a.u. for the different multiplets, where we have obtained average values of $\Gamma = 50.34 \times 10^{-4}$ a.u. for Auger transitions from the ^2P states and $\Gamma = 61.96 \times 10^{-4}$ a.u. for the ^4P state transitions, respectively. For the experimentally observed Auger transitions total rates of

Table 4.47. Relative intensities and total Auger rates for the KLL Auger transitions of oxygen. The eight initial states of the $[1s^{-1}2p^4]$ multiplet are shown in the head of the table. The Auger energies and rates are given in atomic units. Transition energies are listed with respect to the lowest $^4P_{5/2}$ level of the initial state multiplet. ~ 0 denotes an absolute rate $< 10^{-4}$. Note, that the absolute rates for columns 3–10 must be multiplied by 10^{-4}. Occupied orbitals are displayed only as necessary to classify the state. Data from Lohmann and Fritzsche (1996)

		O KLL Auger transitions							
Final	Auger	Relative intensities							
states	energies	$^4P_{5/2}$	$^4P_{3/2}$	$^4P_{1/2}$	$^2D_{5/2}$	$^2D_{3/2}$	$^2P_{3/2}$	$^2P_{1/2}$	$^2S_{1/2}$
$\lvert 2p^4\,^1S_0\rangle$	16.619	~ 0	~ 0	~ 0	0.39	0.39	~ 0	~ 0	11.89
$\lvert 2p^4\,^1D_2\rangle$	16.840	~ 0	~ 0	~ 0	16.55	16.47	0.16	~ 0	0.62
$\lvert 2p^4\,^3P_0\rangle$	16.940	0.13	0.02	11.81	~ 0	~ 0	0.13	7.30	~ 0
$\lvert 2p^4\,^3P_1\rangle$	16.941	0.28	14.92	6.00	~ 0	0.02	3.94	14.60	~ 0
$\lvert 2p^4\,^3P_2\rangle$	16.941	17.82	3.42	0.58	~ 0	0.09	18.21	0.63	~ 0
$\lvert 2s2p^3\,^1P_1^o\rangle$	17.273	~ 0	~ 0	~ 0	5.80	5.83	4.34	4.63	25.97
$\lvert 2s2p^3\,^3S_1^o\rangle$	17.344	5.35	5.40	5.44	~ 0	0.12	21.71	21.62	~ 0
$\lvert 2s2p^3\,^1D_2^o\rangle$	17.374	~ 0	~ 0	~ 0	17.49	17.49	7.34	7.16	~ 0
$\lvert 2s2p^3\,^3P_2^o\rangle$	17.636	9.72	6.40	1.85	2.57	0.35	0.19	0.08	7.32
$\lvert 2s2p^3\,^3P_1^o\rangle$	17.636	3.25	3.25	10.08	0.53	1.76	0.04	0.11	4.45
$\lvert 2s2p^3\,^3P_0^o\rangle$	17.636	~ 0	3.64	1.52	~ 0	0.86	~ 0	0.04	1.51
$\lvert 2s2p^3\,^3D_3^o\rangle$	17.737	18.34	3.08	~ 0	6.53	0.73	0.20	~ 0	~ 0
$\lvert 2s2p^3\,^3D_2^o\rangle$	17.737	3.41	14.13	5.43	2.17	4.25	0.16	0.19	~ 0
$\lvert 2s2p^3\,^3D_1^o\rangle$	17.737	0.24	4.70	16.36	0.30	3.99	0.11	0.18	~ 0
$\lvert 2s2p^3\,^5S_2^o\rangle$	18.070	5.54	5.48	5.45	~ 0	~ 0	~ 0	~ 0	~ 0
$\lvert 2s^22p^2\,^1S_0\rangle$	18.113	~ 0	~ 0	~ 0	5.23	5.17	0.06	~ 0	18.37
$\lvert 2s^22p^2\,^1D_2\rangle$	18.195	~ 0	~ 0	~ 0	42.47	42.24	0.38	0.01	29.91
$\lvert 2s^22p^2\,^3P_2\rangle$	18.296	15.76	21.72	28.00	~ 0	0.12	20.63	31.29	~ 0
$\lvert 2s^22p^2\,^3P_1\rangle$	18.297	13.75	12.79	4.21	~ 0	0.08	16.35	10.13	~ 0
$\lvert 2s^22p^2\,^3P_0\rangle$	18.298	6.07	1.02	3.30	~ 0	0.03	6.09	2.00	~ 0
Abs. tot.	Σ	100	100	100	100	100	100	100	100
rates $\times 10^{-4}$	487.98	62.24	61.85	61.80	69.61	69.41	50.44	50.24	62.39

$\Gamma \sim 50.34 \times 10^{-4}$ a.u. and $\Gamma \sim 61.97 \times 10^{-4}$ a.u. have been obtained for the ^{2}P and ^{4}P initial states, respectively. Thus, our data are in good agreement with the experimental results.

The absolute Auger rate for all O KLL Auger transitions has been obtained as $\Gamma = 487.98 \times 10^{-4}$ a.u. This value is larger compared to a previous value we found for laser excited Na KLL transitions, $\Gamma = 379.67 \times 10^{-4}$ a.u. (Lohmann *et al.* 1996). This is caused by a larger number of initial states for the O KLL Auger transitions. Comparing the normalized total rates, i.e. dividing by the number of initial states, we get $\Gamma_n(\text{O}) = 61.00 \times 10^{-4}$ a.u. and $\Gamma_n(\text{Na}^*) = 94.92 \times 10^{-4}$ a.u., confirming the well-known fact that the absolute Auger rate increases with atomic number Z (e.g. see Chattarji 1976).

Generally, the intensity of the different Auger transitions varies depending on the initial and final states, respectively. This behavior is discussed in the following.

KL$_1$L$_1$ Transitions

The ^{4}P KL$_1$L$_1$ Auger transitions cover $\sim 18\%$ of the total intensity. Transitions to the final singlet states are not allowed if extreme LS coupling is assumed. This is well confirmed by our calculations which show the intensity distributed between the ^{3}P fine structure states, where the $(J \rightarrow J_f)$ $5/2 \rightarrow 2$, $3/2 \rightarrow 1$, and $1/2 \rightarrow 0$ Auger transitions cover most of the group intensity. A similar behaviour occurs for the ^{2}P$_J \rightarrow {}^3$P$_{J_f}$ transitions where the main contribution is by the $3/2 \rightarrow 2$, and $1/2 \rightarrow 1$ transitions. Thus, the highest intensity is by Auger transitions from the highest possible J value of the initial state multiplet to the highest possible value J_f of the final state multiplet; then, by recursively subtracting 1 from both quantum numbers, we get the Auger line with the highest intensity for the next initial fine structure state, and so on. This behaviour can be seen as a propensity rule for the initial P state transitions.

The ^{2}P KL$_1$L$_1$ group intensity is calculated to be 22%.

The ^{2}D group intensity is slightly decreased to 17%, where the ^{2}S group intensity is predicted to be 12% of the total intensity. For the first, almost all intensity stems from the ^{2}D $\rightarrow {}^1$D transitions where for the latter the ^{2}S $\rightarrow {}^1$S transition gives the main contribution.

Thus, for all KL$_1$L$_1$ Auger transitions, the angular momentum L is preferably conserved during the Auger transition clearly showing the LS coupling-type character of the transitions.

KL$_1$L$_{2,3}$ Transitions

The ^{4}P KL$_1$L$_{2,3}$ Auger transitions cover $\sim 46\%$ of the total intensity. Again, transitions to the final singlet states are not allowed in the LS coupling case which is confirmed by our calculations. The intensity distribution of the fine structure transitions shows more variation than in the previous case. However, the propensity rule of the previous Sect. can still be applied, i.e. the largest intensities have been calculated for the $5/2 \to 3$, $3/2 \to 2$, and $1/2 \to 1$ fine structure terms of the $^4P \to {}^3D$ Auger transitions.

However, a different behaviour occurs for ^2P transitions where we obtained a group intensity of 34% for the $KL_1L_{2,3}$ Auger transitions. Here, the main contribution stems from the $^2P \to {}^3S_1$ transition. The other part is from the $^2P \to {}^1P_1$ and $^1D_2$ Auger lines, where the intensity of the remaining lines is at least by a factor of 70 lower. Due to selection rules Auger transitions from doublet to quintet states are not allowed as can be seen from the data.

For the ^2D transitions we get a group intensity of 35% which is similar to the ^2P transitions. The main contribution is from the $^2D \to {}^1P_1$ and $^1D_2$ Auger lines where the intensity of the latter is by a factor of 3 larger. In contrast to the ^2P transitions, the $^2D \to {}^3D_{J_f}$ Auger transitions show a non-zero intensity, too.

The group intensity for the ^2S transitions is further increased to 39% where the major contribution stems from the $^2S \to {}^1S_0$ Auger transition. The rest is distributed between the fine structure states of the $^3P^o_{J_f}$ final states. All other transitions show an almost vanishing intensity.

$KL_{2,3}L_{2,3}$ Transitions

The group intensity for the ^4P transitions has been calculated to be $\sim 35\%$. Transitions to the final singlet states are not allowed. In contrast to the KL_1L_1 Auger transitions the propensity rule no longer applies. The main contribution stems from the $^3P_2$ final state and is increasing with a decreasing total angular momentum of the initial state.

The same holds for the ^2P state transitions. However, we calculated a larger group intensity of $\sim 43\%$.

The behaviour of the ^2D and ^2S transitions is similar to the case of the KL_1L_1 Auger transitions. The group intensities have been obtained as 47% and 48%, respectively. For the first, almost all of the intensity stems from the $^2D \to {}^1D$ transition, where for the latter both, the $^2S \to {}^1S$ and $^2S \to{}^1 D$ transitions, contribute.

The O KLL Spectrum

As has been pointed out in Sect. 4.9.1 only the KLL Auger spectra from the ^2P and ^4P initial states are experimentally accessible via photoionization techniques. The combined, numerical oxygen KLL spectrum stemming from the ^2P and ^4P and initial states has been plotted in Fig. 4.15 using equations (4.14) and (4.35). A numerical FWHM of $\Gamma = 0.215\,\text{eV}$ has been assumed for generating the Lorentz profiles of the Auger lines. The full curves refer to the Auger decay from the ^4P state whereas the broken curves originate from the ^2P state. As has been discussed, oxygen is well described in the LS coupling scheme. Therefore, each line consists of a number of unresolved fine structure states. Our spectroscopic identification of the Auger lines has been included in Fig. 4.15 resulting in an 11 peak spectrum. The

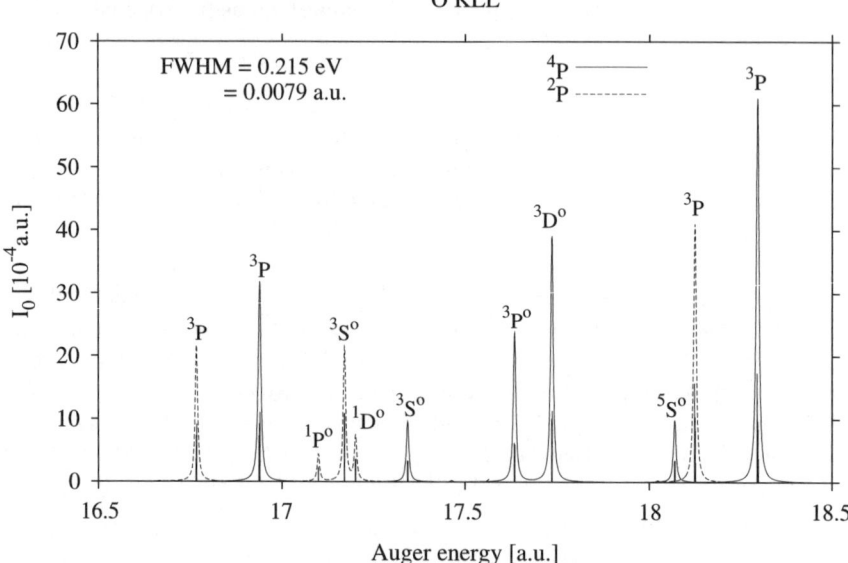

Fig. 4.15. The KLL Auger spectrum of atomic oxygen after photoionization of the 1s shell. The Auger decay propagates from the generated ^2P and ^4P states, only. Full curve: lines originate from ^4P state. Broken curve: lines from ^2P state. Odd parity multiplets are related to $KL_1L_{2,3}$ transitions, even parity lines stem from KL_1L_1 (left side, lower Auger energies) and $KL_{2,3}L_{2,3}$ (right side, higher Auger energies) Auger transitions, respectively

Auger peaks with odd parity refer to the $KL_1L_{2,3}$ Auger decay, whereas the two ^3P peaks with even parity on the left- (lower Auger energies) and right-hand side (higher Auger energies) of Fig. 4.15 are related to the KL_1L_1 and $KL_{2,3}L_{2,3}$ Auger transitions, respectively. The small line stemming from the $1s^{-1}2p \rightarrow (2s2p^3)\,^1P^o$ transition has not been observed in the experiment (see the next section).

For completeness, we show the ^2D and ^2S KLL spectra in Figs. 4.16 and 4.17 together with their spectroscopic Auger line identification. As has been discussed in the previous Sects. almost all prominent lines of the ^2D case stem from $^2D \rightarrow {}^{1,3}D$ transitions whereas Auger decay into S and P states reveals lower intensities. The Auger decay into the KL_1L_1 ^1S state is strongly suppressed which leaves us with a 7 peak spectrum. The ^2S KLL spectrum is even simpler revealing $^2S \rightarrow {}^1S, {}^1D$ transitions for the even parity states and decays solely into $^{1,3}P^o$ states for the odd parity case. Having the Auger decay into the KL_1L_1 ^1D state strongly suppressed eventually results in a 5 peak spectrum.

Comparison with Other Data

In Table 4.48 our results are compared to the 10 lines experimentally resolved by Caldwell and Krause (1993), and to the theoretical MCHF calculations by Saha

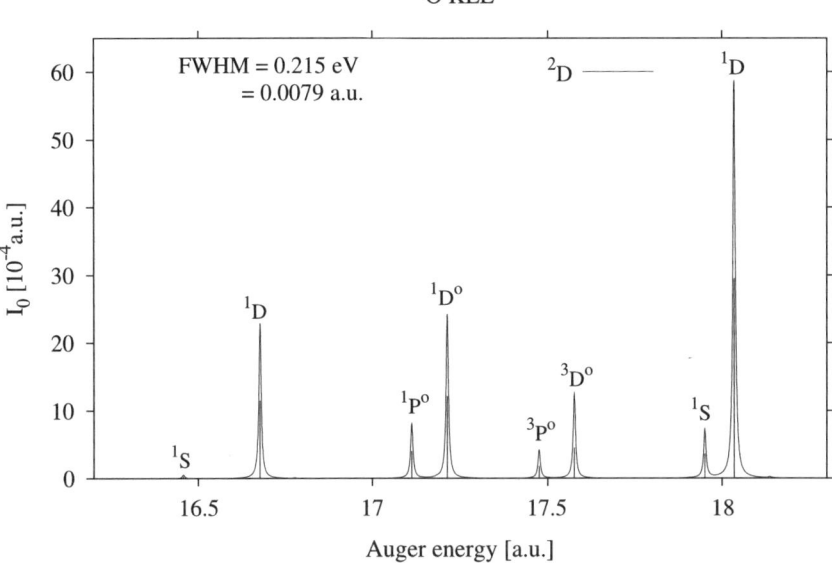

Fig. 4.16. The KLL Auger spectrum of atomic oxygen after photoionization of the 1s shell. The Auger decay from the ^2D state is possible only after a primary excitation from the ground state. Odd parity lines refer to $KL_1L_{2,3}$, even parity lines to KL_1L_1 (left, lower Auger energies) and $KL_{2,3}L_{2,3}$ (right, higher Auger energies) Auger transitions, respectively

(1994) as well as to the calculations by Petrini and Araújo (1994) using the eigenchannel method, and to the MCDF data by Chen (1994). Both have been cited in the paper by Saha and numbers have been taken from there.

In contrast to the energy results, the relative intensities are less sensitive to the numerical approach, at least as long a summation over individual transitions has to be performed to compare with the experiment.

Our results are close to the calculation by Chen (1994). Where generally agreement with the experimental data and the theoretical calculations is good there are some discrepancies, mainly for the $KL_{2,3}L_{2,3}$ $^{2,4}P \rightarrow {}^3P$ Auger transitions. As we have pointed out in Sect. 4.9.2, we cannot compete with the calculations by Saha (1994) due to the small basis set we used. Therefore, some of the experimental data are better reproduced by the calculations of Saha (1994).

4.9.4 Angular Distribution of the KLL Auger Lines

Our main attention has been paid to the analysis of the angular anisotropy which can be found in the spectrum. In Table 4.49 the anisotropy coefficients α_2 for the KLL Auger transitions of oxygen from the initial $^4P_{5/2,3/2}$, $^2P_{3/2}$, and $^2D_{5/2,3/2}$ states are given. Auger transitions from the initial $^2S_{1/2}$ and $^2P_{1/2}$ fine structure states must be isotropic since no alignment can be generated. Large anisotropy parameters have

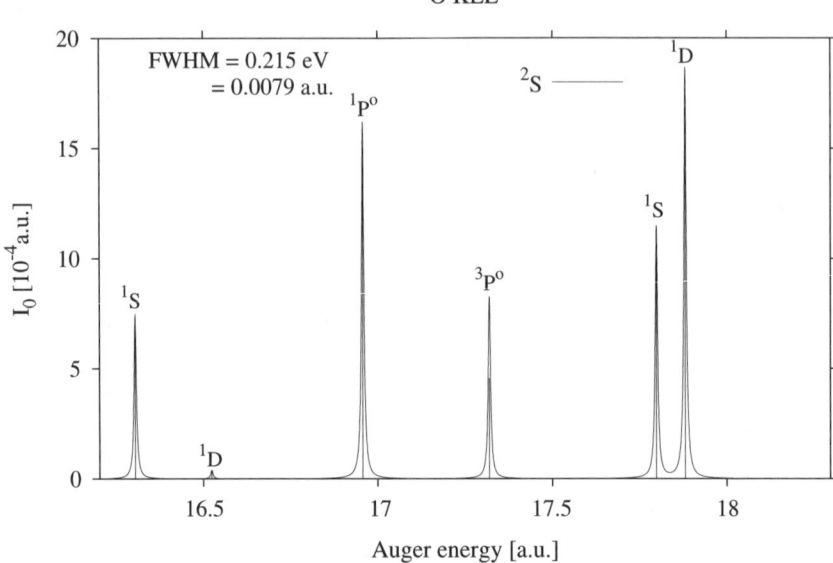

Fig. 4.17. The KLL Auger spectrum of atomic oxygen after photoionization of the 1s shell. The Auger decay from the ^{2}S state is possible only after a primary excitation from the ground state. Odd parity lines refer to $KL_1L_{2,3}$, even parity lines to KL_1L_1 (left, lower Auger energies) and $KL_{2,3}L_{2,3}$ (right, higher Auger energies) Auger transitions, respectively

been found for most of the Auger transitions. As has been demonstrated in earlier publications (e.g. Lohmann 1990) Auger transitions to a final fine structure state with $J_f = 0$ are independent of the Auger transition matrix elements. This result is independent of the specific structure of the atomic shell. The anisotropy parameters are expected to be $\alpha_2 = -1.0$ or -1.069 for Auger transitions from initial $J = 3/2$ or $5/2$ states, respectively, which is confirmed by our calculations.

Some of the Auger transitions become independent of the transition matrix elements if extreme LS coupling is applied. Particularly, considering transitions to final S states, Lohmann and Fritzsche (1996) obtained

$$\alpha_2^{(LS)}\left(^{2S+1}L_J \rightarrow \, ^{2S_f+1}S_{J_f}\right) = \delta_{S_f,J_f}(-1)^{J+S}\sqrt{5(2J+1)}$$
$$\times (2L+1)\begin{pmatrix} L & L & 2 \\ 0 & 0 & 0 \end{pmatrix}\begin{Bmatrix} J & J & 2 \\ L & L & S \end{Bmatrix}. \quad (4.43)$$

The anisotropy parameter α_2 appears to be independent of its final spin and total angular momentum for these transitions. Inserting the appropriate quantum numbers we obtain $\alpha_2 = -\sqrt{14/25} \sim -0.7483$ for the $^4P_{5/2} \rightarrow$ S Auger transitions, $\alpha_2 = 4/5 = 0.8$ for the $^4P_{3/2} \rightarrow$ S transitions, and $\alpha_2 = -1.0$ for the $^2P_{3/2} \rightarrow$ S Auger transitions which is in perfect agreement with our calculations. $^2D_J \rightarrow \, ^3S_1$ transitions are not allowed in the LS coupling due to parity violation.

Table 4.48. Relative intensities of the KLL Auger spectrum of atomic oxygen. Comparison with experiment and other calculations. (a): Lohmann and Fritzsche (1996); (b): Saha (1994); (c): Petrini and Araújo (1994); (d): Caldwell and Krause (1993); †: The values obtained in this work are close to a MCDF calculation by Chen (1994) which is referred to in Saha

O KLL Auger transitions				
	Relative intensity			
Transition	(a)†	(b)	(c)	(d) Exp.
$1s\,^4P \rightarrow 2p2p\,^3P$	1.63	3.050	2.000	3.0(7)
$1s\,^2P \rightarrow 2p2p\,^3P$	1.17	1.290	1.743	1.4(2)
$1s\,^4P \rightarrow 2s2p\,^5S^o$	0.25	0.141	0.157	0.18(6)
$1s\,^4P \rightarrow 2s2p\,^3D^o$	1.00	1.000	1.000	1.00
$1s\,^4P \rightarrow 2s2p\,^3P^o$	0.60	0.485	0.545	0.44(4)
$1s\,^4P \rightarrow 2s2p\,^3S^o$	0.25	0.174	0.218	0.20(2)
$1s\,^2P \rightarrow 2s2p\,^1D^o$	0.17	0.130	0.210	0.16(3)
$1s\,^2P \rightarrow 2s2p\,^3S^o$	0.53	0.506	0.812	0.58(5)
$1s\,^4P \rightarrow 2s2s\,^3P$	0.83	0.636	0.683	0.64(1)
$1s\,^2P \rightarrow 2s2s\,^3P$	0.55	0.586	0.941	0.50(5)

Generally, we point out that, due to coupling rules of the $3nj$ symbols and conservation of spin, α_2 should vanish for all quartet → singlet, and doublet → quintet Auger transitions if LS coupling is assumed. Though we have obtained large anisotropy coefficients for these transitions their discussion is useless because our calculation of the Auger rates show them as strongly forbidden.

A resolved final state fine structure can hardly be realized experimentally. We present the anisotropy parameters α_2 for the unresolved final state fine structure in Table 4.50, where we have considered the lines observed by Caldwell and Krause (1993), only. As can be seen, one still obtains non-zero anisotropy parameters. Large values of α_2 have been obtained for Auger transitions to the final ^5S and ^3S states.

However, as discussed in Sect. 4.9.2, the line widths of the initial state multiplets are usually larger than their fine structure splittings. Thus, as a first attempt, we assume a statistically distributed multiplet $M(LS)$ for an intermediate $^{2S+1}L_J$ state after the primary ionization process, i.e.

$$\alpha_2\left(^{2S+1}L\right) = \left[(2S+1)(2L+1)\right]^{-1} \sum_{J \in M} (2J+1)\,\alpha_2\left(^{2S+1}L_J\right). \qquad (4.44)$$

This is only a crude assumption, e.g. see the discussion in Sect. 4.8 or Lohmann *et al.* (1996), Kabachnik *et al.* (1994). A coherent summation, i.e. (2.282), over the

Table 4.49. The anisotropy coefficients α_2 for the KLL Auger transitions of oxygen. †: Values independent of the Auger transition matrix elements. ‡: Values independent of matrix elements in the LS coupling approximation. Data from Lohmann and Fritzsche (1996)

	O KLL Auger transitions				
	α_2 parameters				
	Initial states				
Final states	$^4P_{5/2}$	$^4P_{3/2}$	$^2D_{5/2}$	$^2D_{3/2}$	$^2P_{3/2}$
$\lvert 2p^4\,^1S_0\rangle$ †	-1.069	-1.0	-1.069	-1.0	-1.0
$\lvert 2p^4\,^1D_2\rangle$	0.153	0.089	0.026	0.024	0.023
$\lvert 2p^4\,^3P_0\rangle$ †	-1.069	-1.0	-1.069	-1.0	-1.0
$\lvert 2p^4\,^3P_1\rangle$	-0.534	0.009	-0.862	-0.067	-0.074
$\lvert 2p^4\,^3P_2\rangle$	-0.006	0.090	0.122	-0.008	-0.010
$\lvert 2s2p^3\,^1P_1^o\rangle$	-0.806	0.339	-0.748	-0.771	0.660
$\lvert 2s2p^3\,^3S_1^o\rangle$ ‡	-0.748	0.800	-0.749	-1.0	-1.0
$\lvert 2s2p^3\,^1D_2^o\rangle$	0.398	0.722	0.750	0.672	-0.003
$\lvert 2s2p^3\,^3P_2^o\rangle$	0.751	-0.154	-0.748	0.799	0.800
$\lvert 2s2p^3\,^3P_1^o\rangle$	-0.748	-0.214	-0.749	-0.933	0.741
$\lvert 2s2p^3\,^3P_0^o\rangle$ †	-1.069	-1.0	-1.069	-1.0	-1.0
$\lvert 2s2p^3\,^3D_3^o\rangle$	-0.217	-0.200	0.856	-0.201	-0.200
$\lvert 2s2p^3\,^3D_2^o\rangle$	0.754	0.046	0.627	0.750	0.467
$\lvert 2s2p^3\,^3D_1^o\rangle$	-0.748	0.376	-0.748	0.800	0.800
$\lvert 2s2p^3\,^5S_2^o\rangle$ ‡	-0.748	0.800	-0.748	0.803	0.778
$\lvert 2s^2 2p^2\,^1S_0\rangle$ †	-1.069	-1.0	-1.069	-1.0	-1.0
$\lvert 2s^2 2p^2\,^1D_2\rangle$	0.005	0.169	0.208	0.194	0.172
$\lvert 2s^2 2p^2\,^3P_2\rangle$	0.155	0.635	-0.026	-0.003	-0.004
$\lvert 2s^2 2p^2\,^3P_1\rangle$	-0.534	-0.039	-0.874	-0.766	-0.755
$\lvert 2s^2 2p^2\,^3P_0\rangle$ †	-1.069	-1.0	-1.069	-1.0	-1.0

fine structure of the initial states has to be carried out to deal satisfactorily with this matter.

The anisotropy coefficients α_2 are given in Table 4.51 for the unresolved final state, and for both, the statistically averaged and the coherently excited initial fine structure states, respectively. Note, that the two results can differ in magnitude and even in sign.

Table 4.50. The anisotropy coefficients α_2 for the KLL Auger transitions of oxygen. The data are given for the unresolved final state fine structure. We have only considered the lines observed by Caldwell and Krause (1993). Data from Lohmann and Fritzsche (1996)

	O KLL Auger transitions		
	α_2 parameters		
	Initial states		
Final states	$^4P_{5/2}$	$^4P_{3/2}$	$^2P_{3/2}$
$\left\vert 2p^4\ {}^3P\right\rangle$	-0.022	0.022	-0.062
$\left\vert 2s2p^3\ {}^1P^o\right\rangle$	†-0.806	†0.339	‡0.660
$\left\vert 2s2p^3\ {}^3S^o\right\rangle$	-0.748	0.800	-1.0
$\left\vert 2s2p^3\ {}^1D^o\right\rangle$	†0.398	†0.722	-0.003
$\left\vert 2s2p^3\ {}^3P^o\right\rangle$	0.375	-0.401	†0.783
$\left\vert 2s2p^3\ {}^3D^o\right\rangle$	-0.072	0.082	†0.252
$\left\vert 2s2p^3\ {}^5S^o\right\rangle$	-0.748	0.800	†0.778
$\left\vert 2s^22p^2\ {}^3P\right\rangle$	-0.320	0.346	-0.430

† Although large anisotropy parameters have been calculated, these lines show almost vanishing intensity.

‡ This line has not been observed. Though we calculated a *measurable non-zero* intensity of approximately 5%.

The Auger lines for which we predict *sufficiently large intensities in conjunction with measurable anisotropy coefficients* are marked by an asterisk. Inspecting our results, we predict a measurable anisotropy for the $^4P \rightarrow {}^3S^o$, $^5S^o$ and $^2P \rightarrow {}^3S^o$, ^3P Auger transitions. The large value of $\alpha_2 = -0.871$ for the $^2P \rightarrow {}^3S^o$ transition should be noted. The Auger intensity for this line has been calculated to be $\sim 22\%$ of the whole ^2P Auger transitions. Therefore, we would suggest this line to be most interesting for an experimental proof of an anisotropic angular distribution of the oxygen KLL Auger lines. We also obtained a measurable anisotropy parameter of $\alpha_2 = -0.303$ for the $^2P \rightarrow {}^1P^o$ Auger transition. Though we calculated a non-zero intensity of approximately 5% this line has not been observed in the experiment by Caldwell and Krause (1993).

The Auger energies, rates and anisotropy coefficients for the KLL Auger transitions of atomic oxygen have been calculated. The energies and rates have been found similar to recent data by Chen (1994) and close enough to the experimental data by Caldwell and Krause (1993) to ensure that our model is able to predict reliable data for the angular distribution parameters. Large anisotropy coefficients for most of the oxygen KLL Auger transitions have been found. Due to the fact that the initial state fine structure splitting is smaller than their widths these numbers are hardly to determine within an experiment. Therefore, a coherent excitation of

Table 4.51. The anisotropy coefficients α_2 for the KLL Auger transitions of oxygen. The data are given for the unresolved final, and for the statistically averaged or coherently summed excited initial fine structure states, respectively. We have only considered the lines observed by Caldwell and Krause (1993). The Auger lines with sufficiently large intensities in conjunction with measurable anisotropy coefficients have been marked by an asterisk. Data from Lohmann and Fritzsche (1996)

	O KLL Auger transitions			
	α_2 parameters			
	Initial states			
Final states	$^4P_{av}$	$^4P_{co}$	$^2P_{av}$	$^2P_{co}$
$\lvert 2p^4\,^3P \rangle$	-0.004	-0.101	-0.041	$* -0.303$
$\lvert 2s2p^3\,^1P^o \rangle$	†	†	0.440	$* ‡0.471$
$\lvert 2s2p^3\,^3S^o \rangle$	-0.107	$* -0.334$	-0.667	$* -0.871$
$\lvert 2s2p^3\,^1D^o \rangle$	†	†	-0.002	-0.056
$\lvert 2s2p^3\,^3P^o \rangle$	0.054	0.119	†	†
$\lvert 2s2p^3\,^3D^o \rangle$	-0.009	0.008	†	†
$\lvert 2s2p^3\,^5S^o \rangle$	-0.107	$* -0.255$	†	†
$\lvert 2s^22p^2\,^1D \rangle$	†	†	0.115	0.109
$\lvert 2s^22p^2\,^3P \rangle$	-0.045	-0.015	-0.287	-0.044

av Averaged initial fine structure.

co Coherent summation over the initial fine structure.

† Lines have not been observed in the experiment.

‡ This line has not been observed. Though we calculated a measurable non-zero intensity of approximately 5%.

* Auger lines for which a measurable anisotropy is predicted.

the initial fine structure states has been considered. The anisotropy parameters have been calculated by averaging and by coherently summing over the initial fine structure states, respectively. It has been shown that the results obtained can differ by magnitude and sign.

For an unresolved initial and final state fine structure we still obtain non-zero anisotropy parameters for some Auger transitions. A measurable anisotropy has been predicted for the $^4P \rightarrow {}^3S^o, {}^5S^o$ and the $^2P \rightarrow {}^3S^o, {}^3P$ Auger lines of the atomic oxygen KLL spectra.

So far, there are almost no angular distribution measurements for open-shell systems. We hope that our work will encourage experimental studies on the angular distribution of open shell KLL Auger transitions.

5 Molecular Auger Processes

While the previous chapters dealt with the Auger emission from free atoms a number of experiments focused on Auger transitions from free molecules. This is an active molecules. This is an active field of theoretical and experimental research. We will concentrate our discussion on diatomic and small polyatomic molecules.

Most of the experimental and theoretical studies of molecular Auger processes have concentrated on determining the energy and probability of a given Auger transition (Aksela *et al.* 1988b; Carravetta and Ågren 1987; Ferrett *et al.* 1988; Larkins 1990; Schimmelpfennig *et al.* 1995; Zähringer *et al.* 1993). Additional information can be obtained by investigating angle and spin resolved Auger processes. For atoms this has been shown within a number of experimental and theoretical investigations (e.g. see Chen 1992; Kuntze *et al.* 1994; Lohmann *et al.* 1993; 1996; Merz and Semke 1990; Müller *et al.* 1995, and the discussion in the previous chapters). Further references may be found in the cited papers. A general theory for the angular distribution of Auger electrons following photoabsorption or photoionization of molecules has been developed by Dill *et al.* (1980) and Chandra and Chakraborty (1992). These authors have shown that vacancies produced in molecular photoionization behave differently than their atomic counterparts. In particular, Dill *et al.* (1980) showed that Auger electrons, emitted in the decay of a K-shell vacancy in a diatomic molecule produced by photons, can have a non-isotropic angular distribution, contrary to the atomic case. This is due to the anisotropic nature of the photon-molecule interaction. Both, photoabsorption as well as photoionization, create in general a non-isotropic axis distribution of the excited or ionized molecules. This anisotropy gives then rise to a non-isotropic angular distribution of the Auger electrons. However, in addition to the anisotropic axis distribution, for non-Σ vacancies, one has to consider the influence of the shape and spatial orientation of the electronic orbitals of the ionized molecules. An interesting step into this direction has been made by Zähringer *et al.* (1992). This discussion is however restricted to Auger emission from molecules with a sharp axis orientation (for example from molecules adsorbed at surfaces). The results do therefore not apply to our present case of interest where we consider the Auger emission from freely rotating molecules.

Meanwhile, several groups have become interested in angle and spin resolved molecular Auger processes and experimental results, applying photoionization techniques, have been obtained (Hemmers *et al.* 1993; Menzel *et al.* 1993; Menzel 1994). Experimental results for the Auger angular distribution using electron impact ionization have been reported for a fixed molecular axis by Zheng *et al.* (1995a) and a theoretical interpretation has been given (Zheng *et al.* 1995b). On the other hand, we investigated the angular distribution of Auger electrons emitted by diatomic molecules in the gas phase after electron impact ionization (Bonhoff *et al.* 1996).

For an analysis of experimental results for freely rotating molecules and also for numerical calculations it is necessary to develop a general theoretical framework. In this chapter we derive the general formulas for the angular distribution and spin polarization of Auger emission after electron impact ionization.

Although the general structure of this chapter is related to the treatments given by Dill *et al.* (1980) and Chandra and Chakraborty (1992), there are important differences:

i) In case of photoionization (or photofragmentation) as initial process the number of independent parameters characterizing the Auger angular distribution is limited by the dipole selection rules. No such restriction exists in case of ionization by electron impact, and additional parameters must be introduced.

ii) We will consider in some detail the coherence properties of the primary ionization process for non-Σ vacancies. The importance of this point seems to have not been discussed in the literature before. As a consequence, the possibility of a coherent superposition of Λ-doublet states results in a further increase in the number of independent parameters, even for photoionization as initial process. From a physical point of view, shape and spatial orientation of the coherently excited electronic orbitals must be considered. In order to take the additional parameters into account the previously derived theories must be further developed.

In the following section we will introduce a convenient set of parameters and derive the general expression for the angular distribution of the Auger electrons. Analogously to the atomic case, we describe the Auger decay dynamics with anisotropy parameters while the molecular ensemble M^+ will be parameterized in terms of *order parameters* which reflect the anisotropy produced during the primary ionization.

The theory will be further extended in Sect. 5.2. From the state multipoles of the emitted Auger electrons we will derive the general expressions for the Cartesian component of the spin polarization vector. The general equations of angular distribution and spin polarization can be simplified for certain polarization states of the ionizing electron beam. Some special cases of interest will be discussed in more detail.

In Sect. 5.3 we will again focus on the angular distribution introducing relative parameters and discuss its general features. Particularly, we discuss the problem of factorization within the two-step model. The importance of the coherencies produced in the primary ionization process will be considered within a simple example. As has been discussed (Bonhoff *et al.* 1996; Lohmann 1996c), we will demonstrate that a coherent ionization requires the introduction of additional parameters, even in the case of photoionization as initial process, which are therefore necessary to correctly interpret the experimental data.

In Sect. 5.4 a brief discussion of the numerical methods will be given. Results for the HF Auger spectrum and for their angular distribution parameters are also presented.

The main focus of this chapter is on diatomic molecules. However, the theory can be applied to polyatomics in a similar way. A discussion of the angular distribution of Auger electrons emitted from non-linear polyatomic molecules may be found in Lehmann *et al.* (1997) where results for the H_2O molecule are presented. A more detailed investigation on the angular distribution of Auger electrons emitted from polyatomics applying group-theoretical methods has been given by Lehmann and Blum (1997). We will adopt their formalism in Sect. 5.5 and investigate the anisotropy of the Auger decay for the case of non-degenerate and degenerate point groups. The consequences of molecular symmetries on the anisotropy parameters of angular distribution are pointed out for some examples.

5.1 Angular Distribution of Molecular Auger Electrons

5.1.1 Basic Framework

We will now derive a general theory for the angular distribution of Auger electrons for processes where the initial vacancy is created by electron impact, and define the relevant anisotropy parameters. We will focus on freely rotating diatomic molecules. Our theoretical approach is based on two assumptions:

i) As in the atomic case, we will consider the molecular Auger emission as a two-step process, that is, ionization (5.1a) and decay (5.1b) are considered as independent processes (e.g. see Mehlhorn 1990):

$$e^- + M \longrightarrow M^+ + e_i^- + e_s^- \qquad (5.1a)$$
$$\quad \longrightarrow M^{++} + e_{Auger}^- \qquad (5.1b)$$

ii) The rotation of the molecules between ionization (5.1a) and decay (5.1b) can be neglected (Dill *et al.* 1980).

The process will be described in a space-fixed coordinate system XYZ where Z coincides with the incoming electron beam axis. A molecule-fixed system xyz will be introduced where z is parallel to the internuclear axis $\mathbf{n}$. The two systems are related by the Euler angles $\alpha\beta\gamma$ as defined by Zare (1988): β and α are the polar and azimuthal angle of $\mathbf{n}$ in the laboratory system (see Fig. 5.1) and γ refers to rotations around the internuclear axis ($\mathbf{n}\|z$).

We will focus on experiments where only the Auger electron e_{Auger} is observed, but not the electrons e and e'. It will be assumed that the initial electrons and molecules M are unpolarized, and that the final spin states are not detected. Further we assume the molecules M initially in Σ-states. Thus, the initial axis distribution is isotropic, and therefore the ensemble of ionized molecules M^+ has two symmetry properties

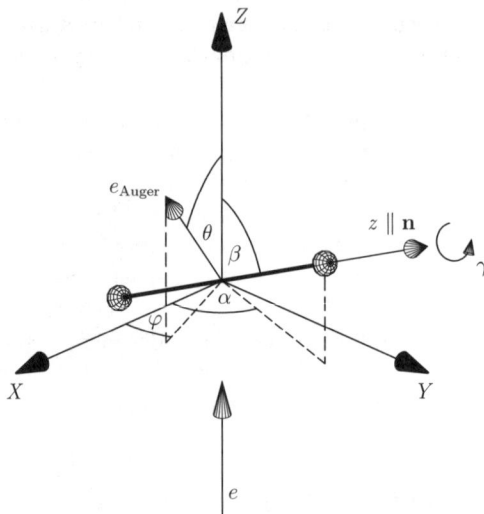

Fig. 5.1. Direction of Auger emission, molecular (**n**), and space-fixed coordinate frame (Lohmann *et al.* 1997)

a) The ensemble is axially symmetric around the incoming electron beam axis (which is $\parallel Z$).
b) For a fixed orientation of the molecular axis **n**, the plane containing **n** and the Z axis is a symmetry plane.

For diatomic molecules the angle γ can be omitted. The direction of the emitted Auger electrons will be characterized by the angles ($\varphi \theta \chi$) with respect to the XYZ system. Generally, the angle χ can be chosen as zero.

5.1.2 Characterization of the Molecular Ensemble M^+

As discussed in the introduction the axis distribution of the molecules M^+ can be expected to be anisotropic. In addition, the molecules will generally be found in electronically excited states, which usually show an anisotropic spatial distribution. The key problem of the theory is then to relate the non-isotropy of the Auger angular distribution to the anisotropy of the molecular ensemble M^+.

Immediately after the ionization the molecules M^+ will be characterized in terms of the states $|\Omega\,\mathbf{n}\rangle$ where **n** is the direction of the internuclear axis and Ω is the sum of the electronic angular momentum component Λ and spin Σ along the molecular axis, suppressing the dependence on the vibrational quantum number for brevity. Using these states as basis we can completely characterize the molecular ensemble M^+ in terms of the relevant density matrix $\hat{\rho}$ with elements $\langle \Omega'\mathbf{n}|\hat{\rho}|\Omega\,\mathbf{n}\rangle$. These elements contain the information on the process (5.1a) under the assumed experimental conditions. We normalize in such a way that the diagonal element $\langle \Omega\,\mathbf{n}|\hat{\rho}|\Omega\,\mathbf{n}\rangle$ is equal to the probability of finding a molecule M^+ in the state with quantum number Ω and axis pointing in the direction **n**. The off-diagonal elements characterize the coherence between states with different Ω, and they can be related to the shape and orientation of the molecular orbitals (see for example the discussion in Sect. 5.3 or Wöste *et al.* 1994), which will influence the Auger angular distribution (5.1b).

In order to develop an effective formalism we will now introduce a parameterization for the elements of $\hat{\rho}$ which will allow us to characterize the non-isotropy of the axes and orbital distributions in a convenient way. The density matrix elements $\langle \Omega' \mathbf{n} | \hat{\rho} | \Omega\, \mathbf{n} \rangle$ are functions of the three Euler angles $(\alpha\beta\gamma)$ and can therefore be expanded in terms of the complete set of rotation matrix elements $\mathcal{D}^{(K)}_{Q'Q}(\alpha\beta\gamma)$. Hence, we write

$$\langle \Omega' \mathbf{n} | \hat{\rho} | \Omega\, \mathbf{n} \rangle = \sum_{KQ'Q} \frac{2K+1}{8\pi^2} \langle \mathcal{D}^{(K)*}_{Q'Q}(\Omega'\Omega) \rangle \mathcal{D}^{(K)}_{Q'Q}(\alpha\beta\gamma), \qquad (5.2)$$

using the conventions of Zare (1988). The third Euler angle γ is redundant for diatomic molecules, which requires

$$Q = \Omega - \Omega'. \qquad (5.3)$$

In the following we will put γ equal to zero. Note that this choice fixes the position of the molecular x- and y-axes in such a way that x always lies in the zZ-plane. For example, x would be parallel to the negative Z-axis for $\beta = \pi/2$.

The expansion coefficients $\langle \mathcal{D}^{(K)*}_{Q'\Omega-\Omega'}(\Omega'\Omega) \rangle$ are called *order parameters* and are defined by

$$\left\langle \mathcal{D}^{(K)*}_{Q'Q}(\Omega'\Omega) \right\rangle = 2\pi \int_0^{2\pi} d\alpha \int_0^{\pi} d\beta \sin\beta \, \langle \Omega' \mathbf{n} | \hat{\rho} | \Omega\, \mathbf{n} \rangle \, \mathcal{D}^{(K)*}_{Q'Q}(\alpha\beta 0), \qquad (5.4)$$

where the asterisk denotes the complex conjugate element. Their properties have been discussed (e.g. see Lehmann *et al.* 1997), and we will illustrate their symmetries and their geometrical importance in more detail using special examples in Sect. 5.5.3. Here, we only point out that the order parameters are determined by the spin polarization of the incoming electron beam via

$$Q' = m_{s_0} - m'_{s_0}, \qquad (5.5)$$

where m_{s_0} denotes the magnetic spin quantum number of the incoming polarized electron beam, and by the transition matrix elements of the primary ionization process. For the considered case of electron impact ionization, and applying tedious but straightforward transformations (Bonhoff *et al.* 1996; Nahrup 1995), the order parameters can be expressed as

$$
\left\langle \mathcal{D}_{Q'Q}^{(K)*}(\Omega'\Omega) \right\rangle = \sum_{\substack{L_0 k_0 \ell_0 \ell_0' \\ M_0 M_0' \\ M_{S_0} M_{S_0}'}} \frac{2\pi}{|\mathbf{p}_e|} i^{\ell_0 + \ell_0'} (-1)^{1/2 - M_{S_0}' + \ell_0' - M_0' + Q'}
$$

$$
\times (2L_0 + 1)\sqrt{(2k_0 + 1)(2\ell_0 + 1)(2\ell_0' + 1)} \left\langle t_{k_0 Q'} \right\rangle
$$

$$
\times \left\langle f(\Omega', \ell_0' M_0' M_{S_0}'), f^*(\Omega, \ell_0 M_0 M_{S_0}) \right\rangle
$$

$$
\times \begin{pmatrix} \ell_0' & \ell_0 & L_0 \\ 0 & 0 & 0 \end{pmatrix} \begin{pmatrix} \ell_0' & \ell_0 & L_0 \\ M_0' & -M_0 & M_0 - M_0' \end{pmatrix}
$$

$$
\times \begin{pmatrix} k_0 & L_0 & K \\ -Q' & 0 & Q' \end{pmatrix} \begin{pmatrix} k_0 & L_0 & K \\ M_{S_0}' - M_{S_0} & M_0' - M_0 & Q \end{pmatrix}
$$

$$
\times \begin{pmatrix} 1/2 & 1/2 & k_0 \\ M_{S_0}' & -M_{S_0} & M_{S_0} - M_{S_0}' \end{pmatrix} \left\langle \Omega_0' | \hat{\rho}_{0,mol} | \Omega_0 \right\rangle. \tag{5.6}
$$

Here, $\mathbf{p}_e$ denotes the electronic momentum, ℓ_0 refers to the partial waves angular momenta of the ionizing electron, whereas the magnetic quantum numbers M_0 and M_{S_0} are the angular momentum and spin projections onto the molecular axis. The state multipoles $\left\langle t_{k_0 Q'} \right\rangle$ are related to the spin polarization state of the ionizing electron beam. The molecular ensemble is described by the elements of the density matrix $\left\langle \Omega_0' | \hat{\rho}_{0,mol} | \Omega_0 \right\rangle$. Supposing an unpolarized molecular ensemble and the molecules to be in $^1\Sigma$ states before the electronic ionization, the projection of the total angular momenta vanish, $\Omega_0 = \Omega_0' = 0$, and thus, the density matrix becomes a constant number

$$
\left\langle \Omega_0' | \hat{\rho}_{0,mol} | \Omega_0 \right\rangle = \left\langle 0 | \hat{\rho}_{0,mol} | 0 \right\rangle, \tag{5.7}
$$

independent of Ω_0. The summation over the contributing partial waves ℓ_1 and ℓ_2 and all possible angular momentum, M_1 and M_2, and spin projections, M_{S_1} and M_{S_2}, of the unobserved scattered and emitted electrons has been abbreviated as

$$
\left\langle f(\Omega', \ell_0' M_0' M_{S_0}'), f^*(\Omega, \ell_0 M_0 M_{S_0}) \right\rangle
$$

$$
= \sum_{\substack{\ell_1 M_1 M_{S_1} \\ \ell_2 M_2 M_{S_2}}} \left\langle \Omega' \mathbf{n}, \ell_1 M_1 M_{S_1} \ell_2 M_2 M_{S_2} | \mathbf{T} | 0 \mathbf{n}, \ell_0' M_0' M_{S_0}' \right\rangle
$$

$$
\times \left\langle \Omega \mathbf{n}, \ell_1 M_1 M_{S_1} \ell_2 M_2 M_{S_2} | \mathbf{T} | 0 \mathbf{n}, \ell_0 M_0 M_{S_0} \right\rangle^*, \tag{5.8}
$$

where conservation of angular momentum in the $\mathbf{T}$ matrix elements yields the selection rules

$$
\Omega + M_1 + M_{S_1} + M_2 + M_{S_2} = M_0 + M_{S_0}, \tag{5.9a}
$$

and

$$
\Omega' + M_1 + M_{S_1} + M_2 + M_{S_2} = M_0' + M_{S_0}'. \tag{5.9b}
$$

Subtraction of the latter from the first expression and using (5.3) yields

$$Q = \Omega - \Omega' = M_0 - M_0' + M_{S_0} - M_{S_0}',$$ (5.10)

and for electron impact ionization we have the general restriction

$$|Q'| \le 1,$$ (5.11)

for the order parameters.

Because of the axial symmetry of the molecular ensemble M^+ around Z the density matrix elements must be independent of α which requires that only terms with

$$Q' = 0$$ (5.12)

contribute to the expansion (5.2). Taking the symmetry conditions (5.3) and (5.12) into account we can rewrite (5.2) as

$$\langle \Omega' \mathbf{n} | \hat{\rho} | \Omega \, \mathbf{n} \rangle = \sum_K \frac{2K+1}{8\pi^2} \Big\langle \mathcal{D}_{0\Omega-\Omega'}^{(K)*}(\Omega'\Omega) \Big\rangle \mathcal{D}_{0\Omega-\Omega'}^{(K)}(\alpha\beta 0).$$ (5.13)

This equation can be further simplified by using the relations

$$\mathcal{D}_{0\Omega-\Omega'}^{(K)}(\alpha\beta 0) = d_{0\Omega-\Omega'}^{(K)}(\beta) = \sqrt{\frac{4\pi}{2K+1}} Y_{K,\Omega-\Omega'}(\beta, 0),$$ (5.14)

and

$$d_{00}^{(K)}(\beta) = P_K(\cos \beta),$$ (5.15)

where $d_{0\,\Omega'-\Omega}^{(K)}(\beta)$ are the *reduced* rotation matrix elements, $Y_{K,\Omega'-\Omega}$ the spherical harmonics, and $P_K(\cos \beta)$ denote the Legendre polynomials.

We point out that the probability density $W(\beta, \alpha)$ of finding a molecule in the ensemble M^+ with axis pointing in the direction (β, α) independently of Ω is given by

$$W(\beta, \alpha) = \sum_\Omega \langle \Omega \, \mathbf{n} | \hat{\rho} | \Omega \, \mathbf{n} \rangle = \sum_{K\Omega} \frac{2K+1}{8\pi^2} \Big\langle \mathcal{D}_{00}^{(K)*}(\Omega\Omega) \Big\rangle P_K(\cos \beta),$$ (5.16)

where we have used (5.15). So, the order parameters $\langle \mathcal{D}_{00}^{(K)*}(\Omega\Omega) \rangle$ with $\Omega = \Omega'$ characterize axis distributions of the molecules M^+. This is completely similar to the characterization of angular momentum distributions by state multipoles (Blum 1996). For an isotropic ensemble all order parameters vanish except that one with $K = 0$. If at least one of the order parameters with K even $(K \neq 0)$ is different from zero, the axis distribution is called *aligned*. If at least one parameter with K odd is non-vanishing, then the axis distribution shows a *head versus tail* distinction and we call such a distribution *oriented*. For further details see for example Michl and Thulstrup (1986) or Blum (1996).

The order parameters with $\Omega \neq \Omega'$ characterize the coherence between the corresponding states. The physical significance of these coefficients will be discussed using simple examples in Sect. 5.3.

Finally, we note that from the symmetry property (b), given in Sect. 5.1.1, the condition

$$\langle \Omega' \mathbf{n} | \hat{\rho} | \Omega \, \mathbf{n} \rangle = (-1)^{\Omega - \Omega'} \langle -\Omega' \mathbf{n} | \hat{\rho} | -\Omega \, \mathbf{n} \rangle \tag{5.17}$$

can be derived for any direction of the molecular axis $\mathbf{n}$. Equation (5.17) shows in particular that states with Ω and $-\Omega$ are equally populated. Equations (5.4) and (5.17) yield the symmetry condition

$$\left\langle \mathcal{D}^{(K)*}_{0\Omega - \Omega'}(\Omega' \Omega) \right\rangle = \left\langle \mathcal{D}^{(K)*}_{0\Omega - \Omega'}(-\Omega' - \Omega) \right\rangle \tag{5.18}$$

for the order parameters.

5.1.3 Angular Distribution of Molecular Auger Electrons

We will now consider the Auger decay (5.1b). The angular distribution $I(\theta, \varphi)$ of the Auger electrons is given in general by the expression

$$I(\theta, \varphi) = \sum_{\Omega_1 \Omega' \Omega m_s} \int d\mathbf{n} \, \langle \Omega_1 \, \mathbf{n} \, \mathbf{p} m_s | \mathbf{T} | \Omega' \mathbf{n} \rangle$$
$$\times \langle \Omega' \mathbf{n} | \hat{\rho} | \Omega \, \mathbf{n} \rangle \langle \Omega_1 \, \mathbf{n} \, \mathbf{p} m_s | \mathbf{T} | \Omega \, \mathbf{n} \rangle^* \tag{5.19}$$

averaging over all unobserved quantities and using the completeness relation. Here $\mathbf{p}$ is the momentum of the Auger electron (with polar angle θ and azimuth φ in the space-fixed system, see Fig. 5.1) and m_s its spin component with respect to Z. $\mathbf{T}$ is the relevant transition operator, and the elements $\langle \Omega_1 \mathbf{n} \, \mathbf{p} m_s | \mathbf{T} | \Omega \, \mathbf{n} \rangle$ are the amplitudes for the indicated Auger transition. In first order Born approximation the operator $\mathbf{T}$ is replaced by the Coulomb operator $\mathbf{V}$.

It is worth noting that for an incoherent superposition of states $|\Omega \mathbf{n}\rangle$ (5.19) reduces to the expression

$$I(\theta, \varphi) = \sum_{\Omega} \int d\mathbf{n} \, \langle \Omega \, \mathbf{n} | \hat{\rho} | \Omega \, \mathbf{n} \rangle \sum_{\Omega_1 m_s} |\langle \Omega_1 \, \mathbf{n} \, \mathbf{p} m_s | \mathbf{T} | \Omega \, \mathbf{n} \rangle|^2. \tag{5.20}$$

There is a simple interpretation for this expression. The last sum in (5.20) can be interpreted as angular distribution of Auger electrons emitted from molecules with sharp $\mathbf{n}$ and sharp Ω. Multiplying with the probability $\langle \Omega \, \mathbf{n} | \hat{\rho} | \Omega \, \mathbf{n} \rangle$ of finding a molecule in this state and integrating over all angles gives the total function $I(\theta, \varphi)$. Equation (5.20) seems to be intuitively obvious and has often been used as a basic equation of departure. However, we stress the point that coherence between the basic molecular states is not included, and therefore (5.20) applies only to special cases. Examples are given in Sect. 5.3.

Returning now to (5.19) we perform a partial wave expansion of the transition amplitudes. Expansion of $|\mathbf{p}\rangle$ in partial waves with respect to the space-fixed system yields

$$|\mathbf{p}\rangle = \frac{1}{\sqrt{k}} \sum_{\ell m} i^{\ell} Y_{\ell m}^{*}(\theta\varphi) |\ell m\rangle. \qquad (5.21)$$

Transformation of orbital and spin states to the molecular system xyz (with $\mathbf{n}$ as z-axis) gives

$$|\mathbf{p}\rangle = \frac{1}{\sqrt{k}} \sum_{\ell m M} i^{\ell} Y_{\ell m}^{*}(\theta\varphi) \, \mathcal{D}_{m\,M}^{(\ell)}(\alpha\beta 0) |\ell M\rangle, \qquad (5.22a)$$

and

$$|m_s\rangle = \sum_{M_s} \mathcal{D}_{m_s\,M_s}^{(1/2)}(\alpha\beta 0) |M_s\rangle. \qquad (5.22b)$$

Here, M and M_s are the components of orbital angular momentum and spin, respectively along $\mathbf{n}$.

Substitution of (5.13) and (5.22) into (5.19) allows to perform the integrals. Making use of the relations (B.12) and (B.13a) we obtain

$$\sum_{m_s} \mathcal{D}_{m_s\,M_s'}^{(1/2)}(\alpha\beta\gamma) \, \mathcal{D}_{m_s\,M_s}^{(1/2)}(\alpha\beta\gamma)^{*} = \delta_{M_s' M_s}. \qquad (5.23a)$$

Applying (B.23a) and using (A.9) gives

$$\int d\alpha \int d\beta \sin\beta \, \mathcal{D}_{m'\,M'}^{(\ell')}(\alpha\beta\gamma) \, \mathcal{D}_{0\,Q}^{(K)}(\alpha\beta\gamma) \, \mathcal{D}_{m\,M}^{(\ell)}(\alpha\beta\gamma)^{*}$$

$$= (-1)^{m-M} \, 4\pi \begin{pmatrix} \ell' & \ell & K \\ m' & -m & 0 \end{pmatrix} \begin{pmatrix} \ell' & \ell & K \\ M' & -M & Q \end{pmatrix}, \qquad (5.23b)$$

and applying (B.18a) while using (B.28) eventually yields

$$\sum_{m} \begin{pmatrix} \ell' & \ell & K \\ m & -m & 0 \end{pmatrix} \mathcal{D}_{-m\,0}^{(\ell')}(\varphi\theta 0) \, \mathcal{D}_{m\,0}^{(\ell)}(\varphi\theta 0) = \begin{pmatrix} \ell' & \ell & K \\ 0 & 0 & 0 \end{pmatrix} P_K(\cos\theta). \qquad (5.23c)$$

Using (5.23a)–(5.23c) for reducing (5.19) we finally obtain

$$I(\theta,\varphi) = \sum_{K\Omega'\Omega} \left\langle \mathcal{D}_{0\,\Omega-\Omega'}^{(K)*}(\Omega'\Omega) \right\rangle A_K(\Omega',\Omega) \, P_K(\cos\theta), \qquad (5.24)$$

for the angular distribution of the Auger electrons. From its structure (5.24) is similar to (2.131) derived for the atomic case. Though, the dependencies of the anisotropy coefficients and the order parameters are different. Here, the anisotropy coefficients $A_K(\Omega',\Omega)$ are defined by the expression

$$A_K(\Omega',\Omega) = \frac{2K+1}{2\pi} \sum_{\substack{\Omega_1 M M' M_s \\ \ell\ell'}} i^{\ell+\ell'}(-1)^{M+\ell'} \frac{\sqrt{(2\ell+1)(2\ell'+1)}}{4\pi k}$$

$$\times \begin{pmatrix} \ell' & \ell & K \\ M' & -M & \Omega - \Omega' \end{pmatrix} \begin{pmatrix} \ell' & \ell & K \\ 0 & 0 & 0 \end{pmatrix}$$

$$\times \langle \Omega_1 \, \ell' M' M_s | \mathbf{T} | \Omega' \rangle \langle \Omega_1 \, \ell M M_s | \mathbf{T} | \Omega \rangle^{*}. \qquad (5.25)$$

Reflection invariance in the molecular xz-plane requires

$$A_K(\Omega', \Omega) = A_K(-\Omega', -\Omega). \tag{5.26}$$

5.2 Spin Polarization of Molecular Auger Electrons

5.2.1 General Expressions

Generally, applying statistical tensor methods, the state multipoles $\langle t_{kq} \rangle$ describing the emitted Auger electrons can be written as

$$I \langle t_{kq} \rangle = \sum_{\substack{K Q' \\ \Omega' \Omega}} \langle \mathcal{D}^{(K)*}_{Q' \Omega - \Omega'}(\Omega' \Omega) \rangle A_{Kkq}(\Omega', \Omega)\, \mathcal{D}^{(K)}_{Q'q}(\varphi \theta \chi), \tag{5.27}$$

where Ω denotes the total magnetic angular momentum component of the molecules (see Sect. 5.1.2).

The intensity I of the Auger electrons and the Cartesian components of the spin polarization vector $\mathbf{P}$, with respect to the Auger system, can be obtained from the state multipoles by applying the relations

$$\frac{1}{\sqrt{2}} = \langle t_{00} \rangle, \tag{5.28a}$$

$$P_x = \left(\langle t_{1-1} \rangle - \langle t_{11} \rangle \right), \tag{5.28b}$$

$$P_y = i \left(\langle t_{1-1} \rangle + \langle t_{11} \rangle \right), \tag{5.28c}$$

$$P_z = \sqrt{2} \langle t_{10} \rangle, \tag{5.28d}$$

and inserting the relevant state multipoles $\langle t_{kq} \rangle$ in (5.27).

The anisotropy parameters $A_{Kkq}(\Omega', \Omega)$ are defined in analogy to the atomic case and characterize the dynamics and geometry of the Auger decay

$$
\begin{aligned}
A_{Kkq}(\Omega', \Omega) = \frac{1}{4\pi k_0} (2K+1) \sum_{\substack{\Omega_f L \ell \ell' \\ m m' m_s m'_s}} & (-1)^{m' + m'_s - q - 1/2}\, i^{\ell - \ell'} \\
\times\ (2L+1) & \sqrt{(2\ell+1)(2\ell'+1)(2k+1)} \\
\times \begin{pmatrix} \frac{1}{2} & \frac{1}{2} & k \\ m'_s & -m_s & m_s - m'_s \end{pmatrix} & \begin{pmatrix} \ell' & \ell & L \\ m' & -m & m - m' \end{pmatrix} \begin{pmatrix} \ell' & \ell & L \\ 0 & 0 & 0 \end{pmatrix} \\
\times \begin{pmatrix} k & L & K \\ -q & 0 & q \end{pmatrix} & \begin{pmatrix} k & L & K \\ m'_s - m_s & m' - m & \Omega - \Omega' \end{pmatrix} \\
\times\ \langle \Omega_f\, \ell' m' m'_s | \mathbf{T} | \Omega' \rangle & \langle \Omega_f\, \ell m m_s | \mathbf{T} | \Omega \rangle^*.
\end{aligned}
\tag{5.29}
$$

The partial wave quantum numbers ℓ, m of the Auger electron as well as its spin quantum number m_s refer to the molecular system.

From the definition of the anisotropy parameters we obtain the symmetry relation

$$A_{Kkq}(\Omega', \Omega)^* = (-1)^{\Omega - \Omega'} A_{Kk-q}(\Omega', \Omega) . \tag{5.30}$$

Using the fact that the transition operator $\mathbf{T}$, describing the Auger decay, is invariant under reflection in the molecular xz-plane, we obtain an additional symmetry

$$A_{Kk-q}(-\Omega', -\Omega) = (-1)^k A_{Kkq}(\Omega', \Omega) . \tag{5.31}$$

The angular dependency of intensity and spin polarization is expressed in terms of rotation matrix elements $\mathcal{D}^{(K)}_{Q'q}(\varphi\theta\chi)$. The quantum numbers Q' and q are connected with the degree of spin polarization of the incoming beam and the Auger electrons, respectively.

Each rotation matrix element has the same rank K and quantum number Q' as its associated order parameter. This shows, that the anisotropy of intensity and spin polarization is a direct consequence of the anisotropy in the molecular ensemble M^+, and the anisotropy parameters $A_{Kkq}(\Omega', \Omega)$ can be interpreted as the different weight factors for the state multipoles $\langle t_{kq} \rangle$; e.g. the intensity I of the Auger electrons is weighted by the anisotropy parameters A_{K00}. Note, that, besides the usual angular momentum coupling rules, the sum over K is not restricted.

Inserting (5.29) into (5.28) we obtain expressions for the Cartesian components of the spin polarization vector as

$$IP_x = \sum_{\substack{KQ' \\ \Omega'\Omega}} \left\langle \mathcal{D}^{(K)*}_{Q'\Omega-\Omega'}(\Omega'\Omega) \right\rangle \left(A_{K1-1}(\Omega', \Omega) \, \mathcal{D}^{(K)}_{Q'-1}(\varphi\theta\chi) \right.$$
$$\left. - A_{K11}(\Omega', \Omega) \, \mathcal{D}^{(K)}_{Q'1}(\varphi\theta\chi) \right), \tag{5.32}$$

$$IP_y = \mathrm{i} \sum_{\substack{KQ' \\ \Omega'\Omega}} \left\langle \mathcal{D}^{(K)*}_{Q'\Omega-\Omega'}(\Omega', \Omega) \right\rangle \left(A_{K1-1}(\Omega', \Omega) \, \mathcal{D}^{(K)}_{Q'-1}(\varphi\theta\chi) \right.$$
$$\left. + A_{K11}(\Omega', \Omega) \, \mathcal{D}^{(K)}_{Q'1}(\varphi\theta\chi) \right), \tag{5.33}$$

and

$$IP_z = \sqrt{2} \sum_{\substack{KQ' \\ \Omega'\Omega}} \left\langle \mathcal{D}^{(K)*}_{Q'\Omega-\Omega'}(\Omega'\Omega) \right\rangle A_{K10}(\Omega', \Omega) \, \mathcal{D}^{(K)}_{Q'0}(\varphi\theta\chi) . \tag{5.34}$$

It is of interest to note the identity

$$P_x(\varphi, \theta, \chi + \pi/2) = P_y(\varphi, \theta, \chi), \tag{5.35}$$

which can be derived from the symmetry relations of the rotation matrices.

The general equations of angular distribution and spin polarization can be simplified for certain polarization states of the incoming ionizing electron beam. In the following we will consider some special cases.

5.2.2 Unpolarized Electron Beam

Assuming an unpolarized electron beam for the primary ionization, from (5.27) we have $Q' = 0$. The angular distribution becomes independent of the angle φ because of the axially symmetric distribution of the molecular ensemble M^+. Thus, the intensity may be written as

$$I(\theta) = \sum_{K\Omega'\Omega} \left\langle \mathcal{D}_{0\Omega-\Omega'}^{(K)*}(\Omega'\Omega) \right\rangle A_K(\Omega', \Omega) \, P_K(\cos\theta). \qquad (5.36)$$

This is the same expression as has been derived in Sect. 5.1 (see also Bonhoff *et al.* 1996) where, using (5.29), the anisotropy parameters have been abbreviated as

$$A_K(\Omega', \Omega) = \sqrt{2} \, A_{K00}(\Omega', \Omega). \qquad (5.37)$$

The angular dependence is given by the Legendre polynomials $P_K(\cos\theta)$. The factors describing the primary ionization and the Auger transition are clearly separated. The coherencies produced in the primary process are taken implicitly into account by the order parameters. They are missing in other theories (Chandra and Chakraborty 1992; Dill *et al.* 1980).

The system is invariant under reflection in the reaction plane defined by the incoming beam axis and the direction of Auger emission. Thus, the polarization vector $\mathbf{P}$ must be perpendicular to the reaction plane. Choosing the coordinate frame of the Auger system in such a way that $\chi = 0$, the x- and z-component of the spin polarization vector vanish

$$P_x = P_z = 0. \qquad (5.38)$$

I.e., only the y-component can be expected to be non-zero. It may be expressed as

$$I P_y(\theta) = 2i \sum_{K\Omega\Omega'} \left\langle \mathcal{D}_{0\Omega-\Omega'}^{(K)*}(\Omega'\Omega) \right\rangle A_{K11}(\Omega', \Omega) \, d_{01}^{(K)}(\theta), \qquad (5.39)$$

where $d_{Q'q}^{(K)}(\theta)$ denotes the reduced rotation matrices.

5.2.3 Transversely Polarized Electron Beam

Using a transversely polarized electron beam, the Auger intensity shows a dependence on the azimuth φ which is caused by the spin-orbit interaction

$$I(\theta, \varphi) = I(\theta)_{unpol} - B(\theta) \sin\varphi. \qquad (5.40)$$

Here, I_{unpol} is given by (5.36) and the function $B(\theta)$ is defined as

$$B(\theta) = 2i \sum_{K\Omega\Omega'} \left\langle \mathcal{D}_{1\Omega-\Omega'}^{(K)*}(\Omega'\Omega) \right\rangle A_K(\Omega', \Omega) \, d_{10}^{(K)}(\theta). \qquad (5.41)$$

Thus, measuring the intensity of the Auger electrons in the reaction plane, we obtain the same result as for the unpolarized case.

5.2.4 Neglect of Spin–Orbit Interaction

$B(\theta)$ can be expected as small for light molecules since the spin-orbit interaction is weak. Without proof we note that $B(\theta) = 0$ for a vanishing spin-orbit interaction. Thus, the intensity of the Auger electrons becomes independent on the polarization of the electron beam,

$$I(\theta, \varphi) = I(\theta)_{unpol} \,. \tag{5.42}$$

Further, using an unpolarized electron beam and neglecting spin-orbit interaction, we always end up with unpolarized Auger electrons,

$$\mathbf{P} = 0. \tag{5.43}$$

5.3 Application and Examples

5.3.1 General Features of the Auger Angular Distribution

Equations (5.24) and (5.25) are our main results. The angular distribution of the Auger electrons is expressed in terms of the Legendre polynomials $P_K(\cos\theta)$. The independence on φ reflects the axial symmetry of the problem. The factors describing the primary ionization (5.1a) and the factors describing the Auger decay (5.1b) are clearly separated. The anisotropy of the molecules immediately after the ionization is characterized by the relevant order parameters $\left\langle \mathcal{D}^{(K)*}_{0,\,\Omega-\Omega'} \right\rangle$ which contain all information on the geometry and dynamics of the ionization (5.1a). The information on the dynamics of the Auger decay is contained in the anisotropy coefficients A_K, which are independent of the molecular orientation since all quantum numbers in the transition matrix elements refer to the molecular axis $\mathbf{n}$. The numerical calculation of $I(\theta)$ requires therefore

i)   the determination of the relevant order parameters,
ii) the calculation of A_K for one particular molecular orientation.

It should be noted that any Legendre polynomial P_K is related to the order parameters with the same rank K. Hence, any order parameter of rank K gives rise to a characteristic angular distribution P_K of the Auger electrons. In this sense one might say that the anisotropy of the Auger angular distribution reflects the anisotropy of the molecular axis and orbital distribution before the Auger decay. The parameters A_K can be considered as weight factors determining the amount by which the Auger emission is influenced by the corresponding order parameters.

5.3.2 The Problem of Factorization

It is useful to re-write (5.24) in terms of relative parameters. Integrating (5.24) over the solid angle $\Omega = (\theta, \phi)$ and applying the orthogonality relation of the Legendre polynomials, the total intensity I_0 is given by

$$I_0 = \int d\Omega\, I(\theta, \varphi) = 4\pi \sum_{\Omega} \left\langle \mathcal{D}^{(0)*}_{00}(\Omega\Omega) \right\rangle A_0(\Omega\Omega). \tag{5.44}$$

The angular distribution can then be rewritten in the following form

$$I(\theta, \varphi) = \frac{I_0}{4\pi} \left(1 + \sum_{K>0} a_K P_K(\cos\theta) \right), \tag{5.45}$$

where the angular distribution parameters a_K are given by

$$a_K = \left(\sum_{\Omega} \left\langle \mathcal{D}_{00}^{(0)*}(\Omega, \Omega) \right\rangle A_0(\Omega\Omega) \right)^{-1}$$
$$\times \sum_{\Omega'\Omega} \left\langle \mathcal{D}_{0\Omega-\Omega'}^{(K)*}(\Omega', \Omega) \right\rangle A_K(\Omega', \Omega). \tag{5.46}$$

Equations (5.45) and (5.46) are generally valid. It is important to note, that the angular distribution parameters a_K can in general not be factorized into a product of two factors, the first one describing the excitation and the second one the Auger decay. This is due to two reasons. The first one is related to the coherent excitation of states $|\Omega\rangle$. This is similar to the atomic case of an unresolved fine structure state, see Sect. 2.7.4, which has been discussed by Mehlhorn and Taulbjerg (1980) and more explicitly by Kabachnik *et al.* (1994). Secondly, even for the case of an incoherent excitation process the summation in (5.46) includes generally more than one term and the angular distribution can therefore not be factorized.

In the case of a primary photoionization with unpolarized photons, only the term with $K = 2$ contributes to (5.45) because of the dipole selection rules and the Auger angular distribution simplifies as follows

$$I(\theta, \varphi) = \frac{I_0}{4\pi} \left(1 + a_2 P_2(\cos\theta) \right). \tag{5.47}$$

This expression is similar to equation (7) given by Dill *et al.* (1980), but even in the case of photoionization it is in general not possible to factorize the angular distribution parameter a_2, as discussed above.

5.3.3 Significance of the Coherence Terms

Our results show in particular the importance of the coherence terms. In order to get some physical insight and to point out several novel features of our parameterization, we will consider some simple examples. In the remainder of this section we will write $\Omega = \Lambda + \Sigma$ and assume that total spin and total orbital angular momentum will be separately conserved during the Auger decay. From (5.3) follows then

$$Q = \Omega - \Omega' = \Lambda - \Lambda', \tag{5.48}$$

since the final spin states are not detected. The corresponding order parameters $\left\langle \mathcal{D}_{0\Lambda-\Lambda'}^{(K)*} \right\rangle$ characterize the molecules M^+.

Let us assume that states with different $|\Lambda|$ are energetically separated so that these states overlap incoherently and the corresponding off-diagonal elements of $\hat{\rho}$

vanish. States with $\Lambda = \pm|\Lambda|$ are very nearly degenerate and will be coherently excited. Hence, if the initial vacancy is a Σ-state, then only terms with $Q = 0$ contribute to (5.24) in the relevant energy region. For Π-states we have $Q = 0, \pm 2$, for Δ-states $Q = 0, \pm 4$ and so on. Equation (5.24) simplifies then accordingly.

In the remainder of this section we will consider spinless molecules M^+ ($\Omega = \Lambda$) in order to point out as simply as possible the importance of the coherence parameters. Instead of using the states $|\Lambda \, \mathbf{n}\rangle$ as a basis one can also use the *directed* electronic orbitals as a basis set. For example, for $|\Lambda| = 1$ we have (Wöste *et al.* 1994)

$$|\Pi_x \, \mathbf{n}\rangle = -\frac{1}{\sqrt{2}} \left(|\Lambda = 1, \mathbf{n}\rangle - |\Lambda = -1, \mathbf{n}\rangle \right), \qquad (5.49a)$$

$$|\Pi_y \, \mathbf{n}\rangle = \frac{i}{\sqrt{2}} \left(|\Lambda = 1, \mathbf{n}\rangle + |\Lambda = -1, \mathbf{n}\rangle \right). \qquad (5.49b)$$

The corresponding electronic lobes are directed along the molecular x-axis for the Π_x-state and along the y-axis for the Π_y-state. From the symmetry condition (5.17) follows that the interference terms between the two states (5.49) vanish

$$\langle \Pi_x \, \mathbf{n} | \hat{\rho} | \Pi_y \, \mathbf{n} \rangle = 0 \qquad (5.50)$$

and only the diagonal terms of $\hat{\rho}$ are different from zero.

Hence, in this basis (5.20) applies and the angular distribution of the Auger electrons can be written in the form

$$I(\theta, \varphi) = \sum_{i=x,y} \int d\mathbf{n} \, \langle \Pi_i \, \mathbf{n} | \hat{\rho} | \Pi_i \, \mathbf{n} \rangle \sum_{\Omega_1 m_s} \left| \langle \Omega_1 \, \mathbf{n} \, \mathbf{p} \, m_s | \mathbf{T} | \Pi_i \, \mathbf{n} \rangle \right|^2. \qquad (5.51)$$

It is important to note that the coherence terms do not vanish if the basis $|\Lambda \, \mathbf{n}\rangle$ is used – which is the most convenient one for taking selection rules into account –. Equation (5.20) does then not apply and (5.19) must be used. It follows that, before one uses relation (5.19) or (5.20), one has to check the coherencies, and one has always to state explicitly the choice of the basis.

Let us now consider the physical significance of the order parameters. As shown by (5.16) the coefficients $\langle \mathcal{D}_{00}^{(K)*} \rangle$ characterize the distribution of the molecular axes. In order to see the physical importance of the interference terms $\langle \mathcal{D}_{0\Omega-\Omega'}^{(K)*} \rangle$ we will specialize to the case that only terms with $K = 0$ and $K = 2$ contribute to (5.24). This is for instance realized for homo-nuclear diatomic molecules if the initial vacancy is created via photoionization.

Assuming an initial vacancy in a pure Π_x-orbital, we obtain, for instance, for the angular distribution of the Auger electrons

$$I(\theta) = 2\langle \mathcal{D}_{00}^{(0)*}(1,1) \rangle A_0(1,1) + 2\left(\langle \mathcal{D}_{00}^{(2)*}(1,1) \rangle A_2(1,1) \right.$$

$$\left. + \langle \mathcal{D}_{02}^{(2)*}(-1,1) \rangle A_2(-1,1) \right) P_2(\cos\theta). \qquad (5.52)$$

For the opposite case of a pure Π_y-orbital we obtain the same expression except that $\langle \mathcal{D}_{02}^{(2)*} \rangle$ changes its sign. The two orbitals $|\Pi_x\rangle$ and $|\Pi_y\rangle$ differ only in their direction in space. Thus, the order parameters $\langle \mathcal{D}_{02}^{(2)*} \rangle$ can be seen as a direct measure for the influence of the spatial orbital orientation on the Auger emission.

5.4 Numerical Values for the Anisotropy Parameters

5.4.1 Numerical Methods

The calculation of the **T** matrix elements is carried out by means of the Greens operator formalism which is an expansion of the computation of Auger spectra by Schimmelpfennig *et al.* (1995). Recently, it has been extended for the calculation of molecular anisotropy parameters by Bonhoff *et al.* (1997). In the following we will give a brief outline.

The molecular wave functions are optimized in a SCF-CI calculation. In contrast to the method described by Schimmelpfennig *et al.* (1995) a one-center expansion of the basis set is used to obtain the phases of the transition matrix elements, which are necessary for the calculation of the anisotropy parameters.

Schimmelpfennig and co-workers have proposed and applied a method to calculate angular integrated Auger spectra for molecules which is based on the representation of the Greens operator for the free particle in a basis set of Gauss type orbitals. Using this ansatz, the information on the scattering waves and their scattering phases is hidden within the Greens operator and cannot be extracted (Schimmelpfennig 1994; Schimmelpfennig *et al.* 1992, 1995).

Our aim is the calculation of angle and spin resolved spectra. Thus, we change the strategy and calculate the **T**-matrix in a one-center expansion for the basis set. By using this approach, we obtain the required scattering phases which are needed to describe interference effects in the angular resolved spectra correctly (e.g. see the previous discussion or Bonhoff *et al.* 1996; Lohmann 1996c).

For this, the electronic Hamiltonian **H** is split into two parts $\mathbf{H}_0$ and **V**, where $\mathbf{H}_0$ includes the Hamiltonian of the remaining molecular fragment and the kinetic energy of the Auger electron. The potential **V** gives the interaction of the Auger electron with the molecular fragment $|\chi\rangle$, including the long-range Coulomb potential. Starting with an energy-normalized partial wave $|f_{\ell m}^0\rangle$ for a free particle we obtain the scattered wave by applying an effective Lippmann–Schwinger equation

$$|f_{\ell m}^+\rangle = |f_{\ell m}^0\rangle + \mathbf{G}_0\,\mathbf{V}|f_{\ell m}^+\rangle = |f_{\ell m}^0\rangle + \mathbf{G}\,\mathbf{V}|f_{lm}^0\rangle, \tag{5.53}$$

where **G** and $\mathbf{G}_0$ are the Greens operators corresponding to **H** and $\mathbf{H}_0$, respectively.

$$\mathbf{G} = \lim_{\epsilon \to 0_+} \frac{1}{E + \mathbf{H} + i\epsilon} \quad \text{and} \quad \mathbf{G}_0 = \lim_{\epsilon \to 0_+} \frac{1}{E + \mathbf{H}_0 + i\epsilon}. \tag{5.54}$$

The **T** matrix elements are given by

$$T_{\ell m} = \langle \phi_i | \mathbf{T} | \mathcal{A}(\chi\, f_{\ell m}^0) \rangle = \langle \phi_i | \mathbf{V}_{decay} | \mathcal{A}(\chi\, f_{\ell m}^+) \rangle, \tag{5.55}$$

where the right-hand side of the last term denotes the anti-symmetrized product of the scattered partial wave $\left| f_{\ell m}^+ \right\rangle$ and the $n-1$ electron function $\left| \chi \right\rangle$ of the residual molecular ion. In order to avoid confusion with the formerly defined potential $\mathbf{V}$ of the Lippmann–Schwinger equation, we have chosen the notation $\mathbf{V}_{decay}$ for the interaction with the autoionizing initial state $\left| \phi_i \right\rangle$.

All the required operators and wavefunctions can be obtained from basis set expansions in Gauss type orbitals, which has been proven to work efficiently (Schimmelpfennig *et al.* 1995). However, for multi-center molecular basis sets we are not able to assign the quantum numbers ℓ and m correctly. In order to describe the molecular effects correctly, we derive all those operators within a molecular basis set and transfer them afterwards to a one center expansion, which is built up by orthonormal solid harmonics. These solid harmonics have real angular parts given by

$$Y_{\ell m}^+ = \frac{1}{\sqrt{2}} \left(Y_{\ell m} + (-1)^m Y_{\ell -m} \right), \qquad (5.56a)$$

and

$$Y_{\ell m}^- = \frac{1}{i\sqrt{2}} \left(Y_{\ell m} - (-1)^m Y_{\ell -m} \right), \qquad (5.56b)$$

where $m \geq 0$. For the transformation of the basis set representations for the different operators and wavefunctions we need the overlap between the one-center basis functions and the molecular multi-center basis set. In order to simplify the calculation of this overlap, the one-center basis is expanded into Cartesian Gauss type functions. The expansion coefficients can be obtained analytically and are given in the literature (Harnung and Schäffer 1972). As has been discussed by Schimmelpfennig *et al.* (1995), we have to use an effective free Greens operator which takes account of the orbitals occupied in the remaining molecular fragment. These contributions are still calculated within the molecular basis set, but the one-particle part is now calculated in the one-center basis set. This yields the possibility of including optimized basis functions for the higher angular momenta, which cannot be handled in the molecular basis set efficiently.

The energy-normalized partial waves $\left| f_{\ell m}^0 \right\rangle$ are introduced in our calculations as the corresponding solid harmonics, from which they can be re-obtained easily, as discussed above. The expansion in the one-center basis can be obtained by different approaches. An easy way is the diagonalization of the imaginary part of the free one-particle Greens operator and ensuring the correct sign of the derivative of the wavefunctions at the origin. Another, more direct approach, is to fit the analytical solutions, i.e. spherical Bessel functions, by a set of Gauss type orbitals within the region of interaction.

5.4.2 Calculations for HF

As an example of diatomic molecules we consider HF. Choosing the $^2\Sigma^+ (1\sigma^{-1})$ as initial state of the molecules M^+ we calculated the energies and intensities for all substantial Auger transitions. Here, the orbitals optimized in a SCF calculation

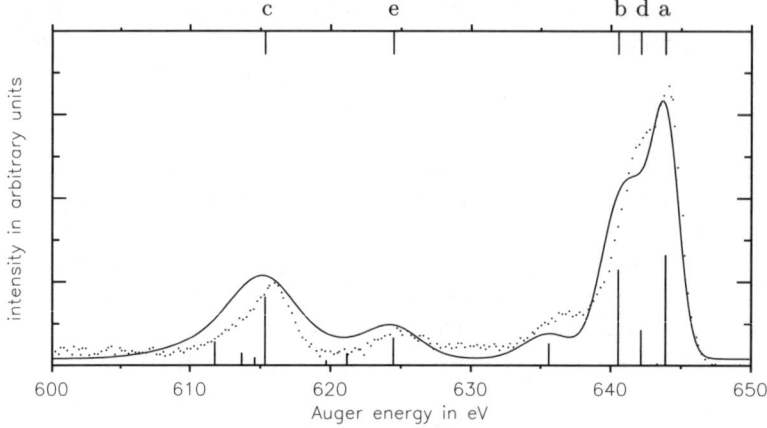

Fig. 5.2. Auger spectrum of HF; our calculations: —; experiment by Shaw *et al.* (1975): · · ·; labels (**a–e**) refer to the calculated anisotropy parameters (see text). Data from Lohmann *et al.* (1997)

have been used for the $^2\Sigma^+$ $(1\sigma^{-1})$ state in order to describe all involved states with correlation-interaction (CI) wavefunctions.

For the multi-center basis set an uncontracted (9s, 5p) basis of primitive Gaussians (Huzinaga 1965) has been employed, where we added four s, four p, and eight d functions, optimized to model Bessel functions in the relevant energy region. For the hydrogen atom we used a (7s, 1p)/[4s, 1p] Huzinaga basis set at the equilibrium geometry with an internuclear distance $R = 1.733$ a.u. The dimensions of the CI spaces in the different CSF are about 20000.

In order to perform the partial wave expansion in a suitable way, a one-center basis expansion has been introduced, where we employed the (9s, 5p) basis for the fluorine atom and added (8s, 10p, 14d, 7f, 6g, 6h, 5i, 5k, 5l, 5m) Gaussian functions up to $\ell = 9$ to model Bessel functions. The continuum wavefunctions are then obtained by solving the effective Lippmann–Schwinger equation (5.53). The resulting spectrum is shown in Fig. 5.2.

The anisotropy parameters were calculated for the final states with the largest total intensity, i.e. $^1\Delta\,(1\pi^{-2})$ (a), $^1\Pi(3\sigma^{-1}1\pi^{-1})$ (b), $^1\Pi(2\sigma^{-1}1\pi^{-1})$ (c), $^1\Sigma^+(1\pi^{-2})$ (d), $^3\Pi(2\sigma^{-1}1\pi^{-1})$ (e). They are given in Table 5.1. As can be seen large values for the anisotropy parameters can be expected for all Auger transitions considered. In contrast to the atomic case of angular distribution, where only even-rank anisotropy parameters α_K can occur (Berezhko and Kabachnik 1977), odd-rank anisotropy parameters can also be non-zero for diatomic molecules. Generally, the magnitude of the anisotropy parameters $A_K(\Omega', \Omega)$ decreases with increasing rank. A detailed discussion of the results for HF has been published by Bonhoff *et al.* (1997).

Table 5.1. Anisotropy parameter $A_K(\frac{1}{2}, \frac{1}{2})$ in 10^{-4} a.u. Data from Lohmann *et al.* (1997)

K	(a) $^1\Delta$	(b) $^1\Pi$	(c) $^1\Pi$	(d) $^1\Sigma^+$	(e) $^3\Pi$
0	1.3252	1.1433	0.8147	0.4264	0.3183
1	0.0000	−0.3622	0.2142	−0.2113	−0.0084
2	−1.8932	0.7525	−0.7846	1.2156	−0.3182
3	0.0000	0.3625	−0.2139	−0.2111	0.0084
4	0.5679	−1.8953	−0.0302	0.8686	−0.0001
5	0.0000	0.0001	−0.0003	0.0078	0.0000
6	0.0000	−0.0004	0.0001	−0.0015	0.0001
7	0.0000	−0.0004	0.0000	0.0064	0.0000
8	0.0000	0.0000	0.0000	0.0000	0.0000

5.5 Angular Distribution of Auger Electrons from Polyatomic Molecules

5.5.1 General Remarks

There have been only few studies performed on polyatomic molecular Auger processes. General equations for the Auger decay of non-linear molecules, following photoabsorption or photoionization, have been derived by Chandra (1989) and Chandra and Chakraborty (1992) where the spin polarization of the ejected electrons has also been considered.

In the following we will develop a general theory for the angular distribution of Auger electrons emitted by non-linear molecules in the gas phase after ionization by electron impact. The general relations will be derived which can be used as a basis for any future numerical calculation and also for an analysis of experimental data. Although the general structure is similar to the discussion in the preceding Sects. and previous treatments (Bonhoff *et al.* 1996; Chandra 1989; Chandra and Chakraborty 1992), there are important differences:

1. In the case of photoionization being the initial process the number of independent parameters is limited by the dipole selection rules. No such restrictions exist in the case of ionization by electron impact and additional parameters are necessary for a correct description. This requirement is identical to the case of diatomic molecules.
2. Analogously to the diatomic case, we introduce a parameterization of the molecular ensemble M^+ in terms of *order parameters*, which characterize the axis distribution of the molecules and reflect the anisotropy produced in the primary ionization. Furthermore, this set of parameters allows us to present the final result in a compact and transparent form.
3. Of particular interest is the case of molecules with higher symmetry. The existence of a threefold or higher rotation axis gives rise to electronic degeneracies.

Degenerate states will be coherently excited by electron impact. The correct description of these coherencies requires the introduction of additional parameters necessary for any future numerical calculation of the process. A detailed discussion of the importance of these parameters has been given in Sect. 5.3 and by Bonhoff *et al.* (1996) for diatomic molecules. However, for nonlinear molecules, these coherence parameters have not been taken properly into account in previous formulations and we will give a detailed treatment here.

In the following section we will introduce the basic notation and discuss the approximations inherent in our treatment. We will then proceed to a characterization of the molecular ensemble after the primary ionization. Because there are important differences between molecules belonging to non-degenerate or degenerate point groups, we will treat both cases separately. In Sect. 5.5.3 we will deal with the non-degenerate case. A set of *order parameters* will be introduced which allows a convenient characterization of the anisotropic axis distribution of the molecules after the primary ionization. The symmetry properties of these parameters will be derived and their geometrical importance will be illustrated by using a special example. The Auger decay of the molecules will be treated in Sect. 5.5.4, leading to a set of *anisotropy parameters* containing all the information on the dynamics of the decay process. Section 5.5.5 covers the Auger decay from molecules with degenerate point groups. This requires, in particular, a generalization of the introduced parameters in order to describe the possible coherencies between degenerate electronic states correctly. In both cases, degenerate and non-degenerate, symmetry properties reduce the number of independent parameters and therefore reduce the computational effort in any numerical calculation. We will give examples of some cases in order to illustrate the methods.

5.5.2 Basic Framework

We will now develop a general theory for Auger processes of polyatomic molecules. It will be assumed that the initial vacancy is created by electron impact. We will concentrate on freely rotating molecules in the gas phase. As has been already discussed for the diatomic case in Sect. 5.1.1, we assume the validity of the two-step model as well as we neglect the rotation of the molecules between ionization and Auger emission. Therefore, (5.1a) and (5.1b) can still be applied.

The process will be described in a space-fixed coordinate system XYZ where Z will be chosen as the beam axis of the incoming electrons in the primary process (5.1a). A second right-handed coordinate system xyz will be chosen to be rigidly connected with the molecular framework. The z-axis will always be assumed to be the principal symmetry axis of the molecule. For planar molecules it will be assumed that the molecular plane is spanned by the y- and z-axes (see Sect. 5.5.3).

The orientation of the molecular system with respect to the space-fixed system will be specified by the three Euler angles $\alpha\beta\gamma$. Here β is the angle between z and Z and α is the azimuthal angle of z with respect to the XYZ system. For example, for $\alpha = 0$ the z-axis would lie within the XZ-plane. The third Euler angle γ specifies

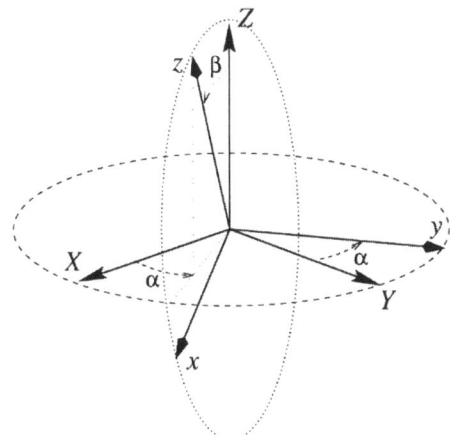

Fig. 5.3. Relation between the space-fixed and the molecular coordinate frame for the special case $\gamma = 0$ (Lehmann and Blum 1997)

a rotation of the xy-plane about the molecular z-axis. For $\gamma = 0$ the x-axis would lie within the zZ-plane and $\gamma \neq 0$ describes a rotation of x out of this plane. The relation between the space-fixed and the molecular system is illustrated in Fig. 5.3 for the special case $\gamma = 0$.

We will concentrate on experiments where only the Auger electrons e_{Auger} will be observed, not the electrons e and e' after the primary process (5.1a). We will further assume that the initial electrons and molecules are unpolarized and that the final spin states are not detected. The initial molecular axes are assumed to be isotropically distributed. Under these conditions the ensemble of molecules M^+ has two main symmetry properties:

i) The ensemble is axially symmetric around the incoming beam axis (Z-axis).
ii) Any plane through Z is a symmetry plane of the molecular ensemble, in particular, any zZ-plane.

Note that these properties are, in general, not valid if, for example, the initial electrons are polarized or the primary ionization is caused by circularly polarized photons.

The initial molecules M are assumed to be in their electronic and vibrational ground state. The states of the ionized molecules will be classified according to their irreducible representations of the relevant point group. We will collectively denote the set of relevant electronic and vibrational quantum numbers of the initial molecules by Γ_0. Similarly the states of M^+ and M^{2+} will be denoted by Γ_1 and Γ_2, respectively.

We will start by first developing the theory for molecules with non-degenerate point groups and postpone the discussion of degenerate point groups to Sect. 5.5.5. There are important differences between these two cases which warrant a separate discussion.

5.5.3 Characterization of the Molecular Ensemble M^+ with Non-Degenerate Point Groups

Description of Anisotropic Axis Distributions

In general, the axes x, y, and z of the molecules M^+ will be anisotropically distributed in space after the primary process (5.1a). We denote the corresponding probability density by $W(\alpha\beta\gamma)$, that is

$$W(\alpha\beta\gamma)\,d\alpha\,\sin\beta d\beta d\gamma \qquad (5.57)$$

is the probability of finding a molecule, in a state with definite Γ_1, averaged over all spins, with the xyz axes lying in the solid angle $d\omega = d\alpha\,\sin\beta d\beta d\gamma$ around the angle $\omega = \alpha\beta\gamma$. Any function of these three Euler angles can be expanded in terms of the complete set of rotation matrix elements $\mathcal{D}^{(K)}_{Q'Q}(\omega)$:

$$W(\omega) = W(\alpha\beta\gamma) = \sum_{KQ'Q} \frac{2K+1}{8\pi^2} \left\langle \mathcal{D}^{(K)*}_{Q'Q} \right\rangle \mathcal{D}^{(K)}_{Q'Q}(\omega), \qquad (5.58)$$

where we have normalized according to

$$\int d\omega\, W(\omega) = 1. \qquad (5.59)$$

The rotation matrices are discussed in Appendix B and are given by (B.7) in the notation of Zare (1988).

The *order parameters* $\left\langle \mathcal{D}^{(K)*}_{Q'Q} \right\rangle$ are defined as expectation values of the corresponding rotation matrix

$$\left\langle \mathcal{D}^{(K)*}_{Q'Q} \right\rangle = \int_0^{2\pi} d\alpha \int_0^{\pi} d\beta\,\sin\beta \int_0^{2\pi} d\gamma\, W(\alpha\beta\gamma)\,\mathcal{D}^{(K)}_{Q'Q}(\omega)^*, \qquad (5.60)$$

where the asterisk denotes the complex conjugate elements. The normalization condition (5.59) requires

$$\left\langle \mathcal{D}^{(0)*}_{00} \right\rangle = 1. \qquad (5.61)$$

The order parameters contain full information on the axis distribution. For example,

$$\left\langle \mathcal{D}^{(1)*}_{00} \right\rangle = \int_0^{2\pi} d\alpha \int_0^{\pi} d\beta\,\sin\beta \int_0^{2\pi} d\gamma\, W(\alpha\beta\gamma)\cos\beta = \langle\cos\beta\rangle, \qquad (5.62)$$

provides us with an effective mean orientation angle between z and Z. For $\left\langle \mathcal{D}^{(2)*}_{00} \right\rangle$ we can write

$$\left\langle \mathcal{D}^{(2)*}_{00} \right\rangle = \left\langle \frac{1}{2}(3\cos^2\beta - 1) \right\rangle, \qquad (5.63)$$

and so on; see Michl and Thulstrup (1986) or Blum (1996) for a more detailed discussion.

In general, we will call an axis distribution *aligned* if at least one order parameter with K even, and $K \neq 0$, contributes to the expansion (5.58), and *oriented* if at least one of the order parameters with K odd is different from zero. The system will be called *polarized*, or simply anisotropic, if at least one order parameter with $K \neq 0$ is non-vanishing. We note that these definitions are completely equivalent to the characterization of rotational orientation and alignment in terms of state multipoles. In fact, for rotating molecules, it has been shown that a close relationship between the two sets of parameters exists (Blum 1996).

Symmetry Properties of the Order Parameters

Equation (5.58) expresses the continuous function $W(\alpha\beta\gamma)$ in terms of an infinite set of discrete order parameters. Such an expansion has particular advantages if only a limited number of parameters contribute. In addition, the symmetry properties of the system are often conveniently characterized in terms of order parameters. We give here some expressions relevant to our present case of interest.

Using the symmetry properties of the rotation matrices (B.12) one obtains from the definition (5.60) the relations

$$\left\langle \mathcal{D}^{(K)*}_{Q'Q} \right\rangle = \left\langle \mathcal{D}^{(K)}_{Q'Q} \right\rangle^*, \tag{5.64a}$$

$$\left\langle \mathcal{D}^{(K)}_{-Q'-Q} \right\rangle = (-1)^{Q'-Q} \left\langle \mathcal{D}^{(K)}_{Q'Q} \right\rangle^*. \tag{5.64b}$$

Furthermore, it follows that order parameters with $Q' = Q = 0$ are real.

According to condition (i) of Sect. 5.5.2 the ensemble of molecules M^+ is axially symmetric around the space-fixed Z-axis. Consequently, all values of α are equally probable so that $W(\alpha\beta\gamma)$ must be independent of α. As follows from the definition of the rotation matrices (B.7) this is only possible if merely order parameters with $Q' = 0$ contribute to the expansion (5.58), which therefore reduces to

$$W(\beta\gamma) = \sum_{KQ} \frac{2K+1}{4\pi} \left\langle \mathcal{D}^{(K)*}_{0Q} \right\rangle \mathcal{D}^{(K)}_{0Q}(0\beta\gamma), \tag{5.65}$$

after the integration over α has been performed. Note, that the element $\mathcal{D}^{(K)}_{0Q}(0\beta\gamma)$ can be expressed in terms of spherical harmonics $Y_{KQ}(\beta\gamma)$ using (B.25) as is discussed in Appendix B.

Now consider condition (ii) of Sect. 5.5.2, which entails that the molecular ensemble must remain invariant under a reflection in any plane through Z. γ is the angle between the xz- and zZ-planes and consequently γ is transformed into $-\gamma$ under a reflection in any zZ-plane. Hence, because of condition (ii), the number of molecules oriented at an angle γ must be equal to the number of molecules oriented at an angle $-\gamma$. This gives the condition

$$W(\beta, \gamma) = W(\beta, -\gamma). \tag{5.66}$$

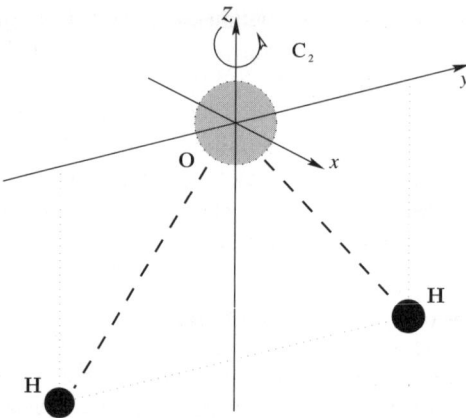

Fig. 5.4. Axis convention of the H_2O molecule. The yz-plane is the molecular plane; after Lehmann and Blum (1997)

Substituting this relation into definition (5.60) we obtain by using (5.64a)

$$
\left\langle \mathcal{D}_{0Q}^{(K)*} \right\rangle = \int_0^{2\pi} d\gamma \int_0^{\pi} d\beta \sin\beta\, W(\beta, -\gamma)\, \mathcal{D}_{0Q}^{(K)}(0\beta\gamma)^{*},
$$

$$
\overset{\gamma \to -\gamma}{=} \int_0^{2\pi} d\gamma \int_0^{\pi} d\beta \sin\beta\, W(\beta, \gamma)\, \underbrace{\mathcal{D}_{0Q}^{(K)}(0\beta - \gamma)^{*}}_{\mathcal{D}_{0Q}^{(K)}(0\beta\gamma)},
$$

$$
= \left\langle \mathcal{D}_{0Q}^{(K)} \right\rangle = \left\langle \mathcal{D}_{0Q}^{(K)*} \right\rangle^{*}. \tag{5.67}
$$

Thus, all order parameters are real if condition (ii) applies.

As a further example of practical importance consider molecules possessing a twofold rotation axis C_2 but no higher rotation axis. Therefore, rotation of the molecules about an angle π around the symmetry axis (z-axis) cannot alter the physical properties of the system. Hence,

$$
W(\alpha, \beta, \gamma) = W(\alpha, \beta, \gamma + \pi), \tag{5.68}
$$

and similar to the derivation of (5.67) we obtain

$$
\left\langle \mathcal{D}_{0Q}^{(K)*} \right\rangle = (-1)^{Q} \left\langle \mathcal{D}_{0Q}^{(K)*} \right\rangle, \tag{5.69}
$$

which shows that the presence of a C_2-axis requires the vanishing of all order parameters with Q odd. Condition (5.69) holds, in particular, for molecules with C_{2v} symmetry, e.g. H_2O shown in Fig. 5.4 to clarify the axis convention.

Examples and Geometrical Interpretations

In order to illuminate further the geometrical meaning of the order parameters we will consider ionization by photon absorption instead of the primary process (5.1a):

$$\gamma + M \longrightarrow M^+ + e. \tag{5.70}$$

Because of dipole selection rules only order parameters with $K \leq 2$ contribute to the expansion (5.58). Assuming linearly polarized light, and choosing the direction of the electric field vector as the Z-axis, it has been shown that no axis orientation can be produced and that only order parameters with $K = 0$ and $K = 2$ are non-vanishing (Michl and Thulstrup 1986; Blum 1996).

Let us consider this case in more detail. Under the assumed experimental conditions relations (5.65) and (5.67) apply and the ensemble M^+ is completely characterized by the *real* order parameters $\langle D_{00}^{(0)} \rangle = 1$, $\langle D_{00}^{(2)} \rangle$, $\langle D_{0\pm1}^{(2)} \rangle$, and $\langle D_{0\pm2}^{(2)} \rangle$. In order to illustrate the concept we will specialize to molecules with C_{2v} symmetry, for example H_2O shown in Fig. 5.4, where the C_2-axis is taken as the z-axis and the molecular plane as the yz-plane.

In this case the order parameters with $Q = \pm 1$ vanish because of condition (5.69). Hence, besides the normalization constant $\langle D_{00}^{(0)} \rangle$, we are left with two independent real alignment parameters $\langle D_{00}^{(2)} \rangle$ and $\langle D_{02}^{(2)} \rangle = \langle D_{0-2}^{(2)} \rangle$. Equation (5.65) reduces then to the expansion

$$W(\beta, \gamma) = \frac{1}{4\pi} \left(1 + \frac{5}{2} \langle D_{00}^{(2)} \rangle [3 \cos^2 \beta - 1] + 5 \sqrt{\frac{3}{2}} \langle D_{02}^{(2)} \rangle \sin^2 \beta \cos 2\gamma \right), \tag{5.71}$$

where explicit expressions for the relevant rotation matrix elements have been inserted. Thus, the axis distribution is completely specified by two order parameters, the usual alignment parameter $\langle D_{00}^{(2)} \rangle$, given by (5.63), and the alignment component $\langle D_{02}^{(2)} \rangle$, which is given by the relation

$$\langle D_{02}^{(2)} \rangle = \int_0^{2\pi} d\gamma \int_0^{\pi} d\beta \sin \beta \, W(\beta, \gamma) \, D_{02}^{(2)}(0\beta\gamma),$$

$$= \sqrt{\frac{3}{8}} \langle \sin^2 \beta \cos 2\gamma \rangle. \tag{5.72}$$

In order to gain further insight into the geometrical importance of the order parameters we relate the set $\langle D_{q'Q}^{(K)} \rangle$ to the direction cosines of the molecular axes. A formalism, based on direction-cosine factors, has previously been outlined by Michl and Thulstrup (1986). Let us denote by $\cos nZ$ the cosine of the angle between the molecular n-axis ($n = x, y, z$) and the space-fixed Z-axis. The Euler angles α, β, γ are related to the angles xZ, yZ, and zZ by the relations

$$\cos zZ = \cos \beta, \tag{5.73a}$$

$$\cos xZ = -\sin \beta \cos \gamma, \tag{5.73b}$$

$$\cos yZ = \sin \beta \sin \gamma. \tag{5.73c}$$

The direction cosines are not independent because

$$\cos^2 zZ + \cos^2 xZ + \cos^2 yZ = 1. \tag{5.74}$$

Using the relations (5.73) we can express the two relevant order parameters in terms of mean values of the squares of the direction cosines

$$\langle \mathcal{D}_{00}^{(2)} \rangle = \frac{3}{2} \langle \cos^2 zZ \rangle - \frac{1}{2}, \tag{5.75a}$$

$$\langle \mathcal{D}_{02}^{(2)} \rangle = \sqrt{\frac{3}{8}} [\langle \cos^2 xZ \rangle - \langle \cos^2 yZ \rangle]. \tag{5.75b}$$

From these *orientation factors* we can read off the restrictions

$$-\frac{1}{2} \le \langle \mathcal{D}_{00}^{(2)} \rangle \le 1, \tag{5.76a}$$

$$-\sqrt{\frac{3}{8}} \le \langle \mathcal{D}_{02}^{(2)} \rangle \le \sqrt{\frac{3}{8}}. \tag{5.76b}$$

For example, if all molecular z-axes were aligned parallel to Z, then $\cos zZ = 1$, $\cos xZ = \cos yZ = 0$ for any molecule and the corresponding distribution would be characterized by $\langle \mathcal{D}_{00}^{(2)} \rangle = 1$ and $\langle \mathcal{D}_{02}^{(2)} \rangle = 0$. If all molecules were to have their x-axes parallel to Z, then $\langle \cos^2 xZ \rangle = 1$, $\langle \cos^2 yZ \rangle = \langle \cos^2 zZ \rangle = 0$, and $\langle \mathcal{D}_{00}^{(2)} \rangle = -\frac{1}{2}$, $\langle \mathcal{D}_{20}^{(2)} \rangle = \sqrt{\frac{3}{8}}$. These relations can be visualized by constructing the *orientation triangle*. Following essentially Michl and Thulstrup (1986), but changing and adapting their definitions to our present case of interest we obtain Fig. 5.5.

Any possible distribution given by definite values of the two alignment parameters will be characterized by points located within this triangle. The isotropic distribution is represented by the point $\langle \mathcal{D}_{00}^{(2)} \rangle = \langle \mathcal{D}_{02}^{(2)} \rangle = 0$. All points on the line $\langle \mathcal{D}_{00}^{(2)} \rangle = \sqrt{6} \langle \mathcal{D}_{02}^{(2)} \rangle + 1$, which limits the orientation triangle to the left, correspond to distributions where the molecular x-axis is perpendicular to Z. The line $\langle \mathcal{D}_{02}^{(2)} \rangle = 0$ contains all distributions where the x- and y-axes are equally distributed around the molecular z-axis. All possible axis distributions of diatomic molecules in Σ-states lie on this line.

In cases where the primary ionization process is caused by electron impact order parameters with $K > 2$ contribute. The corresponding order parameters can, in principle, be related to orientation factors as well. However, the corresponding relations are complex and the use of the set of order parameters is then more convenient and reduces the computational effort considerably.

5.5.4 Angular Distribution of Auger Electrons from Molecular Non-Degenerate Point Groups

Basic Equations

We will now consider the Auger decay (5.1b), assuming that the axis distribution of the molecules M^+ is characterized by (5.58). Let us denote the amplitude for an

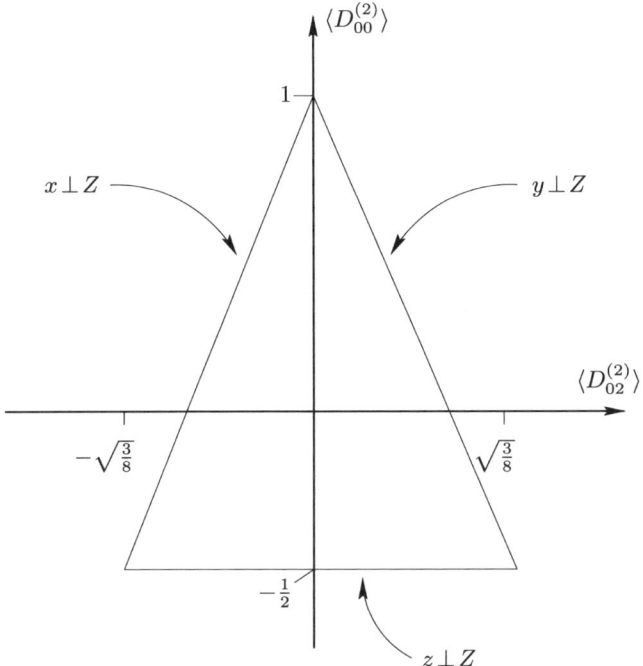

Fig. 5.5. The orientation triangle; after Lehmann and Blum (1997)

Auger transition

$$|\Gamma_1 \omega\rangle \longrightarrow |\Gamma_2 \omega \mathbf{p} m_s\rangle \tag{5.77}$$

by $\langle \Gamma_2 \omega \mathbf{p} m_s | \mathbf{T} | \Gamma_1 \omega\rangle$ where $\mathbf{T}$ is the transition operator, and $\mathbf{p}$ and m_s denote the momentum and spin component of the detected Auger electron, respectively. We normalize in such a way that the squared modulus $|\langle \Gamma_2 \omega \mathbf{p} m_s | \mathbf{T} | \Gamma_1 \omega\rangle|^2$ is the intensity of the emitted Auger electrons in the indicated transitions with definite positions specified by angles $\omega = \alpha\beta\gamma$. The total angular distribution $I(\hat{\mathbf{p}})$ is obtained by multiplying $|\langle \Gamma_2 \omega \mathbf{p} m_s | \mathbf{T} | \Gamma_1 \omega\rangle|^2$ by the probability density $W(\omega)$ of finding a molecule in this particular spatial orientation, summing over the undetected spin components and integrating over the angles α, β, γ. That is

$$I(\hat{\mathbf{p}}) = \sum_{m_s} \int d\omega \, W(\omega) \, |\langle \Gamma_2 \omega \mathbf{p} m_s | \mathbf{T} | \Gamma_1 \omega\rangle|^2. \tag{5.78}$$

Choosing the axis of the direction of the emitted Auger electron $\hat{\mathbf{p}}$ as quantization axis we are able to expand the states of the Auger electrons in the partial wave series

$$|\mathbf{p}\rangle = \frac{1}{\sqrt{k}} \sum_{\ell} i^{\ell} \sqrt{\frac{2\ell + 1}{4\pi}} |\ell 0\rangle. \tag{5.79a}$$

Transforming the partial waves to the molecular system, i.e. choosing the molecular z-axis as the quantization axis, leads to

$$|\mathbf{p}\rangle = \frac{1}{\sqrt{k}} \sum_{\ell M m} i^\ell \sqrt{\frac{2\ell+1}{4\pi}} \mathcal{D}^{(\ell)}_{M0}(\theta\phi0)\, \mathcal{D}^{(\ell)}_{Mm}(\omega)\, |\ell 0\rangle, \qquad (5.79b)$$

where ϕ and θ denote the azimuth and polar angle of $\mathbf{p}$ in the space-fixed coordinate system. Inserting (5.58) and (5.79b) into (5.78) we are now able to perform the integration and after some algebra we obtain

$$I(\theta) = \sum_{KQ} \left\langle \mathcal{D}^{(K)*}_{0Q} \right\rangle A_{KQ}(\Gamma_1)\, P_K(\cos\theta), \qquad (5.80)$$

where we introduced the anisotropy parameters for polyatomic molecules

$$A_{KQ}(\Gamma_1) = \frac{2K+1}{4\pi k} \sum_{\substack{m_s \ell' m' \\ \ell m}} i^{\ell-\ell'}(-1)^m \sqrt{(2\ell+1)(2\ell'+1)}$$

$$\times \langle \Gamma_2\, \ell'm'm_s |\mathbf{T}|\Gamma_1\rangle\langle \Gamma_2\, \ell m\, m_s |\mathbf{T}|\Gamma_1\rangle^*$$

$$\times \begin{pmatrix} \ell' & \ell & K \\ 0 & 0 & 0 \end{pmatrix} \begin{pmatrix} \ell' & \ell & K \\ m' & -m & Q \end{pmatrix}, \qquad (5.81)$$

where m, m' and m_s are related to the molecular z-axis. Equation (5.80) expresses the angular distribution of the Auger electrons in terms of the Legendre polynomials $P_K(\cos\theta)$ and gives a factorization in dynamical and geometrical parts. The anisotropy of the molecules immediately after the ionization is characterized by the relevant order parameters $\langle \mathcal{D}^{(K)*}_{0Q}\rangle$ which contain all information on the geometry and dynamics of the ionization (5.1a) and have been discussed in detail in the preceding sections. The information on the dynamics of the Auger decay (5.1b) is contained in the anisotropy parameters A_{KQ} which are independent of the spatial orientation of the molecules since all quantum numbers in the transition matrix elements refer to the molecular coordinate system.

It should be noted that any Legendre polynomial P_K is related to the order parameters with the same rank K. Hence, any order parameter of rank K gives rise to a characteristic angular distribution P_K of the Auger electrons. In this sense one might say that the anisotropy of the Auger angular distribution reflects the anisotropy of the molecular distribution before the Auger decay as pointed out first by Dill *et al.* (1980) for diatomic molecules. The parameters A_{KQ} can be considered as weight factors determining the amount by which the Auger emission is influenced by the corresponding order parameters. The independence of the Auger electron distribution of the azimuth angle ϕ which is manifest in (5.80) reflects the axial symmetry of the probability density $W(\omega)$ under the assumed conditions; see (5.65).

Symmetry Properties of Anisotropy Parameters

Profiting by the properties of the Wigner $3j$-symbols one obtains from the definition of the anisotropy parameters

$$A^*_{KQ}(\Gamma_1) = (-1)^Q A_{K-Q}(\Gamma_1). \tag{5.82}$$

In particular, all anisotropy parameters with $Q = 0$ are real.

Equation (5.81) simplifies if the molecules have particular symmetry properties. As an example we consider molecules possessing a C_2-axis. Disregarding any further symmetry properties the molecular states can be classified according to the irreducible representations of the point group C_2, i.e. A and B, where B denotes the states which change their sign under C_2-rotation, while states belonging to the A representation remain unchanged. Note that this statement holds true in all other symmetry groups possessing one C_2-axis as the highest rotation axis. The partial waves transform according to

$$C_2|\ell m\rangle = (-1)^m|\ell m\rangle. \tag{5.83}$$

Inserting rotation operators in the **T** matrix elements we find that

$$\langle \Gamma_2\, \ell' m' m_s |\mathbf{T}| \Gamma_1 \rangle \langle \Gamma_2\, \ell m\, m_s |\mathbf{T}| \Gamma_1 \rangle^*$$
$$= (-1)^{m-m'} \langle \Gamma_2\, \ell' m' m_s |\mathbf{T}| \Gamma_1 \rangle \langle \Gamma_2\, \ell m\, m_s |\mathbf{T}| \Gamma_1 \rangle^*, \tag{5.84}$$

which holds for both A and B representations. This gives the condition that $m - m'$ must be even and consequently – taking the second $3j$-symbol in (5.81) into account – all anisotropy parameters with Q odd vanish.

5.5.5 Angular Anisotropy of Auger Electrons from Molecular Ensembles with Degenerate Point Groups

General Theory

We will now discuss the processes (5.1a) and (5.1b) for molecules whose symmetry group is degenerate. That is, the molecular states can be degenerate and will therefore be *coherently* excited in general. The main difference from the discussions in Sects. 5.5.3 and 5.5.4 is that now coherence terms between degenerate states $|\Gamma_1\rangle$ and $|\Gamma'_1\rangle$ must be taken into account, described by off-diagonal elements of the relevant density matrix. Hence, (5.58) no longer suffices and density matrix methods must be applied.

We denote the density matrix describing the molecules M^+ immediately after the ionization process (5.1a) by ρ and its elements by $\langle \Gamma'_1\, \omega | \rho | \Gamma_1\, \omega \rangle$. We normalize in such a way that the diagonal elements with $\Gamma'_1 = \Gamma_1$ are equal to the probability density of finding a molecule in the state Γ_1 with an axis orientation specified by $\omega = \alpha\beta\gamma$. Summing over all degenerate states Γ_1 one obtains the corresponding axis distribution function $W(\omega)$. The off-diagonal elements characterize the coherence between states $|\Gamma_1\rangle$ and $|\Gamma_1'\rangle$ produced during the process (5.1a). In full analogy to (5.2) and (5.58) any matrix element $\langle \Gamma_1' \, \omega | \rho | \Gamma_1 \, \omega \rangle$ can be expanded in terms of rotation matrices

$$\langle \Gamma_1' \, \omega | \rho | \Gamma_1 \, \omega \rangle = \sum_{K Q' Q} \frac{2K+1}{8\pi^2} \langle \mathcal{D}_{0Q}^{(K)*}(\Gamma_1', \Gamma_1) \rangle \mathcal{D}_{0Q}^{(K)}(\omega), \qquad (5.85)$$

for any combination Γ_1, Γ_1'. Note, that the terms with $\mathcal{D}_{Q'Q}^{(K)}(\omega)$, $Q' \neq 0$ are ruled out because of the axial symmetry of the system. Here, the same argument applies as in Sect. 5.3.2. The order parameters $\langle \mathcal{D}_{0Q}^{(K)*}(\Gamma_1', \Gamma_1) \rangle$ in (5.85) are defined by the relation

$$\left\langle \mathcal{D}_{0Q}^{(K)*}(\Gamma_1', \Gamma_1) \right\rangle = \int d\omega \, \langle \Gamma_1' \, \omega | \rho | \Gamma_1 \, \omega \rangle \mathcal{D}_{0Q}^{(K)}(\omega)^*. \qquad (5.86)$$

Equations (5.85) and (5.86) generalize (5.58) and (5.60). For example, for doubly degenerate states $|\Gamma_1\rangle$, $|\Gamma_1'\rangle$ we have to consider the set of order parameters

$$\left\langle \mathcal{D}_{0Q}^{(K)*}(\Gamma_1) \right\rangle, \left\langle \mathcal{D}_{0Q}^{(K)*}(\Gamma_1') \right\rangle, \left\langle \mathcal{D}_{0Q}^{(K)*}(\Gamma_1', \Gamma_1) \right\rangle, \left\langle \mathcal{D}_{0Q}^{(K)*}(\Gamma_1, \Gamma_1') \right\rangle, \qquad (5.87)$$

where the latter order parameters are related by

$$\left\langle \mathcal{D}_{0Q}^{(K)*}(\Gamma_1', \Gamma_1) \right\rangle = (-1)^Q \left\langle \mathcal{D}_{0Q}^{(K)*}(\Gamma_1, \Gamma_1') \right\rangle^*, \qquad (5.88)$$

which can be simply derived from the properties of the rotation matrices; see Appendix B. However, for any fixed values of K and Q we find three independent order parameters for doubly degenerate states. Generally, we can say that by taking the coherencies into account the number of order parameters increases considerably compared to the non-degenerate case. The importance of these coherence parameters for the special case of diatomic molecules has been already discussed in Sect. 5.1.2 and in more detail by Bonhoff et al. (1996). It has been shown that the coherence parameters are particularly required in order to describe the shape and the spatial orientation of the coherently excited orbitals, which in turn will influence the molecular Auger decay (5.1b) in addition to the anisotropy in the axis distribution. It is therefore essential to consider the full set of order parameters for all relevant sets of quantum numbers Γ_1 and Γ_1'.

The angular distribution of Auger electrons, emitted in the process (5.1b), is now given by the expression

$$I(\hat{\mathbf{p}}) = \sum_{\substack{\Gamma_1' \Gamma_1 \\ \Gamma_2 m_s}} \int d\omega \, \langle \Gamma_2 \, \omega \, \mathbf{p} \, m_s | \mathbf{T} | \Gamma_1' \, \omega \rangle \langle \Gamma_1' \, \omega | \rho | \Gamma_1 \, \omega \rangle \langle \Gamma_1 \, \omega | \mathbf{T}^\dagger | \Gamma_2 \, \omega \, \mathbf{p} \, m_s \rangle. \qquad (5.89)$$

For an incoherent superposition of molecular states the off-diagonal elements of the density matrix ρ vanish and (5.89) reduces to the expression

$$I(\hat{\mathbf{p}}) = \sum_{\Gamma_1 \Gamma_2 m_s} \int d\omega \, \left| \langle \Gamma_2 \, \omega \, \mathbf{p} \, m_s | \mathbf{T} | \Gamma_1 \, \omega \rangle \right|^2 \langle \Gamma_1 \, \omega | \rho | \Gamma_1 \, \omega \rangle, \qquad (5.90)$$

which is the same as (5.78) summed over all degenerate states of the molecules M^+ and M^{2+}. – Also compare to the diatomic case (5.20). – This equation has often been used as a starting point for further discussion. However, we stress the point that *the use of (5.90) will be wrong in general since the coherence parameters are not included.*

The further calculations are completely similar to the discussion in Sect. 5.5.4. We insert the expansion (5.85) into (5.89), apply (5.79) and repeat essentially the derivations of Sect. 5.5.4. Eventually, we obtain

$$I(\theta) = \sum_{\substack{KQ \\ \Gamma_1 \Gamma_1'}} \left\langle \mathcal{D}_{0Q}^{(K)*}(\Gamma_1', \Gamma_1) \right\rangle A_{KQ}(\Gamma_1', \Gamma_1) \, P_K(\cos\theta), \qquad (5.91)$$

where the anisotropy parameters $A_{KQ}(\Gamma_1', \Gamma_1)$ for the degenerate case are given by the expression

$$
\begin{aligned}
A_{KQ}(\Gamma_1', \Gamma_1) = \frac{2K+1}{4\pi k} \sum_{\substack{\ell' m' \ell m \\ \Gamma_2 m_s}} & \mathrm{i}^{\ell-\ell'}(-1)^m \sqrt{(2\ell+1)(2\ell'+1)} \\
& \times \left\langle \Gamma_2 \, \ell' m' m_s |\mathbf{T}| \Gamma_1' \right\rangle \left\langle \Gamma_2 \, \ell m \, m_s |\mathbf{T}| \Gamma_1 \right\rangle^* \\
& \times \begin{pmatrix} \ell' & \ell & K \\ 0 & 0 & 0 \end{pmatrix} \begin{pmatrix} \ell' & \ell & K \\ m' & -m & Q \end{pmatrix}.
\end{aligned} \qquad (5.92)
$$

Once more we have to emphasize the importance of the coherencies. The coherence parameters with $\Gamma_1' \neq \Gamma_1$ contribute indispensably to the angular distribution of the Auger electrons. Equations (5.91) and (5.92) are therefore the basis for future calculations of angular distributions which require:

1. The calculation of the order parameters.
2. The determination of the anisotropy parameters for one molecular orientation.

In addition, the compact and transparent form of (5.91) is particularly useful for an analysis of experimental data.

Application of Symmetry Arguments

The use of symmetry arguments often reduces the number of independent anisotropy parameters in (5.89) and hence reduces the computational effort in numerical calculations of Auger angular distributions. We will outline here the arguments for degenerate molecular states. In order to point out the essentials we will concentrate on molecules belonging to the symmetry group C_{3v} as an example and assume that the molecules are spinless for simplicity. First, we consider the case where all the molecular states involved in the process (5.1b) belong to the doubly degenerate E-representation of the point group. – For the different classifications and symmetries of the point groups and their related representations we refer to the literature, e.g. Atkins (1970); Haken (1991); Tinkham (1964). – There are several possibilities for choosing symmetry adapted wavefunctions. Application of the well known projection operator technique leads to a basis set in which one of the σ_v-reflections is diagonal (see, e.g., Douglas and Hollingsworth 1985; Kettle 1995). However, for our present case of interest, it will turn out to be advantageous to use eigenfunctions of the C_3-operator instead. The eigenvalues of the C_3-rotation in any two-dimensional representation must be $\lambda_\pm = \exp(\pm 2\pi i/3)$ so that it is suggested to denote the eigenvectors belonging to $\lambda_\pm$ by $|\Lambda = \pm 1\rangle$. They are defined by the symmetry property

$$C_3|\Lambda\rangle = \exp\left(\frac{2\pi i}{3}\Lambda\right)|\Lambda\rangle \quad \text{with} \quad \Lambda = \pm 1. \tag{5.93}$$

In the Λ-representation we can readily derive selection rules for the matrix elements occurring in (5.92). Since the $\mathbf{T}$ operator is required to be invariant under all symmetry operations we have, in particular, the condition

$$\mathbf{C}_3^\dagger\mathbf{T}\mathbf{C}_3 = \mathbf{T}, \tag{5.94}$$

where now the operator $\mathbf{C}_3$ applies to the total system, i.e. molecule plus projectile. Substituting the symbol $|\Lambda\rangle$ for $|\Gamma\rangle$ and inserting (5.94) into the matrix elements of (5.92) we obtain

$$\langle\Lambda_2\,\ell m\,m_s|\mathbf{T}|\Lambda_1\rangle = \langle\Lambda_2\,\ell m\,m_s|\mathbf{C}_3^\dagger\mathbf{T}\mathbf{C}_3|\Lambda_1\rangle$$
$$= \exp\left(-\frac{2\pi i}{3}[\Lambda_2 + m + m_s - \Lambda_1]\right)\langle\Lambda_2\,\ell m\,m_s|\mathbf{T}|\Lambda_1\rangle, \tag{5.95}$$

where we used (5.93) and similar expressions for the transformation of the projectile partial wave states. From (5.95) it follows that the matrix elements vanish unless the phase factor is equal to $+1$, which requires that the condition

$$\Lambda_2 + m + m_2 - \Lambda_1 = 3k \quad k = 0, \pm 1, \pm 2, \ldots \tag{5.96a}$$

is satisfied. We obtain a similar relation for the second matrix element in (5.92):

$$\Lambda_2 + m' + m_2 - \Lambda_1' = 3k' \quad k' = 0, \pm 1, \pm 2, \ldots. \tag{5.96b}$$

Combining the relations (5.96) we obtain the symmetry condition for the component Q

$$Q = m - m' = (\Lambda_1 - \Lambda_1') + 3N \quad N = (k - k') = 0, \pm 1, \pm 2, \ldots. \tag{5.97}$$

Hence the anisotropy parameters $A_{KQ}(\Lambda_1', \Lambda_1)$ are subject to the following restrictions:

(i) $\Lambda_1 = \Lambda_1' = +1$ $\Rightarrow Q = 3N = 0, \pm 3, \pm 6, \ldots,$

(ii) $\Lambda_1 = \Lambda_1' = -1$ $\Rightarrow Q = 3N = 0, \pm 3, \pm 6, \ldots,$

(iii) $\Lambda_1 = +1, \Lambda_1' = -1 \Rightarrow Q = 3N + 2 = \ldots, -7, -4, -1, +2, +5, \ldots,$

(iv) $\Lambda_1 = -1, \Lambda_1' = +1 \Rightarrow Q = 3N + 1 = \ldots, -5, -2, +1, +4, +7, \ldots.$

For any other value of Q the anisotropy parameters will vanish due to the symmetry of the system.

Equation (5.97) limits the number of non-vanishing anisotropy parameters which must be calculated in order to get numerical results for the angular distribution (5.91). In particular, if the molecules M^+ have been produced by linearly polarized photons as in the example of Sect. 5.5.3, we have $K = 0$ and $K = 2$ and the non-vanishing anisotropy parameters $A_{KQ}(\Gamma_1', \Gamma_1) = A_{KQ}(\Lambda_1', \Lambda_1)$ for the two degenerate states with $\Lambda = \pm 1$ are the following:

$$
\begin{aligned}
&A_{00}(+1,+1)\,, && A_{00}(-1,-1)\,, && A_{20}(+1,+1)\,, && A_{20}(-1,-1)\,, \\
&A_{2+1}(-1,+1)\,, && A_{2-1}(+1,-1) = -A_{2+1}^*(-1,+1)\,, \\
&A_{2+2}(+1,-1)\,, && A_{2-2}(-1,+1) = A_{2+2}^*(+1,-1)\,.
\end{aligned} \tag{5.98}
$$

Similar relations can be derived for other symmetries. For example, if reflection operators exist, which transform $|\Lambda\rangle$ into $|-\Lambda\rangle$, further symmetry relations can be derived.

Finally, we consider the important case where only the final molecular state $|\Gamma_2\rangle$ belongs to the degenerate species whereas $|\Gamma_1\rangle$ belongs to the A_1- or A_2-representation of the C_{3v} point group, respectively. A very common situation in Auger processes of light molecules is that the inner-shell vacancy produced in the primary process is of A_1-symmetry. For example the ground state configuration of NH_3 is

$$
(1a_1)^2(2a_1)^2(1e)^4(3a_1)^2 \tag{5.99}
$$

where the $1a_1$ orbital corresponds to the atomic 1s orbital of nitrogen and $2a_1$, $1e$, and $3a_1$ are bonding orbitals combined from the three hydrogen 1s orbitals and the 2s and 2p orbitals of the nitrogen. The only vacancy that gives rise to an Auger decay is the $(1a_1^{-1})\,{}^2A_1$ state of the molecular ion. Normally the ionization process is fast compared to the molecular vibration. So the states $|\Gamma_1\rangle$ are not vibrationally excited and no vibrational degeneracies occur (Siegbahn $et\ al.$ 1969). However, the final states of the doubly charged molecular ions can be degenerate. In this case (5.97) reduces to

$$
Q = 3N = 0, \pm 3, \pm 6, \ldots . \tag{5.100}
$$

Hence, only parameters $A_{KQ}(\Gamma_1 = A_{1,2})$ with $Q = 0, \pm 3, \pm 6, \ldots$ contribute to the expansion (5.91). This result holds true even if the final state $|\Gamma_2\rangle$ is also non-degenerate.

6 Conclusion and Outlook

We discussed the general theory and numerical methods necessary for the calculation of angle and spin resolved Auger processes. In the second chapter, the theoretical formalism has been developed within the two-step model. Applying the framework of density matrices and state multipoles we investigated the primary ionization or excitation of an inner shell electron via electron or photon impact and discussed the subsequent Auger decay. The general equations of angular distribution and spin polarization have been derived and we discussed in detail which information can be obtained in such experiments using either electron or photon impact for the generation of the inner shell hole. Here, different polarization states of the electrons or photons have been considered where we stressed the physically important process of generation of spin polarization out of alignment. We have derived non-linear interrelations between the angular distribution and spin polarization parameters of the Auger decay. Resonant Auger transitions have been extensively discussed considering the cases of either linearly, circularly, or unpolarized light for the primary photoexcitation process. For concluding the theory we discussed some special cases of Auger transitions in more detail. Like spin polarization of isotopic multiplets, the angle dependent intensities and spin polarizations for specific intermediate ionic hole states, or the Auger emission from an unresolved intermediate fine structure state. Eventually, we considered relations for the asymmetry parameters of the Auger decay and for the linear dichroism of Auger emission.

In Chap. 3, we described the numerical methods developed for the evaluation of the relevant observables. We pointed out that numerical calculations for angle and spin resolved Auger transitions must not only provide the transition amplitudes but the scattering phases, too. Our calculations have been performed within an MCDF approach which accounts for the large fine structure splitting in heavy atoms. The used program packages ANISO and RATR and their specific approximations have been discussed. Both programs have been developed in the context of scattering theory and Δ-SCFCI methods have been applied.

We have given selected results for the angular distribution and spin polarization of Auger electrons emitted from free atoms and molecules in the last two chapters.

In Chap. 4 we focused on angular distribution and spin polarization parameters and their comparison to experimental and other theoretical data where most data have been available for the rare gases. In Sect. 4.1, we compared the theoretical and experimental data of the angular anisotropy of Auger electrons emitted from noble gas atoms. The spin polarization of Auger electrons emitted after photoionization with circularly polarized light has been discussed for Auger transitions in Ar, Kr and Xe in Sect. 4.2 where the effect of polarization transfer has been considered in more detail. Spin polarization parameters for isotropic Auger multiplets have been discussed for the Ar L_2MM Auger transitions in Sect. 4.3. A comparison between recent theoretical and experimental data for the angle and spin resolved Xe $M_{4,5}N_{4,5}N_{4,5}$ Auger spectrum has been given in Sect. 4.4. Correlation effects in the angle and spin resolved Hg $N_{6,7}O_{4,5}O_{4,5}$ Auger spectrum have been investigated in Sect. 4.5 by focusing on initial and final state configuration interaction, respectively. We discussed the angular distribution of resonant Auger transitions which has been investigated applying a spectator model in Sect. 4.6.1.

The investigation of angle and spin resolved Auger transitions is still an active area of research. Though, for the atomic case, the mechanisms causing an anisotropic angular distribution of the emitted Auger electrons have been essentially understood – at least for the case of a primary photoionization/excitation – there is still much research to do with respect to spin polarization. A deeper insight into the mechanism why a specific spin polarization is produced can be obtained from the propensity rules derived in Sect. 4.6.2. However, these rules need further generalization and development.

A further step into this direction has been achieved by investigating the resonantly excited $Ar^*(2p_{1/2,3/2}^{-1}4s_{1/2})_{J=1}$ $L_{2,3}M_{2,3}M_{2,3}$ Auger spectrum, both, experimentally and numerically, which has been discussed in Sect. 4.7. The surprisingly large dynamic spin polarization for certain unresolved groups of Auger lines, found in the experiment, has been in contrast to the hitherto understanding of the resonantly excited $Ar^*L_{2,3}M_{2,3}M_{2,3}$ Auger spectrum as a showcase for a large transferred but vanishing dynamic spin polarization. This effect has been explained as configuration interaction induced, caused by internal selection rules based on the propensity rules derived in Sect. 4.6.2.

The Auger spectra of open-shell systems have been investigated by several groups (e.g., Carre *et al.* 1990; Dorn *et al.*, 1995). With respect to angular distribution, this is a relatively new field of research. For the rare gases, Auger transitions from an initial ns^{-1} state have to be isotropic since no alignment can be produced. As has been shown, this can be different for open-shell systems where the outer shell valence electron couples its angular momentum with the inner shell ns hole which could result in an anisotropic angular distribution. We performed theoretical calculations for the Auger angular distribution of the alkalis in Sect. 4.8, and for atomic oxygen in Sect. 4.9 where large anisotropy parameters for KLL Auger transitions in conjunction with large Auger intensities have been predicted for selected Auger lines.

While the previous chapters concentrated on angle and spin resolved Auger emission from free atoms, Chap. 5 dealt with the Auger electron emission from free molecules. Here, we concentrated on diatomic and small polyatomic molecules. In Sect. 5.1, we have presented the mathematical framework needed to calculate the anisotropy parameters for the angular distribution and spin polarization of molecular Auger electrons after ionization by electron impact. Analogously to the atomic case applying a two-step model, we introduced parameters $A_{Kkq}(\Omega', \Omega)$ describing the Auger decay, whereas all information on geometry and dynamics of the ionization process is contained in the order parameters $\langle \mathcal{D}_{0\Omega-\Omega'}^{(K)*} \rangle$. The general expressions for the spin polarization vector have been derived in Sect. 5.2. They can be simplified for certain polarization states of the incoming ionizing electron beam which has been discussed within a few examples. Introducing relative parameters in Sect. 5.3, we have shown that generally a factorization of the angular distribution and spin polarization parameters is not possible which is in contrast to the atomic case.

It has been shown, that the coefficients $\langle \mathcal{D}_{00}^{(K)*} \rangle$ with $\Omega = \Omega' = 0$ characterize the axis distribution of the molecules M^+. Using a special example we have shown that the order parameters with $\Omega \neq \Omega'$ are related to the shape and spatial orientation of the orbitals. This latter set of parameters is closely related to the coherencies produced during the primary ionization. If these coherencies are not explicitly taken into account, then the terms $\langle \mathcal{D}_{0Q}^{(K)*} \rangle$ with $Q \neq 0$ would be missing in the expression for the angular distribution. This would not change the overall structure of the general relation for the angular distribution of the molecular Auger emission. However, it is of course essential for any numerical calculation to include all relevant parameters. Equations (5.24) and (5.25) are therefore the necessary basis for any numerical work.

In Sect. 5.4, the numerical method used for the calculations has been briefly reviewed, and the angular distribution data and the Auger spectrum of HF, respectively, have been discussed.

A general theory for the angular distribution of Auger electrons emitted from polyatomic molecules after electron impact has been derived in Sect. 5.5, taking into account the anisotropy produced in the primary ionization process as well as the dynamics of the Auger decay. The general relations for the angular distribution and the related anisotropy parameters have been given for, both, the non-degenerate and the degenerate case, respectively. In both cases the angular distribution of the Auger electrons is expressed in terms of Legendre polynomials describing the angular dependence. As for the diatomic case the information on the primary ionization process is provided by the order parameters $\langle \mathcal{D}_{Q'Q}^{(K)*}(\Gamma_1', \Gamma_1) \rangle$ whereas the dynamics of the Auger emission process is contained in the anisotropy parameters $A_{KQ}(\Gamma_1', \Gamma_1)$. The use of both sets of parameters clearly separates the influence of the two processes and allows for presenting the final results in a condensed and transparent form.

The importance of the coherence parameters for the theory of angular distribution has been emphasized. Considering the C_{2v} and C_{3v} point groups as examples, it has been shown how the use of symmetry principles reduces the number of independent parameters and therefore the efforts in any numerical calculation of the angle resolved molecular Auger process. The ideas can easily be applied to other symmetry groups.

Our aim is to further extend the theory of angle and spin resolved Auger emission and its numerical application to more complicated open shell systems. Since the

RATR program, respectively its successor RATIP (Fritzsche *et al.* 2000; Fritzsche 2001), allows for a full inclusion of relativistic and exchange effects it will be applied to heavy open-shell atoms. A good candidate might be the $N_{6,7}O_{4,5}O_{4,5}$ Auger spectrum of gold. Here, the initial hole state is in an f-shell which can couple with the open d-shell in various ways. Since gold is a 6th-row element its fine structure splitting should be large enough to resolve its fine structure levels energetically. The Au $N_{6,7}O_{4,5}O_{4,5}$ Auger spectrum has been measured by Aksela *et al.* (1983b) showing a large number of Auger lines.

Of particular interest is the calculation of the Au $N_1O_{4,5}O_{4,5}$ Auger transitions. Our previous calculation of the Cs KLL Auger spectra showed an almost vanishing fine structure splitting, which is caused by the small overlap of the wavefunctions of the inner shell $1s^{-1}$ hole with the outer shell 6s electron. Due to this, the Cs spectra showed a Xe like, and therefore isotropic structure. This can be different for the $(4s^{-1}5d^9)$ initial state configuration of gold. The data would yield information on the influence of the fine structure splitting on the angular distribution and of the strength of correlation effects for such type of transitions.

Until today, there are only few investigations or data on spin polarized Auger emission from open shell atoms. Our own attempt on mercury is, from a theoretical point of view, similar to the case of the rare gases due to the closed sub-shell ground state of mercury; e.g., see Sect. 4.5. An experiment has been performed by Kuntze *et al.* (1993) investigating Auger transitions on barium. This would be another task for the future.

The angular distribution and spin polarization of the emitted Auger electrons are functions of anisotropy and spin polarization parameters, describing the decay dynamics, and of the general tensors of alignment and orientation, $\mathcal{A}_{KQ}$ and $\mathcal{O}_{KQ}$, respectively, containing information about the primary ionization process. While a number of theoretical data are available for the first set of parameters, numerical data for alignment and orientation have been obtained analytically for the case of photoexcitation. Numerical and experimental data for valence shell orientation and alignment may be found in the literature. The Auger emission process, however, generally requires knowledge of alignment and orientation after a deep inner shell ionization and excitation, respectively.

The first investigation on deep inner shell alignment and orientation has been performed by Berezhko *et al.* (1978b) for the case of photoionization, almost 30 years ago. The theoretical formalism has been discussed in Sect. 2.3.1. Only recently, a large variety of numerical deep inner shell alignment and orientation data for closed shell atoms and cations have been published (Kleiman and Lohmann 2003; Kleiman and Becker 2005). Lohmann and Kleiman (2006) developed a theoretical model for calculating alignment and orientation of photoionized open shell atoms and provided first numerical open shell data and predictions for inner shell alignment and orientation, respectively.

In contrast to photoexcitation, alignment and orientation after electron impact excitation are no longer analytical expressions, but depend on the electron excitation matrix elements; e.g., see Sect. 2.3.3. Here, alignment tensor and orientation parameters have been investigated comparatively detailed in the light of coincidence and non-coincidence experiments (e.g., Andersen *et al.* 1997; Srivastava *et al.* 1996a, 1996b) but only limited research has been performed in the context of studying the subsequent Auger emission processes (see, e.g., Mehlhorn 1990; Kaur and Srivastava 1999; Feuerstein *et al.* 1999; Theodosiou 1987). However, even such studies have been confined to electron excitation from the more outer sub-valence atomic shells, only and their subsequent decay via Auger emission has been to a final state with total angular momentum $J_f = 0$ in which case the angular distribution parameter of Auger emission becomes a constant number independent of the matrix elements; e.g., see Sect. 2.5.2.

A more extended investigation, including Auger transitions to final states with $J_f \neq 0$, has been performed by Lohmann *et al.* (2002) who calculated the primary excitation cross section, as well as the related alignment and orientation parameters, for the electronically excited $Ar^*(2p_{3/2}^{-1}4s_{1/2})_{J=1}$ $L_3M_{2,3}M_{2,3}$ Auger transitions (e.g., Kleinpoppen *et al.* 2005).

This kind of research is even more important realizing the fact that, in some sense, the emission of an Auger electron after electron impact excitation can be seen as a special case of present $(e, 2e)$ experiments (e.g., Paripás *et al.* 1997; Balashov and Bodrenko 1999, 2000; Taouil *et al.* 1999; Weigold *et al.* 2002; Bartschat 2003; Knyr *et al.* 2003; Lower *et al.* 2004), and particularly the work by Birgit Lohmann (1996a). I.e., coincidence experiments observing the scattered and the emitted Auger electrons simultaneously,

$$
e^- + A \longrightarrow A^* + e_s^-
$$
$$
\hookrightarrow A^{+*} + e_{Auger}^-. \tag{6.1}
$$

Due to their Fermi character the two emitted electrons are indistinguishable. Thus, within the limit of the two-step model, the investigation of Auger emission experiments after electron impact excitation can yield complementary information on $(e, 2e)$ experiments, too

It is a task for the future to calculate alignment and orientation parameters after deep inner shell electron impact ionization. The case of electron impact ionization is more complicated since, in a general approach, we are here dealing with a three-body problem. With respect to current multi-coincidence studies, the Auger emission after electron impact ionization must be interpreted as a special case of an $(e, 3e)$ experiment, observing, both, the scattered and the ionized electron, as well as the emitted Auger electron in coincidence,

$$
e^- + A \longrightarrow A^+ + e_i^- + e_s^-
$$
$$
\hookrightarrow A^{2+} + e_{Auger}^-. \tag{6.2}
$$

Attempts into this direction have been performed (e.g., Coplan *et al.* 2002; Bartschat 2003; Knyr *et al.* 2003).

Another active field of research are experiments of the type

$$\gamma_{Syn} + A \longrightarrow A^*$$
$$\longrightarrow A^{+*} + e^-_{Auger_1}$$
$$\longrightarrow A^{2+} + e^-_{Auger_2}, \tag{6.3}$$

where the two Auger electrons, emitted from the resonantly excited and from the singly ionized intermediate state, respectively, are observed in coincidence. The investigation of this $(\gamma, 2e)$ coincidence process is important for the interpretation of a variety of experiments (e.g., Viefhaus $et\ al.$ 1996, 1998; Wehlitz $et\ al.$ 1999). Theoretical investigations have been published (Kabachnik and Schmidt 1995) for some special cases of interest. However, there is need for a general theory, assuming either the photon or the atom or both in an arbitrarily polarized state. This should enable experimentalists to systematically study photoexcited targets through the periodic table and completely analyze the relevant amplitudes to a larger degree of completeness. A more complete theoretical analysis and investigation into this field has been published by Zimmermann (2000).

The research of $(\gamma, 2e)$ coincidence processes also covers the area of Auger–photoelectron coincidence spectroscopy (APECS) which is a still growing field (e.g., Bartynski 2003; Bolognesi $et\ al.$ 2003).

$$\gamma_{Syn} + A \longrightarrow A^{+*} + e^-_{Phot}$$
$$\longrightarrow A^{2+} + e^-_{Auger}. \tag{6.4}$$

Alignment effects after autoionization decay of atomic states have been observed in ions (see, e.g. Zimmermann $et\ al.$ 2000), furthermore, interference effects in the alignment and orientation in Kr II ion states have been analyzed (Lagutin $et\ al.$ 2003); interestingly, the direction of electron emission from photoionized atoms can be controlled by varying the phase of the ionizing laser field (Paulus $et\ al.$ 2003).

The case of observing the spin polarization of the emitted electrons in coincidence has not been considered so far. The theoretical analysis of spin resolved $(\gamma, 2e)$ experiments of the type of (6.3) leads to so-called $tensor\ polarization$ parameters. Though an electron spin filter has not yet been experimentally realized, the theoretical investigation of the tensor polarization can provide additional information which may result in new experimental approaches for such type of experiments. In particular, observing the tensor polarization could yield a $new\ method$ of indirectly determining the electronic spin polarization. Additional information can be obtained by using polarized targets.

This field of research is complemented by so-called $(\gamma, e\gamma)$ coincidence experiments observing the emitted photoelectron and the fluorescence photon simultaneously

$$\gamma_{Syn} + A \longrightarrow A^{+*} + e^-_{Phot}$$
$$\longrightarrow A^+ + h\nu. \tag{6.5}$$

The general formalism for describing this process has been developed (Lohmann *et al.* 2003a), and detailed examples with respect to complete photoionization experiments have been published (e.g., Beyer *et al.* 1995; Schmidt 1997; Ueda *et al.* 1998; Becker and Crowe 2001). However, specific applications with respect to angle and spin resolved coincidence studies for both, fluorescence photon and the photoelectron, still need to be performed.

Related to this field of research are resonant Auger studies with a subsequent fluorescence polarization analysis

$$\gamma_{Syn} + A \longrightarrow A^*$$
$$\hookrightarrow A^{+*} + e^-_{Auger}$$
$$\hookrightarrow A^+ + h\nu. \tag{6.6}$$

Here, experiments utilizing Auger resonant Raman techniques have been performed (Meyer *et al.* 2001; O'Keeffe *et al.* 2003, 2004). However, a polarization analysis has been achieved for the emitted fluorescence photons, only.

Furthermore, experiments for triple coincidence processes, like $(\gamma, e\gamma e)$ which may be described in a double two-step model are under discussion.

Another field of research are molecular Auger processes. Numerical investigations have been performed for the Auger emission from HCl (Bonhoff 1998b), while the experimental spectrum has been obtained by Aksela *et al.* (1983a). Experiments for the angular distribution of molecular Auger electrons have been done for resonantly excited CO by Hemmers *et al.* (1993), while numerical investigations for the angular distribution parameters of CO have been performed by Bonhoff (1998a).

Numerical results for the molecular spectra and anisotropy parameters of angular distribution have been obtained for freely rotating HF (see Sect. 5.4) and H_2O molecules (Lehmann *et al.* 1997). For the latter, further extensions to polyatomic molecules are of interest. Here, point group symmetries can be applied in order to derive selection rules for the relevant transition amplitudes (see Sect. 5.5).

While a theory for the angular distribution of Auger electrons emitted in the decay of molecular vacancies created by electron impact has been developed (see Sect. 5.1), the spin polarization of molecular Auger electrons still needs further development. First attempts for diatomics have been made (see Sect. 5.2). For the case of an electron impact ionization the number of independent parameters is not restricted by dipole selection rules. Here, it is necessary to determine the maximum number of parameters, that can be obtained from such kind of experiments and to investigate whether and what additional information can be obtained if polarized electron or photon beams are used for the primary inner shell ionization of the molecules. For atoms such additional parameters are known to provide information about, for instance, shape, size and spatial orientation of the electronic charge cloud.

A current field of molecular research are angle resolved photoelectron–photoion coincidence studies and experiments utilizing photoionization techniques using either linearly or circularly polarized light (Golovin 1991; Motoki *et al.* 2000; Heiser *et al.* 1997; Takahashi *et al.* 2000; Landers *et al.* 2001; Lafosse *et al.* 2000)

$$\gamma_{Syn} + M(AB) \longrightarrow M(AB)^{+*} + e^-$$
$$\longrightarrow A^{+*} + B. \tag{6.7}$$

The photoemission of the molecule $M(AB)$ results in an excited molecular ion $M(AB)^{+*}$ which subsequently dissociates into an ion state A^+ and a neutral atom B. The photoelectron and the ion are then observed in coincidence. An angle resolved coincidence experiment has been performed by Geßner *et al.* (2002) for the photoemission of NO, observing the angular distribution of the N^+ ion relative to the photoelectron emission direction. Changing the helicity of the incoming circularly polarized light a so-called circular dichroism in the angular distribution has been observed. Related calculations which employed a multi-channel Schwinger configuration interaction calculation were found in good agreement with the experiment (Geßner *et al.* 2002).

The outlined research is complemented by investigations of the emission of spin polarized Auger electrons from a solid surface. As has been pointed out by Kessler (1985), large degree of spin polarization has been observed for such transitions. This has been confirmed in an experiment by Müller *et al.* (1995). Numerically, such investigations could be realized using Bloch waves for modeling the surface effects. Attempts into that direction have been done by van der Laan and Thole (1995). Experiments have been performed detecting a solid state electronic spin effect by utilizing spin polarized electron energy loss spectroscopy with ultrathin Co films on copper (001)-crystals (Vollmer *et al.* 2003); the connection of this solid state effect to electron spin effects in collision physics has still to be worked out.

A good survey of the present state of the art angle and spin resolved Auger emission, and more generally, all areas of such type of coincidence studies and related experiments may be found in the *Conference Proceedings* of the bi-annual *International Symposium on Polarization and Correlation in Electronic and Atomic Collisions* held together with the *(e, 2e), Double Photoionization and Related Topics Symposium* (e.g., Hanne *et al.* 2003) and the *International Conference on Photonic, Electronic and Atomic Collisions* (the so-called ICPEAC, e.g., Schuch *et al.* 2004).

In 1975 the *Stirling Symposium on Electron and Photon Interactions with Atoms* took place, finding myself still in high school working as a baccalaureate for my graduation during that time. The Stirling Symposium was dedicated to honor Ugo Fano and a *Festschrift* for him was published (Kleinpoppen and McDowell 1976). Bederson and Miller (1976) presented an updated summary on spin polarization in electron–atom scattering, and a large part of invited papers were devoted to angular correlation, $(e, 2e)$, and electron-spin effects. Accordingly, the Stirling meeting can be considered as a predecessor of the present bi-annual Symposia on both, $(e, 2e)$, Double Photoionization and Related Topics, and Polarization and Correlation in Electronic and Atomic Collisions, and as the head-waters of most of nowadays state of the art research in the fields of coincident angle and spin resolved atomic and molecular physics.

A Vector Addition Coefficients of Angular Momenta

In the following, we will give an overview of the vector addition coefficients and their properties together with a variety of contraction and orthogonality relations most frequently used in angular momentum coupling theory. Concerning a more complete description and introduction to the theory of angular momentum coupling we refer to the books by Brink and Satchler (1962); Edmonds (1974) or Zare (1988). For a more extensive compendium of angular momentum coupling equations and relations we refer to the book by Varshalovich *et al.* (1988).

The vector addition coefficients have been numerically calculated and a variety of tables for the Clebsch–Gordan and the nj-symbols exist. We only mention the extended tables by Rotenberg *et al.* (1959) and Varshalovich *et al.* (1988).

A.1 Clebsch–Gordan Coefficients and $3j$-Symbols

The Clebsch–Gordan coefficients, or vector addition coefficients, are defined by the unitary transformation of the coupling of two angular momenta

$$|(ab)c\gamma\rangle = \sum_{\alpha\beta}|a\alpha, b\beta\rangle(a\alpha, b\beta|c\gamma), \tag{A.1}$$

and vanish unless the selection rule

$$\alpha + \beta = \gamma, \tag{A.2}$$

and the triangular condition

$$|a - b| \leq c \leq a + b \tag{A.3}$$

are fulfilled.

The Clebsch–Gordan coefficients are real quantities

$$\left(a\alpha, b\beta|c\gamma\right)^* = \left(a\alpha, b\beta|c\gamma\right), \tag{A.4}$$

which yields the inverse transformation

$$|a\alpha, b\beta\rangle = \sum_{c\gamma}|(ab)c\gamma\rangle(a\alpha, b\beta|c\gamma). \tag{A.5}$$

Symmetry properties:

$$\left(a\alpha, b\beta | c\gamma\right) = (-1)^{a+b-c}\left(a-\alpha, b-\beta | c-\gamma\right), \tag{A.6a}$$

$$= (-1)^{a+b-c}\left(b\beta, a\alpha | c\gamma\right), \tag{A.6b}$$

$$= \sqrt{\frac{2c+1}{2b+1}}(-1)^{a-\alpha}\left(a\alpha, c-\gamma | b-\beta\right), \tag{A.6c}$$

$$= \sqrt{\frac{2c+1}{2a+1}}(-1)^{b+\beta}\left(c-\gamma, b\beta | a-\alpha\right). \tag{A.6d}$$

Special cases:

$$\left(a\alpha, b\beta | 00\right) = \frac{(-1)^{a-\alpha}}{\sqrt{2a+1}}\delta_{ab}\delta_{\alpha-\beta}, \tag{A.7a}$$

$$\left(a\alpha, 00 | c\gamma\right) = \delta_{ac}\delta_{\alpha\gamma}. \tag{A.7b}$$

If c and γ take their maximum values, we have:

$$\left(aa, bb | a+b\ a+b\right) = 1 \tag{A.7c}$$

The orthonormality of $|a\alpha, b\beta\rangle$ and $|(ab)c\gamma\rangle$ yields the orthogonality relations:

$$\sum_{c\gamma}\left(a\alpha', b\beta' | c\gamma\right)\left(a\alpha, b\beta | c\gamma\right) = \delta_{\alpha'\alpha}\delta_{\beta'\beta}, \tag{A.8a}$$

$$\sum_{\alpha\beta}\left(a\alpha, b\beta | c'\gamma'\right)\left(a\alpha, b\beta | c\gamma\right) = \delta_{c'c}\delta_{\gamma'\gamma}. \tag{A.8b}$$

The Wigner $3j$-symbols are defined as

$$\begin{pmatrix} a & b & c \\ \alpha & \beta & \gamma \end{pmatrix} = \frac{(-1)^{a-b-\gamma}}{\sqrt{2c+1}}\left(a\alpha, b\beta | c-\gamma\right). \tag{A.9}$$

Note the appearance of $-\gamma$ on the right, so that now the selection rule

$$\alpha + \beta + \gamma = 0, \tag{A.10}$$

must be fulfilled.

The $3j$-symbol is invariant under cyclic permutations of its columns and is multiplied by $(-1)^{a+b+c}$ by non-cyclic ones, and by changing the signs of its magnetic components α, β, γ. In particular:

$$\begin{pmatrix} a & b & c \\ \alpha & \beta & \gamma \end{pmatrix} = \begin{pmatrix} b & c & a \\ \beta & \gamma & \alpha \end{pmatrix} = \begin{pmatrix} c & a & b \\ \gamma & \alpha & \beta \end{pmatrix}, \tag{A.11a}$$

$$\begin{pmatrix} a & b & c \\ \alpha & \beta & \gamma \end{pmatrix} = (-1)^{a+b+c}\begin{pmatrix} b & a & c \\ \beta & \alpha & \gamma \end{pmatrix}, \tag{A.11b}$$

$$= (-1)^{a+b+c}\begin{pmatrix} a & b & c \\ -\alpha & -\beta & -\gamma \end{pmatrix}. \tag{A.11c}$$

Special cases:

$$\begin{pmatrix} a & b & 0 \\ \alpha & \beta & 0 \end{pmatrix} = \frac{(-1)^{a-\alpha}}{\sqrt{2a+1}} \delta_{ab} \delta_{\alpha-\beta}. \qquad \text{(A.12a)}$$

If $\alpha = \beta = \gamma = 0$, and $a + b + c$ is odd, we have

$$\begin{pmatrix} a & b & c \\ 0 & 0 & 0 \end{pmatrix} = 0, \qquad \text{(A.12b)}$$

and if $2p \equiv a + b + c$ is even, we have

$$\begin{pmatrix} a & b & c \\ 0 & 0 & 0 \end{pmatrix} = (-1)^p \sqrt{\Delta(abc)} \frac{p!}{(p-a)!(p-b)!(p-c)!}, \qquad \text{(A.12c)}$$

where

$$\Delta(abc) \equiv \frac{(a+b-c)!(b+c-a)!(c+a-b)!}{(a+b+c+1)!}. \qquad \text{(A.12d)}$$

If the arguments of β and γ change by 1 an important relation is

$$\begin{pmatrix} a & b & c \\ \frac{1}{2} & \frac{1}{2} & -1 \end{pmatrix} = \frac{-1}{2} \begin{pmatrix} a & b & c \\ \frac{1}{2} & \frac{-1}{2} & 0 \end{pmatrix} \frac{(2b+1) + (-1)^{a+b-c}(2a+1)}{\sqrt{c(c+1)}}. \qquad \text{(A.13)}$$

Orthogonality relations:

$$\sum_{\alpha\beta} \begin{pmatrix} a & b & c \\ \alpha & \beta & \gamma \end{pmatrix} \begin{pmatrix} a & b & c' \\ \alpha & \beta & \gamma' \end{pmatrix} = \frac{1}{2c+1} \delta_{cc'} \delta_{\gamma\gamma'}, \qquad \text{(A.14a)}$$

$$\sum_{c\gamma} (2c+1) \begin{pmatrix} a & b & c \\ \alpha & \beta & \gamma \end{pmatrix} \begin{pmatrix} a & b & c \\ \alpha' & \beta' & \gamma \end{pmatrix} = \delta_{\alpha\alpha'} \delta_{\beta\beta'}. \qquad \text{(A.14b)}$$

A.2 Racah Coefficients and $6j$-Symbols

The coupling of three angular momenta, $|m\,m'\,m''\rangle \equiv |jm\rangle|j'm'\rangle|j''m''\rangle$, allows for the generation of two usually different basis sets in the related Hilbert sub-space

$$|(j'j)g', j''; J'M'\rangle = \sum_{\substack{mm' \\ m''\mu'}} |m\,m'\,m''\rangle(j'm', jm|g'\mu')(g'\mu', j''m''|JM), \quad \text{(A.15a)}$$

or

$$|j', (jj'')g''; JM\rangle = \sum_{\substack{mm' \\ m''\mu''}} |m\,m'\,m''\rangle(jm, j''m''|g''\mu'')(j'm', g''\mu''|JM). $$

$$\text{(A.15b)}$$

The $6j$-symbols are defined by the unitary transformation between the two basis sets

$$\langle j', (jj'')g''; JM | (j'j)g', j''; J'M' \rangle = \delta_{JJ'}\delta_{MM'}(-1)^{j+j'+j''+J}$$
$$\times \sqrt{(2g'+1)(2g''+1)} \left\{ \begin{matrix} j' & j & g' \\ j'' & J & g'' \end{matrix} \right\},$$

(A.16)

that is, we have

$$| (j'j)g', j''; JM \rangle = \sum_{g''} | j', (jj'')g''; JM \rangle \sqrt{(2g'+1)(2g''+1)}$$
$$\times (-1)^{j+j'+j''+J} \left\{ \begin{matrix} j' & j & g' \\ j'' & J & g'' \end{matrix} \right\}.$$

(A.17)

The $6j$-symbols are related to the Racah, or W-coefficients via

$$\left\{ \begin{matrix} a & b & c \\ d & e & f \end{matrix} \right\} = (-1)^{a+b+d+e} W(abed; cf).$$

(A.18)

Triangular conditions: the $6j$-symbol is non-zero only, if the four triangular conditions are fulfilled by the six angular momenta which may be illustrated in the following way

$$\left\{ \begin{matrix} \circ & \\ & \diagdown \\ & \circ - \circ \end{matrix} \right\}, \left\{ \begin{matrix} \circ - \circ - \circ \\ \end{matrix} \right\}, \left\{ \begin{matrix} & & \circ \\ \circ - \circ & \diagup \end{matrix} \right\}, \left\{ \begin{matrix} \circ & \diagup & \\ \circ & & \diagdown \circ \end{matrix} \right\}.$$

(A.19)

Particularly, the sum of its arguments must be integer

$$a+b+c+d+e+f = n, \quad \text{where} \quad n \in \mathbb{N}.$$

(A.20)

Symmetries: the $6j$-symbol is invariant under any permutation of its columns, and also for interchanging of the upper and lower arguments in each of any two columns, resulting in symmetry relations between 24 different $6j$-symbols, i.e.

$$\left\{ \begin{matrix} a & b & c \\ d & e & f \end{matrix} \right\} = \left\{ \begin{matrix} a & c & b \\ d & f & e \end{matrix} \right\} = \left\{ \begin{matrix} b & a & c \\ e & d & f \end{matrix} \right\} = \left\{ \begin{matrix} a & e & f \\ d & b & c \end{matrix} \right\} = \left\{ \begin{matrix} d & c & e \\ a & f & b \end{matrix} \right\}, \quad \text{etc.} \quad \text{(A.21)}$$

Special value:

$$\left\{ \begin{matrix} a & b & 0 \\ c & d & f \end{matrix} \right\} = (-1)^{a+c+f} \frac{\delta_{ab}\delta_{cd}}{\sqrt{(2a+1)(2c+1)}}.$$

(A.22)

Contraction of $3j$-symbols: according to the definition (A.16) the $6j$-symbols may be expressed in terms of four $3j$-symbols

$$\sum_{\substack{\alpha\beta\gamma \\ \alpha'\beta'\gamma'}} (-1)^{A+B+C+\alpha+\beta+\gamma} \begin{pmatrix} A & B & c \\ \alpha & -\beta & \gamma' \end{pmatrix} \begin{pmatrix} B & C & a \\ \beta & -\gamma & \alpha' \end{pmatrix} \begin{pmatrix} C & A & b \\ \gamma & -\alpha & \beta' \end{pmatrix} \begin{pmatrix} a & b & c \\ \alpha' & \beta' & \gamma' \end{pmatrix}$$

$$= \left\{ \begin{matrix} a & b & c \\ A & B & C \end{matrix} \right\}.$$

(A.23)

Note, that the sum runs in fact over two indices, only, as the primed and unprimed magnetic components of the $3j$-symbols are not independent; see (A.10). Omitting one index, e.g. γ' from the summation, the left-hand side of (A.23) must be multiplied by $(2c+1)$ which yields

$$\sum_{\substack{\alpha\beta\gamma \\ \alpha'\beta'}} (-1)^{A+B+C+\alpha+\beta+\gamma} \begin{pmatrix} A & B & c \\ \alpha & -\beta & \gamma' \end{pmatrix} \begin{pmatrix} B & C & a \\ \beta & -\gamma & \alpha' \end{pmatrix} \begin{pmatrix} C & A & b \\ \gamma & -\alpha & \beta' \end{pmatrix} \begin{pmatrix} a & b & c_1 \\ \alpha' & \beta' & \gamma'_1 \end{pmatrix}$$

$$= \delta_{cc_1} \delta_{\gamma'\gamma'_1} \frac{1}{2c+1} \begin{Bmatrix} a & b & c \\ A & B & C \end{Bmatrix}. \tag{A.24}$$

Again, the sum is over two indices, only. Further relations can be obtained applying the orthogonality properties (A.14) of the $3j$-symbols. Multiplying both sides of (A.23) with the last $3j$-symbol yields

$$\sum_{\alpha\beta\gamma} (-1)^{A+B+C+\alpha+\beta+\gamma} \begin{pmatrix} A & B & c \\ \alpha & -\beta & \gamma' \end{pmatrix} \begin{pmatrix} B & C & a \\ \beta & -\gamma & \alpha' \end{pmatrix} \begin{pmatrix} C & A & b \\ \gamma & -\alpha & \beta' \end{pmatrix}$$

$$= \begin{pmatrix} a & b & c \\ \alpha' & \beta' & \gamma' \end{pmatrix} \begin{Bmatrix} a & b & c \\ A & B & C \end{Bmatrix}. \tag{A.25}$$

Due to the same argument, the sum is over one index, only. Continuing in the same manner, we get

$$\sum_{C\gamma} (-1)^{C+\gamma} (2C+1) \begin{pmatrix} a & B & C \\ \alpha' & \beta & -\gamma \end{pmatrix} \begin{pmatrix} b & A & C \\ \beta' & \alpha & \gamma \end{pmatrix} \begin{Bmatrix} a & b & c \\ A & B & C \end{Bmatrix}$$

$$= (-1)^{b+B+c+C} \sum_{\gamma'} (-1)^{c+\gamma'} \begin{pmatrix} a & b & c \\ \alpha' & \beta' & -\gamma' \end{pmatrix} \begin{pmatrix} B & A & c \\ \beta & \alpha & \gamma' \end{pmatrix}. \tag{A.26a}$$

Note, that due to (A.10), the formal sum over γ and γ' is over one term, only. Thus, (A.26) may be re-expressed as

$$\sum_{C} (-1)^{a+b-c+A+B+C-\alpha'-\alpha} (2C+1) \begin{Bmatrix} a & b & c \\ A & B & C \end{Bmatrix} \begin{pmatrix} B & a & C \\ \beta & \alpha' & -\gamma \end{pmatrix} \begin{pmatrix} b & A & C \\ \beta' & \alpha & \gamma \end{pmatrix}$$

$$= \begin{pmatrix} a & b & c \\ \alpha' & \beta' & -\gamma' \end{pmatrix} \begin{pmatrix} A & B & c \\ \alpha & \beta & \gamma' \end{pmatrix}. \tag{A.26b}$$

Continuing with this procedure yields

$$\sum_{BC} (-1)^{a+b-c+A+B+C-\alpha'-\alpha} (2B+1)(2C+1) \begin{Bmatrix} a & b & c \\ A & B & C \end{Bmatrix}$$

$$\times \begin{pmatrix} B & a & C \\ \beta & \alpha' & -\gamma \end{pmatrix} \begin{pmatrix} b & A & C \\ \beta' & \alpha & \gamma \end{pmatrix} \begin{pmatrix} A & B & c \\ \alpha & \beta & \gamma' \end{pmatrix} = \begin{pmatrix} a & b & c \\ \alpha' & \beta' & -\gamma' \end{pmatrix}, \tag{A.27}$$

and

$$\sum_{cBC}(-1)^{a+b-c+A+B+C-\alpha'-\alpha}(2c+1)(2B+1)(2C+1)\begin{Bmatrix} a & b & c \\ A & B & C \end{Bmatrix}$$

$$\times \begin{pmatrix} B & a & C \\ \beta & \alpha' & -\gamma \end{pmatrix}\begin{pmatrix} b & A & C \\ \beta' & \alpha & \gamma \end{pmatrix}\begin{pmatrix} A & B & c \\ \alpha & \beta & \gamma' \end{pmatrix}\begin{pmatrix} a & b & c \\ \alpha' & \beta' & -\gamma' \end{pmatrix} = 1. \tag{A.28}$$

If $a+b+e$ is even a special case is

$$(-1)^{a+b+c+d+1}\sqrt{(2a+1)(2b+1)}\begin{pmatrix} a & b & e \\ 0 & 0 & 0 \end{pmatrix}\begin{Bmatrix} a & b & e \\ d & c & \frac{1}{2} \end{Bmatrix} = \begin{pmatrix} c & d & e \\ -\frac{1}{2} & \frac{1}{2} & 0 \end{pmatrix}. \tag{A.29}$$

Racah–Elliot relation and orthogonality relations:

$$\sum_{x}(-1)^{2x}(2x+1)\begin{Bmatrix} a & b & x \\ a & b & f \end{Bmatrix} = 1. \tag{A.30}$$

$$\sum_{x}(-1)^{a+b+x}(2x+1)\begin{Bmatrix} a & b & x \\ b & a & f \end{Bmatrix} = \delta_{f0}\sqrt{(2a+1)(2b+1)}. \tag{A.31}$$

$$\sum_{x}(2x+1)\begin{Bmatrix} a & b & x \\ c & d & f \end{Bmatrix}\begin{Bmatrix} c & d & x \\ a & b & g \end{Bmatrix} = \delta_{fg}\frac{1}{(2f+1)}. \tag{A.32}$$

$$\sum_{x}(-1)^{f+g+x}(2x+1)\begin{Bmatrix} a & b & x \\ c & d & f \end{Bmatrix}\begin{Bmatrix} c & d & x \\ b & a & g \end{Bmatrix} = \begin{Bmatrix} a & d & f \\ b & c & g \end{Bmatrix}. \tag{A.33}$$

$$\sum_{x}(-1)^{a+b+c+d+e+f+g+h+j+x}(2x+1)\begin{Bmatrix} a & b & x \\ c & d & g \end{Bmatrix}\begin{Bmatrix} c & d & x \\ e & f & h \end{Bmatrix}\begin{Bmatrix} e & f & x \\ b & a & j \end{Bmatrix}$$

$$= \begin{Bmatrix} g & h & j \\ e & a & d \end{Bmatrix}\begin{Bmatrix} g & h & j \\ f & b & c \end{Bmatrix}. \tag{A.34}$$

A.3 9j-Symbols

The coupling of four angular momenta **a**, **b**, **d**, and **e** resulting in a total angular momentum **i** with z-component m leads to two different basis systems for the Hilbert sub-space of the total angular momentum $|im\rangle$:

$$|(ab)c, (de)f; im\rangle \quad \text{and} \quad |(ad)g, (be)h; im\rangle.$$

As in the three vector case, the corresponding eigenfunctions of the basis sets are not independent. They are connected by a linear transformation

$$|(ad)g, (be)h; im\rangle = \sum_{cf}|(ab)c, (de)f; im\rangle$$

$$\times \langle(ab)c, (de)f; im|(ad)g, (be)h; im\rangle. \tag{A.35}$$

The transformation coefficient in (A.35) that changes the coupling defines the $9j$-symbol of Wigner

$$\langle (ab)c, (de)f; im|(ad)g, (be)h; im\rangle$$

$$= \sqrt{(2c+1)(2f+1)(2g+1)(2h+1)} \begin{Bmatrix} a & b & c \\ d & e & f \\ g & h & i \end{Bmatrix}$$

$$\equiv X(abc, def, ghi), \tag{A.36}$$

which is identical to the X-function of Fano.

Triangular conditions: the $9j$-symbol vanishes unless the triangular conditions for the triads (a, b, c), (d, e, f), (g, h, i), (a, d, g), (b, e, h), and (c, f, i) are fulfilled.

Symmetry: the $9j$-symbol is invariant under interchange of rows and columns (reflection about a diagonal) and is multiplied by $(-1)^p$, where $p = a + b + c + d + e + f + g + h + i$, upon interchanging of two adjacent rows or columns, resulting in 72 symmetry relations.

Orthogonality:

$$\sum_{cf} (2c+1)(2f+1) \begin{Bmatrix} a & b & c \\ d & e & f \\ g & h & i \end{Bmatrix} \begin{Bmatrix} a & b & c \\ d & e & f \\ j & k & i \end{Bmatrix} = \frac{\delta_{gj}\delta_{hk}}{(2g+1)(2h+1)}. \tag{A.37}$$

Sum rule:

$$\sum_{jk} (2j+1)(2k+1) \begin{Bmatrix} a & b & c \\ d & e & f \\ j & k & i \end{Bmatrix} \begin{Bmatrix} a & e & j \\ d & b & k \\ g & h & i \end{Bmatrix} = \begin{Bmatrix} a & b & c \\ d & e & f \\ g & h & i \end{Bmatrix}. \tag{A.38}$$

Identical rows or columns: in this case the $9j$-symbol vanishes unless the sum of all arguments is even

$$\begin{Bmatrix} a & b & c \\ a & b & c \\ g & h & j \end{Bmatrix} = 0 \quad \text{if } g + h + j = 2k + 1, \tag{A.39a}$$

and

$$\begin{Bmatrix} a & a & c \\ d & d & f \\ g & g & j \end{Bmatrix} = 0 \quad \text{if } c + f + j = 2k + 1, \tag{A.39b}$$

where $k \in \mathbb{N}$.

Special values of the arguments: if one argument equals zero the $9j$-symbol is reduced to a $6j$-symbol

$$\begin{Bmatrix} a & b & c \\ d & e & f \\ g & h & 0 \end{Bmatrix} = \delta_{cf}\delta_{gh} \frac{(-1)^{b+c+d+g}}{\sqrt{(2c+1)(2g+1)}} \begin{Bmatrix} a & b & c \\ e & d & g \end{Bmatrix}. \tag{A.40}$$

If two arguments are equal to zero we have

$$
\begin{Bmatrix} a & b & c \\ d & 0 & f \\ g & h & 0 \end{Bmatrix} = \delta_{df}\delta_{bh}\delta_{cf}\delta_{gh} \frac{(-1)^{a-b-c}}{(2b+1)(2c+1)}, \tag{A.41}
$$

and for three arguments equal to zero we obtain

$$
\begin{Bmatrix} a & b & c \\ d & e & f \\ 0 & 0 & 0 \end{Bmatrix} = \frac{\delta_{ad}\delta_{be}\delta_{cf}}{\sqrt{(2a+1)(2b+1)(2c+1)}}, \tag{A.42a}
$$

and

$$
\begin{Bmatrix} 0 & b & c \\ d & 0 & f \\ g & h & 0 \end{Bmatrix} = \delta_{bc}\delta_{bd}\delta_{bf}\delta_{bg}\delta_{bh} \frac{(-1)^{2b}}{(2b+1)^2}. \tag{A.42b}
$$

If one arguments equals unity, the $9j$-symbol can be also reduced to a $6j$-symbol

$$
\begin{Bmatrix} a & b & c \\ d & e & c \\ g & g & 1 \end{Bmatrix} = (-1)^{b+c+d+g} \frac{a(a+1) - d(d+1) - b(b+1) + e(e+1)}{\sqrt{(g+1)(2g+1)2g(c+1)(2c+1)2c}} \begin{Bmatrix} a & b & c \\ e & d & g \end{Bmatrix}. \tag{A.43}
$$

Important relations to the $3j$- and $6j$-symbols exist if one of the triads in the $9j$-symbol equals $(1/2, 1/2, 1)$

$$
\begin{Bmatrix} a & b & c \\ d & e & f \\ \frac{1}{2} & \frac{1}{2} & 1 \end{Bmatrix} \begin{Bmatrix} c & f & 1 \\ \frac{1}{2} & \frac{1}{2} & g \end{Bmatrix} = \frac{(-1)^{2g}}{3} \begin{Bmatrix} a & b & c \\ \frac{1}{2} & g & e \end{Bmatrix} \begin{Bmatrix} e & d & f \\ \frac{1}{2} & g & a \end{Bmatrix}
$$

$$
- \delta_{cf} \frac{(-1)^{b+d-g}}{6(2c+1)} \begin{Bmatrix} a & b & c \\ e & d & \frac{1}{2} \end{Bmatrix}. \tag{A.44}
$$

In addition, if c, d, e are integer and $c + d + e$ is even, we get

$$
\sqrt{6(2c+1)(2d+1)(2e+1)} \begin{pmatrix} c & d & e \\ 0 & 0 & 0 \end{pmatrix} \begin{Bmatrix} a & b & c \\ d & e & c \\ \frac{1}{2} & \frac{1}{2} & 1 \end{Bmatrix} = \begin{pmatrix} a & b & c \\ \frac{1}{2} & \frac{1}{2} & -1 \end{pmatrix}. \tag{A.45}
$$

If $c + d + e$ is odd, we have two other relations

$$
\begin{pmatrix} c+1 & d & e \\ 0 & 0 & 0 \end{pmatrix} \begin{Bmatrix} a & b & c \\ d & e & c+1 \\ \frac{1}{2} & \frac{1}{2} & 1 \end{Bmatrix} = (-1)^{b+c+1/2} \begin{pmatrix} a & b & c \\ \frac{1}{2} & -\frac{1}{2} & 0 \end{pmatrix}
$$

$$
\times \frac{[(d-a)(2a+1) + (e-b)(2b+1) + c + 1]}{\sqrt{6(c+1)(2c+1)(2c+3)(2d+1)(2e+1)}}, \tag{A.46a}
$$

and

$$\begin{pmatrix} c-1 & d & e \\ 0 & 0 & 0 \end{pmatrix} \begin{Bmatrix} a & b & c \\ d & e & c-1 \\ \frac{1}{2} & \frac{1}{2} & 1 \end{Bmatrix} = (-1)^{b+c+1/2} \begin{pmatrix} a & b & c \\ \frac{1}{2} & -\frac{1}{2} & 0 \end{pmatrix}$$

$$\times \frac{[(d-a)(2a+1)+(e-b)(2b+1)-c]}{\sqrt{6c(2c+1)(2c-1)(2d+1)(2e+1)}}. \quad \text{(A.46b)}$$

Contraction of 3*j*-symbols: the following equations are continuously applied as a useful tool in angular momentum coupling theory in order to reduce extended expressions by analytically carrying out the sum over the magnetic quantum numbers.

$$\begin{Bmatrix} a & b & c \\ d & e & f \\ g & h & i \end{Bmatrix} = (2a+1) \sum_{\substack{\beta\gamma\delta\epsilon \\ \phi\eta\nu\rho}} \begin{pmatrix} a & b & c \\ \alpha & \beta & \gamma \end{pmatrix} \begin{pmatrix} b & e & h \\ \beta & \epsilon & \eta \end{pmatrix} \begin{pmatrix} c & f & i \\ \gamma & \phi & \nu \end{pmatrix}$$

$$\times \begin{pmatrix} a & d & g \\ \alpha & \delta & \rho \end{pmatrix} \begin{pmatrix} d & e & f \\ \delta & \epsilon & \phi \end{pmatrix} \begin{pmatrix} g & h & i \\ \rho & \eta & \nu \end{pmatrix}. \quad \text{(A.47)}$$

$$\begin{pmatrix} a & b & c \\ \alpha & \beta & \gamma \end{pmatrix} \begin{Bmatrix} a & b & c \\ d & e & f \\ g & h & i \end{Bmatrix} = \sum_{\substack{\delta\epsilon\phi \\ \eta\nu\rho}} \begin{pmatrix} b & e & h \\ \beta & \epsilon & \eta \end{pmatrix} \begin{pmatrix} c & f & i \\ \gamma & \phi & \nu \end{pmatrix} \begin{pmatrix} a & d & g \\ \alpha & \delta & \rho \end{pmatrix}$$

$$\times \begin{pmatrix} d & e & f \\ \delta & \epsilon & \phi \end{pmatrix} \begin{pmatrix} g & h & i \\ \rho & \eta & \nu \end{pmatrix}. \quad \text{(A.48)}$$

$$\sum_{b\beta} (2b+1) \begin{pmatrix} a & b & c \\ \alpha & \beta & \gamma \end{pmatrix} \begin{pmatrix} b & e & h \\ \beta & \epsilon & \eta \end{pmatrix} \begin{Bmatrix} a & b & c \\ d & e & f \\ g & h & i \end{Bmatrix}$$

$$= \sum_{\delta\phi\nu\rho} \begin{pmatrix} c & f & i \\ \gamma & \phi & \nu \end{pmatrix} \begin{pmatrix} a & d & g \\ \alpha & \delta & \rho \end{pmatrix} \begin{pmatrix} d & e & f \\ \delta & \epsilon & \phi \end{pmatrix} \begin{pmatrix} g & h & i \\ \rho & \eta & \nu \end{pmatrix}. \quad \text{(A.49)}$$

Note, that due to the selection rule (A.10), the sum over β on the left side is over one term, only.

$$\sum_{b\beta c\gamma} (2b+1)(2c+1) \begin{pmatrix} a & b & c \\ \alpha & \beta & \gamma \end{pmatrix} \begin{pmatrix} b & e & h \\ \beta & \epsilon & \eta \end{pmatrix} \begin{pmatrix} c & f & i \\ \gamma & \phi & \nu \end{pmatrix} \begin{Bmatrix} a & b & c \\ d & e & f \\ g & h & i \end{Bmatrix}$$

$$= \sum_{\delta\rho} \begin{pmatrix} a & d & g \\ \alpha & \delta & \rho \end{pmatrix} \begin{pmatrix} d & e & f \\ \delta & \epsilon & \phi \end{pmatrix} \begin{pmatrix} g & h & i \\ \rho & \eta & \nu \end{pmatrix}. \quad \text{(A.50)}$$

Due to the same argument, the sums over β and γ on the left, and over δ and ρ on the right side are over one term, only.

Contraction of $6j$-symbols: analogously, the $9j$-symbol may expressed in terms of $6j$-symbols

$$\begin{Bmatrix} a & b & c \\ d & e & f \\ g & h & i \end{Bmatrix} = \sum_x (-1)^{2x}(2x+1) \begin{Bmatrix} a & b & c \\ f & i & x \end{Bmatrix} \begin{Bmatrix} d & e & f \\ b & x & h \end{Bmatrix} \begin{Bmatrix} g & h & i \\ x & a & d \end{Bmatrix}, \qquad (A.51)$$

$$\sum_x (2x+1) \begin{Bmatrix} a & f & x \\ d & q & e \\ p & c & b \end{Bmatrix} \begin{Bmatrix} a & f & x \\ e & b & s \end{Bmatrix} = (-1)^{2s} \begin{Bmatrix} a & b & s \\ c & d & p \end{Bmatrix} \begin{Bmatrix} c & d & s \\ e & f & q \end{Bmatrix}, \qquad (A.52)$$

$$\sum_x (-1)^{R+x}(2x+1) \begin{Bmatrix} a & f & x \\ d & q & e \\ p & c & b \end{Bmatrix} \begin{Bmatrix} a & f & x \\ b & e & s \end{Bmatrix} = (-1)^{2s} \begin{Bmatrix} p & q & s \\ e & a & d \end{Bmatrix} \begin{Bmatrix} p & q & s \\ f & b & c \end{Bmatrix}, \qquad (A.53)$$

where $R = a+b+c+d+e+f+p+q$.

It is sometimes helpful to re-order the coupling of the quantum numbers in the $3nj$-symbols. This can be achieved with the recursion relation

$$(-1)^{a+f+b+i} \sum_x (-1)^{2x}(2x+1) \begin{Bmatrix} a & b & x \\ d & e & f \\ g & h & i \end{Bmatrix} \begin{Bmatrix} a & b & x \\ c & \lambda & a' \end{Bmatrix} \begin{Bmatrix} i & f & x \\ \lambda & c & f' \end{Bmatrix}$$

$$= (-1)^{a'+f'+g+e} \sum_y (-1)^{2y}(2y+1) \begin{Bmatrix} a' & b & y \\ d & e & f' \\ g & h & i \end{Bmatrix} \begin{Bmatrix} f' & e & y \\ d & \lambda & f \end{Bmatrix} \begin{Bmatrix} g & a' & y \\ \lambda & d & a \end{Bmatrix}. \qquad (A.54)$$

B Rotation Matrices and Spherical Harmonics

The behaviour of particles in a scattering system is directly related to its description in the relevant coordinate frame. These relations are dealt with by the theory of rotation matrices. The book of Zare (1988) yields a detailed and broad insight. We are giving a comprehensive overview of the important relations.

B.1 Transformation Properties of Angular Momentum Under Rotation

Any rotation $\mathbf{R}_n(\omega)$ can be specified by giving three parameters, two to fit its rotation axis $\hat{\mathbf{n}}$ and one to fit its rotation angle ω about $\hat{\mathbf{n}}$. For an arbitrary rotation about $\hat{\mathbf{n}}$ by an angle ω, we obtain

$$\mathbf{R}_n(\omega) = \exp(-\mathrm{i}\omega\mathbf{J} \cdot \hat{\mathbf{n}}). \tag{B.1}$$

We note, that an arbitrary rotation cannot change the value of the angular momentum J since $\mathbf{J}^2$ commutes with the rotation operator

$$\left[\mathbf{R}_n(\omega), \mathbf{J}^2\right] = \left[\exp(-\mathrm{i}\omega\mathbf{J} \cdot \hat{\mathbf{n}}), \mathbf{J}^2\right] = \sum_\nu \frac{1}{\nu!}(-\mathrm{i}\omega)^\nu \left[(\mathbf{J} \cdot \hat{\mathbf{n}})^\nu, \mathbf{J}^2\right] = 0. \tag{B.2}$$

Thus, a rotation acting on the angular momentum eigenstates $\left|JM\right\rangle$ of $\mathbf{J}^2$ and J_z can only transform $\left|JM\right\rangle$ into a linear combination of other M values

$$\left|Jm\right\rangle = \mathbf{R}(\alpha, \beta, \gamma)\left|JM\right\rangle = \sum_{M'} \mathcal{D}^{(J)}_{M'M}(\alpha, \beta, \gamma)\left|JM'\right\rangle, \tag{B.3}$$

where the expansion coefficients

$$\mathcal{D}^{(J)}_{M'M}(\alpha, \beta, \gamma) = \left\langle JM'|\mathbf{R}(\alpha, \beta, \gamma)|JM\right\rangle, \tag{B.4}$$

are the elements of a $(2J + 1) \times (2J + 1)$ unitary matrix for $\mathbf{R}$, called the *rotation matrix*, and form the irreducible representation of the rotation group of dimension $2J + 1$ corresponding to an angular momentum J and a rotation around the Euler angles (α, β, γ); see Fig. B.1.

The Euler angles (α, β, γ) are a set of parameters to specify arbitrary rotations by three successive finite rotations; i.e. to make the XYZ space-fixed coordinate frame coincide with the xyz body-fixed frame:

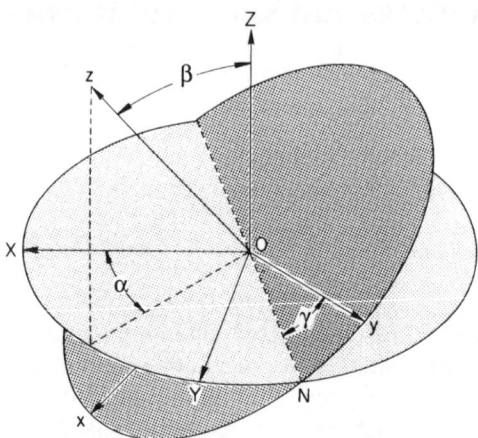

Fig. B.1. The Euler angles α, β, and γ relating the XYZ space-fixed and the xyz body-fixed coordinate frames

1. A counterclockwise rotation α about Z, the vertical axis. This carries the Y axis into the line of nodes N.
2. A counterclockwise rotation β about the line of nodes N. This carries the Z axis into the z axis, i.e. the figure axis of the body.
3. A counterclockwise rotation γ about z, the figure axis. This carries the line of nodes N into the y axis.

Using (B.1), we may express such a rotation as

$$\mathbf{R}(\alpha, \beta, \gamma) = \exp(-i\gamma \mathbf{J} \cdot \hat{\mathbf{n}}_\gamma) \exp(-i\beta \mathbf{J} \cdot \hat{\mathbf{n}}_\beta) \exp(-i\alpha \mathbf{J} \cdot \hat{\mathbf{n}}_\alpha)$$
$$= \exp(-i\gamma J_z) \exp(-i\beta J_N) \exp(-i\alpha J_Z), \tag{B.5}$$

where the form of $\mathbf{R}$ has been chosen to be an active rotation of the physical system. It can be shown, e.g. Zare (1988), that the three Euler angle rotations may all be carried out in the *same* XYZ space-fixed coordinate frame if the order of the rotations is reversed. That is, first a rotation γ about the Z axis, then a rotation β about the Y axis, and finally a rotation α about the same Z axis:

$$\mathbf{R}(\alpha, \beta, \gamma) = \exp(-i\alpha J_Z) \exp(-i\beta J_Y) \exp(-i\gamma J_Z). \tag{B.6}$$

Associated with this is the equivalent rotation of the coordinate frame $(\gamma, \beta, \alpha)^{-1} = (-\alpha, -\beta, -\gamma)$, so their usage results in a different sign convention for the angles of rotation.

Substituting (B.6) into (B.4) and using the fact that the basis vectors $|JM\rangle$ are eigenfunctions of J_Z the rotation matrices simplify as follows

$$\mathcal{D}^{(J)}_{M'M}(\alpha, \beta, \gamma) = \exp(-i\alpha M') d^{(J)}_{M'M}(\beta) \exp(-i\gamma M), \tag{B.7}$$

where

$$d^{(J)}_{M'M}(\beta) = \langle JM'|\exp(-i\beta J_Y)|JM\rangle, \tag{B.8}$$

are the reduced rotation matrices. They are real quantities and explicit expressions as well as extended tables may be found in the literature (Brink and Satchler 1962; Messiah 1979; Varshalovich *et al.* 1988; Zare 1988).

B.2 Symmetry Properties of Rotation Matrices

The reduced rotation matrices $d^{(J)}_{MM'}$ satisfy a number of useful symmetry relations:

$$
\begin{aligned}
d^{(J)}_{MM'}(\beta) &= (-1)^{M-M'} d^{(J)}_{M'M}(\beta) = d^{(J)}_{-M'-M}(\beta) = d^{(J)}_{M'M}(-\beta) \\
&= (-1)^{J-M} d^{(J)}_{M-M'}(\pi - \beta) = (-1)^{J+M'} d^{(J)}_{M-M'}(\pi + \beta).
\end{aligned} \tag{B.9}
$$

Note, that rotation by $-\beta$ is the inverse of rotation by β, that is $d^{(J)}_{M'M}(-\beta) = \left[d^{(J)}_{M'M}(\beta)\right]^{-1}$. As the $d^{(J)}_{MM'}$ are real and are elements of a unitary transformation, we get $\left[d^{(J)}_{M'M}(\beta)\right]^{-1} = \left[d^{(J)}_{M'M}(\beta)\right]^{\dagger} = d^{(J)}_{MM'}(\beta)$.

Using Zare (1988) we obtain special values:

$$d^{(J)}_{M'M}(0) = \delta_{M'M} \quad \text{and} \quad d^{(J)}_{M'M}(\pi) = (-1)^{J+M'}\delta_{M'-M}. \tag{B.10}$$

Combining (B.9) and (B.10) yields the result

$$
\begin{aligned}
d^{(J)}_{M'M}(2\pi) &= (-1)^{J+M'} d^{(J)}_{-M'M}(\pi) \\
&= (-1)^{2J}\delta_{M'M} = (-1)^{2J} d^{(J)}_{M'M}(0).
\end{aligned} \tag{B.11}
$$

The symmetry properties for the rotation matrices $\mathcal{D}^{(J)}_{MM'}$ are obtained as:

$$\mathcal{D}^{(J)}_{MM'}(\alpha\beta\gamma)^* = (-1)^{M-M'}\mathcal{D}^{(J)}_{-M-M'}(\alpha\beta\gamma) = \mathcal{D}^{(J)}_{M'M}(-\gamma-\beta-\alpha), \tag{B.12}$$

where $(-\gamma - \beta - \alpha)$ is the rotation inverse to $(\alpha\beta\gamma)$. In the following we may use the contraction $(\omega = \alpha\beta\gamma)$.

As the rotation matrices are unitary, they satisfy the sum rules:

$$
\begin{aligned}
\sum_{M'}\left[\mathcal{D}^{(J)}_{M'M}(\omega)\right]^{\dagger}\mathcal{D}^{(J)}_{M'N}(\omega) &= \sum_{M'} d^{(J)}_{M'M}(-\beta)d^{(J)}_{M'N}(\beta) \\
&= \sum_{M'} d^{(J)}_{MM'}(\beta)d^{(J)}_{M'N}(\beta) = \delta_{MN},
\end{aligned} \tag{B.13a}
$$

and

$$
\begin{aligned}
\sum_{M}\left[\mathcal{D}^{(J)}_{M'M}(\omega)\right]^{\dagger}\mathcal{D}^{(J)}_{N'M}(\omega) &= \sum_{M} d^{(J)}_{M'M}(-\beta)d^{(J)}_{N'M}(\beta) \\
&= \sum_{M} d^{(J)}_{MM'}(\beta)d^{(J)}_{N'M}(\beta) = \delta_{M'N'}.
\end{aligned} \tag{B.13b}
$$

B.3 The Clebsch–Gordan Series and its Inverse

The connection between the uncoupled $|J_1M_1\rangle|J_2M_2\rangle$ and coupled representations $|JM\rangle$ under a rotational transformation is given by the so-called *Clebsch–Gordan series*.

$$\mathcal{D}^{(J_1)}_{M'_1 M_1}(\omega)\mathcal{D}^{(J_2)}_{M'_2 M_2}(\omega)$$
$$= \sum_J (J_1M_1, J_2M_2|JM)(J_1M'_1, J_2M'_2|JM')\mathcal{D}^{(J)}_{M'M}(\omega). \quad (B.14)$$

In principle, the right-hand side must be summed over M and M', too. Applying the selection rule (A.2), the sum can be omitted. The Clebsch–Gordan series can be re-written in terms of $3j$-symbols

$$\mathcal{D}^{(J_1)}_{M'_1 M_1}(\omega)\mathcal{D}^{(J_2)}_{M'_2 M_2}(\omega)$$
$$= \sum_J (2J+1) \begin{pmatrix} J_1 & J_1 & J \\ M_1 & M_2 & M \end{pmatrix} \begin{pmatrix} J_1 & J_2 & J \\ M'_1 & M'_2 & M' \end{pmatrix} \mathcal{D}^{(J)}_{M'M}(\omega)^*. \quad (B.15)$$

Note the complex conjugate rotation matrix on the right hand side which enters by using (A.9) and (B.12).

The *inverse Clebsch–Gordan series* yields

$$\mathcal{D}^{(J)}_{M'M}(\omega) = \sum_{M_1 M'_1 M_2 M'_2} (J_1M_1, J_2M_2|JM)(J_1M'_1, J_2M'_2|JM')$$
$$\times \ \mathcal{D}^{(J_1)}_{M'_1 M_1}(\omega)\mathcal{D}^{(J_2)}_{M'_2 M_2}(\omega). \quad (B.16)$$

Note, that the summation of the right-hand side is not independent. Applying the selection rule (A.2), either two of the summation indices M_1, M'_1 or M_2, M'_2 can be omitted. Re-writing in terms of $3j$-symbols and omitting the sum over M_2 and M'_2 we obtain

$$\mathcal{D}^{(J)}_{M'M}(\omega)^* = \sum_{M_1 M'_1} (2J+1) \begin{pmatrix} J_1 & J_2 & J \\ M_1 & M_2 & M \end{pmatrix} \begin{pmatrix} J_1 & J_2 & J \\ M'_1 & M'_2 & M' \end{pmatrix}$$
$$\times \ \mathcal{D}^{(J_1)}_{M'_1 M_1}(\omega)\mathcal{D}^{(J_2)}_{M'_2 M_2}(\omega). \quad (B.17)$$

Analogously, we obtain the relations

$$\mathcal{D}^{(J)}_{M'M}(\omega)^* \begin{pmatrix} J_1 & J_2 & J \\ M_1 & M_2 & M \end{pmatrix}$$
$$= \sum_{M'_1 M'_2} \begin{pmatrix} J_1 & J_2 & J \\ M'_1 & M'_2 & M' \end{pmatrix} \mathcal{D}^{(J_1)}_{M'_1 M_1}(\omega)\mathcal{D}^{(J_2)}_{M'_2 M_2}(\omega), \quad (B.18a)$$

and

$$
\mathcal{D}^{(J)}_{M\,M'}(\omega)^* \begin{pmatrix} J_1 & J_2 & J \\ M_1 & M_2 & M \end{pmatrix}
$$
$$
= \sum_{M_1' M_2'} \begin{pmatrix} J_1 & J_2 & J \\ M_1' & M_2' & M' \end{pmatrix} \mathcal{D}^{(J_1)}_{M_1\,M_1'}(\omega)\mathcal{D}^{(J_2)}_{M_2\,M_2'}(\omega). \quad \text{(B.18b)}
$$

Eventually, the contraction of three rotation matrices results in a $3j$-symbol

$$
\sum_{M M_1 M_2} \mathcal{D}^{(J)}_{M\,M'}(\omega)\mathcal{D}^{(J_1)}_{M_1\,M_1'}(\omega)\mathcal{D}^{(J_2)}_{M_2\,M_2'}(\omega) \begin{pmatrix} J_1 & J_2 & J \\ M_1 & M_2 & M \end{pmatrix}
$$
$$
= \begin{pmatrix} J_1 & J_2 & J \\ M_1' & M_2' & M' \end{pmatrix}. \quad \text{(B.19)}
$$

The closure relation for two consecutive rotations gives

$$
\sum_{M''} \mathcal{D}^{(J)}_{M\,M''}(\omega_2)\mathcal{D}^{(J)}_{M''\,M'}(\omega_1) = \mathcal{D}^{(J)}_{M\,M'}(\omega), \quad \text{(B.20)}
$$

where $\omega = (\alpha\beta\gamma)$ is the resultant of first $\omega_1 = (\alpha_1\beta_1\gamma_1)$ and second $\omega_2 = (\alpha_2\beta_2\gamma_2)$. Analogously, the closure relation for the reduced rotation matrices yields

$$
\sum_{M''} d^{(J)}_{M\,M''}(\beta_2) d^{(J)}_{M''\,M'}(\beta_1) = d^{(J)}_{M\,M'}(\beta_1 + \beta_2). \quad \text{(B.21)}
$$

B.4 Integrals Over Rotation Matrices

Integrating over the solid angle element $d\Omega = d\alpha \sin\beta d\beta d\gamma$ and using (B.12) and (B.14) yields the orthogonality relation results in a between two rotation matrices having the same argument

$$
\int d\Omega\, \mathcal{D}^{(J_1)}_{M_1'\,M_1}(\omega)^* \mathcal{D}^{(J_2)}_{M_2'\,M_2}(\omega) = \frac{8\pi^2}{2J_1+1}\delta_{J_1 J_2}\delta_{M_1' M_2'}\delta_{M_1 M_2}. \quad \text{(B.22)}
$$

Applying the Clebsch–Gordan series and (B.22) allows for evaluating the integral over three rotation matrices of the same argument. With (B.14) we obtain

$$
\int d\Omega\, \mathcal{D}^{(J_3)}_{M_3'\,M_3}(\omega)^* \mathcal{D}^{(J_2)}_{M_2'\,M_2}(\omega)\, \mathcal{D}^{(J_1)}_{M_1'\,M_1}(\omega)
$$
$$
= \frac{8\pi^2}{2J_3+1}\big(J_1 M_1, J_2 M_2 | J_3 M_3\big)\big(J_1 M_1', J_2 M_2' | J_3 M_3'\big), \quad \text{(B.23a)}
$$

and using (B.15) yields

$$
\int d\Omega\, \mathcal{D}^{(J_3)}_{M_3'\,M_3}(\omega)\, \mathcal{D}^{(J_2)}_{M_2'\,M_2}(\omega)\, \mathcal{D}^{(J_1)}_{M_1'\,M_1}(\omega)
$$
$$
= 8\pi^2 \begin{pmatrix} J_1 & J_2 & J_3 \\ M_1 & M_2 & M_3 \end{pmatrix} \begin{pmatrix} J_1 & J_2 & J_3 \\ M_1' & M_2' & M_3' \end{pmatrix}. \quad \text{(B.23b)}
$$

B.5 Relation with the Spherical Harmonics

Special cases: if the angular momentum L is integer and one magnetic quantum number is equal to zero the dependence on the angles γ or α becomes obsolete, and the rotation matrices can be expressed in terms of spherical harmonics $Y_{LM}(\beta, \alpha)$. For the spherical harmonics the notation $Y_{LM}(\theta, \phi)$ is more common. In the following, we therefore change the notation of the Euler angles accordingly; i.e. $\alpha \leftrightarrow \phi$, $\beta \leftrightarrow \theta$, and $\gamma \leftrightarrow \chi$. This yields the relation

$$\mathcal{D}_{M0}^{(L)}(\omega) = d_{M0}^{(L)}(\theta)\, e^{-iM\phi} = C_{LM}^{*}(\theta, \phi) = \sqrt{\frac{4\pi}{2L+1}}\, Y_{LM}^{*}(\theta, \phi), \qquad (B.24)$$

where $C_{LM}(\theta, \phi)$ are the *renormalized* spherical harmonics introduced by Condon and Shortley (1935). With the first magnetic quantum number equal to zero we obtain

$$\mathcal{D}_{0M}^{(L)}(\omega) = d_{0M}^{(L)}(\theta)\, e^{-iM\chi} = (-1)^{M} d_{M0}^{(L)}(\theta)\, e^{-iM\chi} = (-1)^{M} C_{LM}^{*}(\theta, \chi)$$

$$= C_{L-M}(\theta, \chi) = \sqrt{\frac{4\pi}{2L+1}}\, Y_{L-M}(\theta, \chi), \qquad (B.25)$$

where we have used the relation $Y_{LM}^{*} = (-1)^{M} Y_{L-M}$. Particularly, for $\chi = \phi = 0$, the reduced rotation matrices may be also expressed in terms of associated Legendre polynomials $P_{L}^{M}(\cos\theta)$ as

$$d_{M0}^{(L)}(\theta) = C_{LM}^{*}(\theta, 0) = \sqrt{\frac{(L-M)!}{(L+M)!}}\, P_{L}^{M}(\cos\theta), \qquad (B.26)$$

if $M \geq 0$, and

$$d_{0M}^{(L)}(\theta) = (-1)^{M} C_{LM}^{*}(\theta, 0) = (-1)^{M} \sqrt{\frac{(L-M)!}{(L+M)!}}\, P_{L}^{M}(\cos\theta). \qquad (B.27)$$

Having both magnetic quantum numbers equal to zero we simply get the Legendre polynomials

$$\mathcal{D}_{00}^{(L)}(\omega) = d_{00}^{(L)}(\theta) = P_{L}(\cos\theta). \qquad (B.28)$$

Some of the relations derived above are often applied using spherical harmonics. Using (B.24), the orthogonality relation (B.22) may be expressed as

$$\int \sin\theta d\theta d\phi\, C_{kq}^{*}(\theta, \phi) C_{KQ}(\theta, \phi) = \frac{4\pi}{2K+1} \delta_{kK} \delta_{qQ}. \qquad (B.29)$$

From the sum rules (B.13a) and (B.13b) we get

$$\sum_{q} |C_{kq}(\theta, \phi)|^{2} = 1, \qquad (B.30)$$

and for $q = 0$ we obtain

$$\sum_k (2k + 1)C_{k0}(\theta, \phi) = 2\delta(\cos\theta - 1). \tag{B.31}$$

From the closure relation (B.20) we obtain the addition theorem for the renormalized spherical harmonics

$$\sum_q C_{kq}(\theta, \phi)C_{kq}^*(\theta', \phi') = P_k(\cos\omega), \tag{B.32}$$

where ω is the angle between the two directions (θ, ϕ) and (θ', ϕ').

From the Clebsch–Gordan series and its inverse, see (B.15) and (B.17), we obtain similar contractions between two renormalized spherical harmonics of the same argument

$$C_{a\alpha}(\theta, \phi)C_{b\beta}(\theta, \phi) = \sum_c C_{c\gamma}(\theta, \phi)(2c + 1)(-1)^\gamma \begin{pmatrix} a & b & c \\ \alpha & \beta & -\gamma \end{pmatrix} \begin{pmatrix} a & b & c \\ 0 & 0 & 0 \end{pmatrix}, \tag{B.33}$$

where the sum over γ on the right-hand side is fixed to one argument due to the selection rule (A.10). Summing over the magnetic components on the left-hand side yields

$$\sum_{\alpha\beta} C_{a\alpha}(\theta, \phi)C_{b\beta}(\theta, \phi) \begin{pmatrix} a & b & c \\ \alpha & \beta & -\gamma \end{pmatrix} = C_{c\gamma}(\theta, \phi)(-1)^\gamma \begin{pmatrix} a & b & c \\ 0 & 0 & 0 \end{pmatrix}. \tag{B.34}$$

Combining (B.22) and (B.33) yields the integral over three renormalized spherical harmonics

$$\int \sin\theta\, d\theta\, d\phi\; C_{a\alpha}(\theta, \phi)C_{b\beta}(\theta, \phi)C_{c\gamma}(\theta, \phi) = 4\pi \begin{pmatrix} a & b & c \\ \alpha & \beta & -\gamma \end{pmatrix} \begin{pmatrix} a & b & c \\ 0 & 0 & 0 \end{pmatrix}. \tag{B.35}$$

For all three magnetic quantum numbers equal to zero, we obtain a similar integral in terms of Legendre polynomials

$$\int \sin\theta\, d\theta\; P_a(\cos\theta)P_b(\cos\theta)P_c(\cos\theta) = 2\begin{pmatrix} a & b & c \\ 0 & 0 & 0 \end{pmatrix}^2. \tag{B.36}$$

B.6 Left–Right Asymmetry of Rotation Matrices

For evaluating asymmetry parameters it is useful to investigate the symmetries of the reduced rotation matrices under the operation $\theta \longrightarrow -\theta$.

Considering the case of electron impact excitation/ionization we need to consider only a few of the reduced rotation matrices $d_{Q'Q}^{(K)}(\theta)$ since the quantum numbers Q and Q' are restricted by the selection rule

$$|Q| \le 1, \tag{B.37}$$

and only particular combinations of the reduced rotation matrices can occur (see Sect. 2.5.3).

Due to the definition of the asymmetry parameters we need to consider the sum and the difference of the occurring combinations of reduced rotation matrices with the argument being θ and $-\theta$, respectively. I.e., we need to consider expressions of the type $d^{(K)}_{Q'Q}(\theta) \pm d^{(K)}_{Q'Q}(-\theta)$.

In order to obtain the relevant results the symmetry relations (B.9) of the reduced rotation matrices must be applied. For the simple most case we obtain

$$d^{(K)}_{00}(\theta) - d^{(K)}_{00}(-\theta) = 0, \tag{B.38}$$

and by using (B.28)

$$d^{(K)}_{00}(\theta) + d^{(K)}_{00}(-\theta) = 2d^{(K)}_{00}(\theta) = 2P_K(\cos\theta). \tag{B.39}$$

Analogously, applying (B.26) and (B.27), we obtain for the difference

$$d^{(K)}_{10}(\theta) - d^{(K)}_{10}(-\theta) = d^{(K)}_{10}(\theta) - d^{(K)}_{01}(\theta) = 2\,C^*_{K1}(\theta, 0)$$

$$= \frac{2}{\sqrt{K(K+1)}}\, P^1_K(\cos\theta). \tag{B.40}$$

The sum yields

$$d^{(K)}_{10}(\theta) + d^{(K)}_{10}(-\theta) = d^{(K)}_{10}(\theta) + d^{(K)}_{01}(\theta)$$

$$= C^*_{K1}(\theta, 0) - C^*_{K1}(\theta, 0) = 0. \tag{B.41}$$

For the other occurring differences we simply obtain

$$\left[d^{(K)}_{11}(\theta) \pm d^{(K)}_{-11}(\theta)\right] - \left[d^{(K)}_{11}(-\theta) \pm d^{(K)}_{-11}(-\theta)\right]$$

$$= \left[d^{(K)}_{11}(\theta) \pm d^{(K)}_{-11}(\theta)\right] - \left[d^{(K)}_{11}(\theta) \pm d^{(K)}_{-11}(\theta)\right] = 0, \tag{B.42}$$

and for the summations we eventually get

$$\left[d^{(K)}_{11}(\theta) \pm d^{(K)}_{-11}(\theta)\right] + \left[d^{(K)}_{11}(-\theta) \pm d^{(K)}_{-11}(-\theta)\right]$$

$$= \left[d^{(K)}_{11}(\theta) \pm d^{(K)}_{-11}(\theta)\right] + \left[d^{(K)}_{11}(\theta) \pm d^{(K)}_{-11}(\theta)\right]$$

$$= 2\left[d^{(K)}_{11}(\theta) \pm d^{(K)}_{-11}(\theta)\right]. \tag{B.43}$$

Note, that the derived relations are valid for any integer rank $K \geq 0$ of the reduced rotation matrices.

Eventually, we give explicit expressions of (B.43) for some cases of interest. For $K = 1$ we obtain

$$d^{(1)}_{11}(\theta) + d^{(1)}_{-11}(\theta) = 1, \tag{B.44}$$

and

$$d_{11}^{(1)}(\theta) - d_{-11}^{(1)}(\theta) = \cos\theta, \tag{B.45}$$

and for $K = 2$ we get

$$d_{11}^{(2)}(\theta) + d_{-11}^{(2)}(\theta) = \cos\theta, \tag{B.46}$$

and

$$d_{11}^{(2)}(\theta) - d_{-11}^{(2)}(\theta) = \cos(2\theta). \tag{B.47}$$

The identity of (B.45) and (B.46) is by chance. For $K = 3$ we obtain

$$d_{11}^{(3)}(\theta) + d_{-11}^{(3)}(\theta) = \frac{1}{4}(5\cos^2\theta - 1), \tag{B.48}$$

which is unequal to the result of (B.47).

Eventually, we give expressions for the angular functions for the *magic angle* $\theta_M = 54.7^o$,

$$\sin\theta_M = \sqrt{\frac{2}{3}} \quad \text{and} \quad \cos\theta_M = \frac{1}{\sqrt{3}}. \tag{B.49}$$

C Irreducible Tensorial Sets

Dealing with tensorial sets is essential in interpreting multi-particle scattering processes. Therefore, its application is an important means for the understanding of Auger emission. Important relations are given below. A good introduction into the field may be found in Brink and Satchler (1962) or Zare (1988).

C.1 Definition and Basic Properties

Definition: Tensor operator $\equiv$ The manifold of operators which, under rotation, linearly transform into each other.

Irreducible tensor operator: The $(2k + 1)$ operators T_{KQ}, $Q = -K, \ldots, +K$, are called the *standard components of an irreducible tensor operator* $\mathbf{T}_K$ *of rank K* if they transform under rotation as

$$\mathcal{D}T_{KQ}\mathcal{D}^{-1} = \sum_q T_{Kq}\mathcal{D}_{qQ}^{(K)}. \tag{C.1}$$

Scalar operator $\equiv$ Irreducible tensor operator of rank zero; i.e.,

$$\mathcal{D}T_{00}\mathcal{D}^{-1} = T_{00}\mathcal{D}_{00}^{(0)} = T_{00}. \tag{C.2}$$

Vector operator $\equiv$ Irreducible tensor operator of first rank; let V_x, V_y, V_z be its components in the Cartesian x, y, z-coordinate frame, then its standard components are

$$V_+ = \frac{1}{\sqrt{2}}\left(V_x + iV_y\right), \tag{C.3a}$$

$$V_0 = V_z, \tag{C.3b}$$

$$V_- = \frac{1}{\sqrt{2}}\left(V_x - iV_y\right). \tag{C.3c}$$

Commutation relations with the angular momentum:

$$\left[J_\pm, T_{kq}\right] = \sqrt{k(k+1) - q(q+1)}T_{kq\pm1}, \tag{C.4a}$$

$$\left[J_z, T_{kq}\right] = qT_{kq}. \tag{C.4b}$$

Hermitian conjugate:

$$\mathbf{S}_K = \mathbf{T}_K^\dagger \quad \text{if} \quad S_{KQ} = (-1)^Q T_{KQ}^\dagger.$$ (C.5)

One of the basic properties of irreducible tensor operators is the *Wigner–Eckart* theorem (see also Sect. 2.2):

$$\langle JM|T_{KQ}|J'M'\rangle = \frac{(-1)^{2K}}{\sqrt{2J+1}} \langle J\|T_K\|J'\rangle \left(J'M', KQ|JM\right)$$

$$= (-1)^{J-M} \langle J\|T_K\|J'\rangle \begin{pmatrix} J & K & J' \\ -M & Q & M' \end{pmatrix},$$ (C.6)

where $\langle J\|T_K\|J'\rangle$ is the so-called reduced matrix element. Its complex conjugate is given (for K integer) as:

$$\langle J\|T_K\|J'\rangle^* = (-1)^{J'-J}\langle J\|T_K^\dagger\|J'\rangle.$$ (C.7)

Note, that the notation of the Wigner–Eckart theorem is not unique in the literature. Our form coincides with the convention used by Blum (1996); Edmonds (1974); Fano and Racah (1959); Messiah (1979); Racah (1942); Varshalovich *et al.* (1988) and Zare (1988). Some authors, including Brink and Satchler (1962); Rose (1957) and Wigner (1959), define the reduced matrix element to be a factor of $\sqrt{2J+1}$ smaller.

Special values of tensor operators are the *identity operator*

$$\langle J\|\mathbf{1}\|J'\rangle = \delta_{JJ'}\sqrt{2J+1},$$ (C.8)

and the total angular momentum

$$\langle J\|\mathbf{J}\|J'\rangle = \delta_{JJ'}\sqrt{J(J+1)(2J+1)}.$$ (C.9)

In the following, some of the multipole expansions useful in physics are given:

$$\exp(i\mathbf{k}\mathbf{r}) = \sum_\ell i^\ell(2\ell+1)j_\ell(kr)\mathbf{C}_\ell\left(\theta_k\phi_k\right)\mathbf{C}_\ell\left(\theta_r\phi_r\right),$$ (C.10)

$$\delta(\mathbf{a}-\mathbf{b}) = \frac{1}{4\pi a^2}\delta(a-b)\sum_\ell(2\ell+1)\mathbf{C}_\ell\left(\theta_a\phi_a\right)\mathbf{C}_\ell\left(\theta_b\phi_b\right),$$ (C.11)

$$\exp\left(-\gamma[\mathbf{a}-\mathbf{b}]^2\right) = \sum_\ell i^{-\ell}(2\ell+1)e^{-\gamma(a^2+b^2)}j_\ell(2i\gamma ab)$$

$$\times\,\mathbf{C}_\ell\left(\theta_a\phi_a\right)\mathbf{C}_\ell\left(\theta_b\phi_b\right),$$ (C.12)

where $j_\ell(kr)$ denote the spherical Bessel functions; and if $\boldsymbol{\rho} = \mathbf{b}-\mathbf{a}$, with $b \geq a$,

$$\frac{1}{\rho} = \sum_\ell\frac{a^\ell}{b^{\ell+1}}\mathbf{C}_\ell\left(\theta_a\phi_a\right)\mathbf{C}_\ell\left(\theta_b\phi_b\right),$$ (C.13)

$$\frac{e^{ik\rho}}{\rho} = ikh_0^{(1)}(k\rho) = ik \sum_\ell (2\ell+1) j_\ell(ka) h_\ell^{(1)}(kb)$$

$$\times \mathbf{C}_\ell (\theta_a \phi_a) \, \mathbf{C}_\ell (\theta_b \phi_b) \,, \tag{C.14}$$

$$\frac{e^{-\alpha\rho}}{\alpha\rho} = -\sum_\ell (2\ell+1) j_\ell(i\alpha a) h_\ell^{(1)}(i\alpha b) \mathbf{C}_\ell (\theta_a \phi_a) \, \mathbf{C}_\ell (\theta_b \phi_b) \,, \tag{C.15}$$

where $h_\ell^{(1)}(kr)$ denote the spherical Hankel functions of the first type; and if $\mathbf{r} = \mathbf{a} + \mathbf{b}$, we get the expansion

$$r^\ell C_{\ell m} (\theta_r, \phi_r) = \sum_{\lambda\mu} \sqrt{\frac{2\ell!}{2\lambda!2(\ell-\lambda)!}} (\ell - \lambda \, m - \mu, \lambda\mu | \ell m)$$

$$\times a^{\ell-\lambda} b^\lambda C_{\ell-\lambda\, m-\mu} (\theta_a \phi_a) \, C_{\lambda\mu} (\theta_b \phi_b) \,. \tag{C.16}$$

C.2 Tensorial Products of Irreducible Tensor Operators

Definition: Let $\mathbf{T}_{k_1}$, $\mathbf{U}_{k_2}$ be two irreducible tensor operators of rank k_1 and k_2, respectively. Then $\mathbf{T}_{k_1} \otimes \mathbf{U}_{k_2} \equiv$ the manifold of the $(2k_1+1)(2k_2+1)$, not necessarily linear independent, operators $T_{k_1 q_1} U_{K_2 q_2}$. It is a (reducible) tensor operator.

$\mathbf{V}_K \equiv \left[\mathbf{T}_{k_1} \otimes \mathbf{U}_{k_2} \right]_K$ is the tensorial product of rank K, which is the irreducible tensor operator of rank K with its components

$$V_{KQ}(k_1, k_2) = \sum_{q_1 q_2} (k_1 q_1, k_2 q_2 | K Q) T_{k_1 q_1} U_{k_2 q_2}, \tag{C.17}$$

where, $|k_1 - k_2| \le K \le k_1 + k_2$ is a necessary condition. Re-writing (C.17) in terms of $3j$-symbols yields

$$V_{KQ}(k_1, k_2) = \sum_{q_1 q_2} (-1)^{k_1 - k_2 + Q} \sqrt{2K+1} \begin{pmatrix} k_1 & k_2 & K \\ q_1 & q_2 & -Q \end{pmatrix} T_{k_1 q_1} U_{k_2 q_2}. \tag{C.18}$$

In case that $k_1 = k_2 = k$, the scalar product is defined as

$$S \equiv \left(\mathbf{T}_k \cdot \mathbf{U}_k \right) = \sum_q (-1)^q T_{kq} U_{k-q} \,. \tag{C.19}$$

Note, that S is not irreducible. It is related to the irreducible tensor of rank zero via

$$S = (-1)^k \sqrt{2k+1} V_{00}(k, k). \tag{C.20}$$

Reduced matrix elements: Suppose a composite quantum system to be generated from its joint sub-systems 1 and 2. Let $\mathbf{J}_1$ and $\mathbf{J}_2$ be the total angular momenta of the related sub-systems, and $\mathbf{J} = \mathbf{J}_1 + \mathbf{J}_2$.

Let $|\tau_1 J_1 M_1\rangle$ and $|\tau_2 J_2 M_2\rangle$ be the basis sets of system 1 and 2, respectively.

$\mathbf{T}_{k_1}$ and $\mathbf{U}_{k_2}$ are irreducible tensor operators which solely interact with the variables in the Hilbert sub-space of system 1 and 2, respectively.

Let $\mathbf{V}_K$ be the tensorial product of rank K according to its definition (C.17). Then, in the standard basis $\{\tau_1\tau_2\mathbf{J}_1^2\mathbf{J}_2^2\mathbf{J}^2 J_z\}$ the matrix elements of $\mathbf{V}_K(k_1, k_2)$ are given as

$$\langle\tau_1\tau_2 J_1 J_2 J\|\mathbf{V}_K\|\tau_1'\tau_2'J_1'J_2'J'\rangle = \sqrt{(2J+1)(2K+1)(2J'+1)}$$

$$\times \begin{Bmatrix} J_1' & J_2' & J' \\ k_1 & k_2 & K \\ J_1 & J_2 & J \end{Bmatrix} \langle\tau_1 J_1\|\mathbf{T}_{k_1}\|\tau_1'J_1'\rangle\langle\tau_2 J_2\|\mathbf{U}_{k_2}\|\tau_2'J_2'\rangle. \quad \text{(C.21)}$$

Special cases occur if the tensor operator acts in one of the sub-systems, only.

Let $\mathbf{U} = \mathbf{1}$ be the identity operator acting on sub-system 2. Then, we have $K = k_1 = k$, and (C.21) can be reduced to

$$\langle\tau_1\tau_2 J_1 J_2 J\|\mathbf{T}_k\|\tau_1'\tau_2'J_1'J_2'J'\rangle = \delta_{\tau_2\tau_2'}\delta_{J_2 J_2'}\langle\tau_1 J_1\|\mathbf{T}_k\|\tau_1'J_1'\rangle$$

$$\times (-1)^{J'+J_1+J_2+k}\sqrt{(2J+1)(2J'+1)}\begin{Bmatrix} J_1 & k & J_1' \\ J' & J_2 & J \end{Bmatrix}. \quad \text{(C.22)}$$

In the opposite case we have $\mathbf{T} = \mathbf{1}$ the identity operator acting in sub-system 1. Then, $K = k_2 = k$, and (C.21) is reduced to

$$\langle\tau_1\tau_2 J_1 J_2 J\|\mathbf{U}_k\|\tau_1'\tau_2'J_1'J_2'J'\rangle = \delta_{\tau_1\tau_1'}\delta_{J_1 J_1'}\langle\tau_2 J_2\|\mathbf{U}_k\|\tau_2'J_2'\rangle$$

$$\times (-1)^{J+J_1+J_2'+k}\sqrt{(2J+1)(2J'+1)}\begin{Bmatrix} J_2 & k & J_2' \\ J' & J_1 & J \end{Bmatrix}. \quad \text{(C.23)}$$

In case of a compound scalar operator $\mathbf{V}_0(k, k)$ we have $K = 0$ and $k_1 = k_2 = k$. Applying (C.21) and (A.40) yields

$$\langle\tau_1\tau_2 J_1 J_2 J\|\mathbf{V}_0(k,k)\|\tau_1'\tau_2'J_1'J_2'J\rangle$$

$$= \langle\tau_1\tau_2 J_1 J_2 J\|\left[\mathbf{T}_k \otimes \mathbf{U}_k\right]_0^{(0)}\|\tau_1'\tau_2'J_1'J_2'J\rangle$$

$$= \delta_{JJ'}(-1)^{k+J+J_2+J_1'}\sqrt{\frac{2J+1}{2k+1}}\begin{Bmatrix} J_1 & k & J_1' \\ J_2' & J & J_2 \end{Bmatrix}$$

$$\times \langle\tau_1 J_1\|\mathbf{T}_k\|\tau_1'J_1'\rangle\langle\tau_2 J_2\|\mathbf{U}_k\|\tau_2'J_2'\rangle. \quad \text{(C.24)}$$

Often, the scalar product $\mathbf{T}_k \cdot \mathbf{U}_k$ is used in place of $\left[\mathbf{T}_k \otimes \mathbf{U}_k\right]_0^{(0)}$ as the scalar operator. Applying the Wigner–Eckart theorem (C.6) and using (C.19) and (C.20), we can re-write (C.24) as

$$\langle\tau_1\tau_2 J_1 J_2 J M|(\mathbf{T}_k \cdot \mathbf{U}_k)|\tau_1'\tau_2'J_1'J_2'J'M'\rangle = \delta_{JJ'}\delta_{MM'}(-1)^k$$

$$\times \sqrt{\frac{2k+1}{2J+1}}\langle\tau_1\tau_2 J_1 J_2 J\|\mathbf{V}_0(k,k)\|\tau_1'\tau_2'J_1'J_2'J\rangle$$

$$= \delta_{JJ'}\delta_{MM'}(-1)^{J+J_2+J_1'}\begin{Bmatrix} J_1 & k & J_1' \\ J_2' & J & J_2 \end{Bmatrix}$$

$$\times \langle\tau_1 J_1\|\mathbf{T}_k\|\tau_1'J_1'\rangle\langle\tau_2 J_2\|\mathbf{U}_k\|\tau_2'J_2'\rangle. \quad \text{(C.25)}$$

Eventually, a more general case is the situation when the system is in a state $|\tau J M\rangle$ with sharp angular momentum J that is not decomposable into sub-system states. Then, the matrix elements of the compound tensor $V_{KQ}(k_1, k_2) = \left[\mathbf{T}_{k_1} \otimes \mathbf{T}_{k_2}\right]_Q^{(K)}$, where $V_{KQ}(k_1, k_2)$ can act only on the set of variables of the system, are obtained as

$$
\begin{aligned}
\langle \tau J \| \mathbf{V}_K \| \tau' J' \rangle &= \langle \tau J \| \left[\mathbf{T}_{k_1} \otimes \mathbf{T}_{k_2}\right]_K \| \tau' J' \rangle \\
&= (-1)^{K+J+J'} \sqrt{2K+1} \sum_{\tau'' J''} \begin{Bmatrix} k_1 & k_2 & K \\ J & J' & J'' \end{Bmatrix} \\
&\quad \times \langle \tau J \| \mathbf{T}_{k_1} \| \tau'' J'' \rangle \langle \tau'' J'' \| \mathbf{T}_{k_2} \| \tau' J' \rangle . \quad \text{(C.26)}
\end{aligned}
$$

A particular application of (C.22) or (C.23) is the reduction of composite systems described in the LSJ coupling scheme. Suppose $\mathbf{L} + \mathbf{S} = \mathbf{J}$ and $\mathbf{L'} + \mathbf{S'} = \mathbf{J'}$. If the tensor operator T_{KQ} acts only on the system with angular momenta L, L' then

$$
\begin{aligned}
\langle (LS)J \| \mathbf{T}_K \| (L'S')J' \rangle &= \delta_{SS'} (-1)^{L+S+J'+K} \sqrt{(2J+1)(2J'+1)} \\
&\quad \times \begin{Bmatrix} L & L' & K \\ J' & J & S \end{Bmatrix} \langle L \| \mathbf{T}_K \| L' \rangle . \quad \text{(C.27)}
\end{aligned}
$$

Basic reduced matrix elements for the identity and the total angular momentum operator have been given in (C.8) and (C.9), respectively. Reduced matrix elements for the spherical harmonics yield

$$
\langle \ell \| \mathbf{Y}_k \| \ell' \rangle = (-1)^\ell \sqrt{\frac{(2\ell+1)(2k+1)(2\ell'+1)}{4\pi}} \begin{pmatrix} \ell & k & \ell' \\ 0 & 0 & 0 \end{pmatrix} . \quad \text{(C.28)}
$$

Using the renormalized spherical harmonics instead gives a more compact relation

$$
\langle \ell \| \mathbf{C}_k \| \ell' \rangle = (-1)^\ell \sqrt{(2\ell+1)(2\ell'1)} \begin{pmatrix} \ell & k & \ell' \\ 0 & 0 & 0 \end{pmatrix} . \quad \text{(C.29)}
$$

Considering the composite angular momentum – electron spin system the renormalized spherical harmonics act in the angular momentum Hilbert sub-space, only. Applying (C.22) and (C.29) gives

$$
\begin{aligned}
\langle (\ell 1/2)j \| \mathbf{C}_k \| (\ell' 1/2)j' \rangle &= \sqrt{(2\ell+1)(2\ell'+1)(2j+1)(2j'+1)} \\
&\quad \times (-1)^{j'+1/2+k} \begin{pmatrix} \ell & k & \ell' \\ 0 & 0 & 0 \end{pmatrix} \begin{Bmatrix} \ell & k & \ell' \\ j' & \frac{1}{2} & j \end{Bmatrix} , \quad \text{(C.30)}
\end{aligned}
$$

and combining with (A.29) eventually yields

$$
\langle (\ell 1/2)j \| \mathbf{C}_k \| (\ell' 1/2)j' \rangle = (-1)^{j'-1/2-k} \sqrt{(2j+1)(2j'+1)} \begin{pmatrix} j & j' & k \\ \frac{1}{2} & -\frac{1}{2} & 0 \end{pmatrix} , \quad \text{(C.31)}
$$

provided that $\ell + \ell' + k$ is even, and zero otherwise.

D Expansion of Dipole Matrix Elements

Following Amusia and Cherepkov (1975, pp. 16) the T matrix elements may be written as

$$\langle JM\mathbf{p}^{(-)}m_s|T_i|J_0M_0\omega\mathbf{n}\lambda\rangle = \sum_{q=1}^{N}\langle JM\mathbf{p}^{(-)}m_s|\exp^{i\mathbf{k}\cdot\mathbf{r}}(\mathbf{e}_\lambda\cdot\mathbf{p}_q^*)|J_0M_0\rangle, \quad (D.1)$$

where N denotes the number of electrons in the atomic shell, $\mathbf{e}_\lambda$ ($\lambda = \pm1$) is the photon polarization vector which is chosen that $\mathbf{e}_\lambda\cdot\mathbf{n} = 0$, and $\mathbf{p}_q$ is the momentum of the qth electron of the atom. For completeness we note, that (D.1) holds for photoionization. In the case of photoemission the operator must be replaced by its complex conjugate, i.e. $\mathbf{e}_\lambda^*\cdot\mathbf{p}_q$.

Applying the long-wavelength limit of the dipole approximation, i.e. $kr_q \ll 1$, the exponential function can be replaced by unity; $\exp(i\mathbf{k}\cdot\mathbf{r}) \approx 1$.

Thus, for an arbitrarily polarized photon beam, the T matrix elements may be written as

$$\langle JM\mathbf{p}^{(-)}m_s|T_i|J_0M_0\omega\mathbf{n}\lambda\rangle = \sum_{q=1}^{N}\langle JM\mathbf{p}^{(-)}m_s|(\mathbf{e}_\lambda\cdot\mathbf{p}_q^*)|J_0M_0\rangle$$

$$= \sum_{q=1}^{N}i\,\omega\langle JM\mathbf{p}^{(-)}m_s|(\mathbf{e}_\lambda\cdot\mathbf{r}_q^*)|J_0M_0\rangle, \quad (D.2)$$

where the first term denotes the "velocity form" and the latter the "length form" of the dipole transition matrix element. The dipole approximation is valid in a rather broad region of energy (Amusia 1990)

$$Z^2 < \omega \ll Z\alpha^{-1}, \quad (D.3)$$

where $\alpha^{-1} = 137$ denotes the fine structure constant. Throughout this book the length form of the dipole matrix elements is used.

In our chosen coordinate frame the polarization vector $\mathbf{e}_\lambda$ can be eliminated by noting that in the helicity system the coordinate system is "spanned" by the three unit vectors $\mathbf{e}_{+1}$, $\mathbf{e}_{-1}$, $\mathbf{n}$, and that the dipole operator $\mathbf{r}$ can be therefore expanded in terms of this basis[1]

$$\mathbf{r} = r_{+1}^*\mathbf{e}_{+1} + r_{-1}^*\mathbf{e}_{-1} + r_0^*\mathbf{n}, \quad (D.4)$$

---

[1] Here, and throughout the following the index q and the summation over q, referring to the qth electron, are suppressed if not causing ambiguities.

where $r_{\pm 1}$ and r_0 are the components of $\mathbf{r}$ along the directions of $\mathbf{e}_{\pm 1}$ and $\mathbf{n}$, respectively. I.e., $r_{\pm 1}$ and r_0 are the spherical components of the vector $\mathbf{r}$. In this system the scalar product of $\mathbf{r}$ and $\mathbf{e}_\lambda$ is given by

$$\mathbf{e}_\lambda \cdot \mathbf{r}^* = r_\lambda . \tag{D.5}$$

The final state electron wavefunction can be expanded into partial waves. Applying the results of Lohmann (1990) we get

$$\left| \mathbf{p}^{(-)} m_s \right\rangle = \psi_{\mathbf{p}m_s}^{(-)}(\mathbf{r}) = \frac{1}{|\mathbf{p}|} \sum_{\substack{\ell m j m_j \\ m' \mu}} i^\ell \, e^{-i\sigma_\ell^j} \, Y_{\ell m}^*(\hat{\mathbf{p}}) \, R_{\varepsilon \ell}^j(r) Y_{\ell m'}(\hat{\mathbf{r}}) \, \chi_\mu$$

$$\times \left(\ell m, 1/2 m_s | j m_j \right) \left(\ell m', 1/2 \mu | j m_j \right), \tag{D.6}$$

which yields for the expansion coefficients

$$a_{\ell m}^j = \left\langle j \ell m | \mathbf{p}^{(-)} \right\rangle = \frac{1}{|\mathbf{p}|} i^\ell \, e^{-i\sigma_\ell^j} \, Y_{\ell m}^*(\hat{\mathbf{p}}) . \tag{D.7}$$

Inserting the partial wave expansions into the transition matrix element we get

$$\left\langle J M \mathbf{p}^{(-)} m_s | r_\lambda | J_0 M_0 \right\rangle = \sum_{\ell m} a_{\ell m}^{j*} \left\langle J M \ell m 1/2 m_s | r_\lambda | J_0 M_0 \right\rangle$$

$$= \sum_{\substack{\ell m j m_j \\ J_1 M_1}} a_{\ell m}^{j*} \left\langle (Jj) J_1 M_1 | r_\lambda | J_0 M_0 \right\rangle$$

$$\times \left(\ell m, 1/2 m_s | j m_j \right) \left(J M, j m_j | J_1 M_1 \right). \tag{D.8}$$

Using the fact that $\hat{r}_\lambda$ is a tensor operator of rank one and applying the Wigner–Eckart theorem (C.6) we get

$$\left\langle J_1 M_1 | r_\lambda | J_0 M_0 \right\rangle = (-1)^{J_1 - M_1} \begin{pmatrix} J_1 & 1 & J_0 \\ -M_1 & \lambda & M_0 \end{pmatrix} \left\langle J_1 \| r \| J_0 \right\rangle. \tag{D.9}$$

With this, we obtain for the dipole matrix element

$$\left\langle J M \mathbf{p}^{(-)} m_s | T_i | J_0 M_0 \omega \mathbf{n} \lambda \right\rangle = \left\langle J M \mathbf{p}^{(-)} m_s | d_\lambda | J_0 M_0 \right\rangle$$

$$= \sum_{\substack{\ell m j m_j \\ J_1 M_1}} a_{\ell m}^{j*} \left\langle (Jj) J_1 \| d \| J_0 \right\rangle$$

$$\times (-1)^{-\ell + 1/2 - m_j - J + j - J_1}$$

$$\times \sqrt{(2j+1)(2J_1+1)} \begin{pmatrix} \ell & 1/2 & j \\ m & m_s & -m_j \end{pmatrix}$$

$$\times \begin{pmatrix} J & j & J_1 \\ M & m_j & -M_1 \end{pmatrix} \begin{pmatrix} J_1 & 1 & J_0 \\ -M_1 & \lambda & M_0 \end{pmatrix}, \tag{D.10}$$

where we introduced the abbreviation $d_\lambda = i\omega r_\lambda$. Inserting the expansion coefficients we finally end up with (2.41).

E Anisotropy Parameters
for Electron Impact Ionization

E.1 Expansion of Matrix Elements

The transition matrix elements of (2.92) can be evaluated applying a triple partial wave expansion (see Appendix D)

$$
\langle J M \mathbf{p}_1^{(-)} m_{s_1} \mathbf{p}_2^{(-)} m_{s_2} | V | J_0 M_0 \mathbf{p}_0^{(+)} m_{s_0} \rangle = \sum_{\substack{\ell_0 \ell_1 m_1 \\ \ell_2 m_2}} a_{\ell_1 m_1}^{j_1 *} b_{\ell_2 m_2}^{j_2 *} c_{\ell_0}^{j_0}
$$

$$
\times \langle J M \ell_1 m_1 1/2 m_{s_1} \ell_2 m_2 1/2 m_{s_2} | V | J_0 M_0 \ell_0 0 1/2 m_{s_0} \rangle
$$

$$
= \sum_{\substack{\ell_0 \ell_1 m_1 \ell_2 m_2 j_0 m_{j_0} \\ j_1 m_{j_1} j_2 m_{j_2} \\ J_1 M_1 J_2 M_2 J_f M_f}} a_{\ell_1 m_1}^{j_1 *} b_{\ell_2 m_2}^{j_2 *} c_{\ell_0}^{j_0}
$$

$$
\times \langle ([J j_1] J_1 j_2) J_f M_f | V | (J_0 j_0) J_2 M_2 \rangle
$$

$$
\times \left(\ell_1 m_1, 1/2 m_{s_1} | j_1 m_{j_1} \right) \left(\ell_2 m_2, 1/2 m_{s_2} | j_2 m_{j_2} \right)
$$

$$
\times \left(\ell_0 0, 1/2 m_{s_0} | j_0 m_{j_0} \right) \left(J M, j_1 m_{j_1} | J_1 M_1 \right)
$$

$$
\times \left(J_1 M_1, j_2 m_{j_2} | J_f M_f \right) \left(J_0 M_0, j_0 m_{j_0} | J_2 M_2 \right). \tag{E.1}
$$

Using the fact that V is a zero-order tensor operator and applying the Wigner–Eckart theorem (C.6) we get

$$
\langle J_f M_f | V | J_2 M_2 \rangle = (-1)^{J_f - M_f} \begin{pmatrix} J_f & 0 & J_2 \\ -M_f & 0 & M_2 \end{pmatrix} \langle J_f \| V \| J_2 \rangle
$$

$$
= \frac{1}{\sqrt{2J_f + 1}} \langle J_f \| V \| J_2 \rangle \, \delta_{J_f J_2} \, \delta_{M_f M_2}. \tag{E.2}
$$

With this, we finally obtain for the transition matrix element

$$\langle JM\mathbf{p}_1^{(-)}m_{s1}\mathbf{p}_2^{(-)}m_{s2}|V|J_0M_0\mathbf{p}_0^{(+)}m_{s0}\rangle = \sum_{\substack{\ell_0\ell_1m_1\ell_2m_2j_0m_{j_0}\\j_1m_{j_1}j_2m_{j_2}\\J_1M_1J_fM_f}} a_{\ell_1m_1}^{j_1\,*}$$

$$\times b_{\ell_2m_2}^{j_2\,*}c_{\ell_0}^{j_0}\langle([Jj_1]J_1j_2)J_f\|V\|(J_0j_0)J_f\rangle$$

$$\times (-1)^{-\ell_1+1/2-m_{j_1}-\ell_2+1/2-m_{j_2}-\ell_0+1/2-m_{j_0}}$$

$$\times (-1)^{-J+j_1-M_1-J_1+j_2-M_f-J_0+j_0-M_f}$$

$$\times \sqrt{(2j_1+1)(2j_2+1)(2j_0+1)(2J_1+1)(2J_f+1)}$$

$$\times \begin{pmatrix} \ell_1 & 1/2 & j_1 \\ m_1 & m_{s_1} & -m_{j_1} \end{pmatrix}\begin{pmatrix} \ell_2 & 1/2 & j_2 \\ m_2 & m_{s_2} & -m_{j_2} \end{pmatrix}\begin{pmatrix} \ell_0 & 1/2 & j_0 \\ 0 & m_{s_0} & -m_{j_0} \end{pmatrix}$$

$$\times \begin{pmatrix} J & j_1 & J_1 \\ M & m_{j_1} & -M_1 \end{pmatrix}\begin{pmatrix} J_1 & j_2 & J_f \\ M_1 & m_{j_2} & -M_f \end{pmatrix}\begin{pmatrix} J_0 & j_0 & J_f \\ M_0 & m_{j_0} & -M_f \end{pmatrix}. \tag{E.3}$$

E.2 Derivation of Anisotropy Parameters

Inserting the derived expression (E.3) for the transition matrix element twice into (2.92) the anisotropy parameter B_e can be written as

$$B_e(K'Q',kq) = \frac{\sqrt{(2k+1)(2K'+1)}}{2J_0+1}\int d\mathbf{p}_1\int d\mathbf{p}_2 \sum_{\substack{MM'm_{s_1}m_{s_2}\\m_{s_0}m'_{s_0}M_0}}$$

$$\times \sum_{\substack{\ell_0\ell_1m_1\ell_2m_2\\j_0m_{j_0}j_1m_{j_1}j_2m_{j_2}\\J_1M_1J_fM_f}}\sum_{\substack{\ell'_0\ell'_1m'_1\ell'_2m'_2\\j_0m'_{j_0}j'_1m'_{j_1}j'_2m'_{j_2}\\J'_1M'_1J'_fM'_f}} (-1)^{J-M+1/2-m_{s_0}}$$

$$\times a_{\ell_1m_1}^{j_1\,*}b_{\ell_2m_2}^{j_2\,*}c_{\ell_0}^{j_0}\langle([Jj_1]J_1j_2)J_f\|V\|(J_0j_0)J_f\rangle$$

$$\times a_{\ell'_1m'_1}^{j'_1}b_{\ell'_2m'_2}^{j'_2}c_{\ell'_0}^{j'_0\,*}\langle([Jj'_1]J'_1j'_2)J'_f\|V\|(J_0j'_0)J'_f\rangle$$

$$\times (-1)^{-\ell_1+1/2-m_{j_1}-\ell_2+1/2-m_{j_2}-\ell_0+1/2-m_{j_0}}$$

$$\times (-1)^{-J+j_1-M_1-J_1+j_2-M_f-J_0+j_0-M_f}$$

$$\times (-1)^{-\ell'_1+1/2-m'_{j_1}-\ell'_2+1/2-m'_{j_2}-\ell'_0+1/2-m'_{j_0}}$$

$$\times (-1)^{-J+j'_1-M'_1-J'_1+j'_2-M'_f-J_0+j'_0-M'_f}$$

$$\times \sqrt{(2j_1+1)(2j_2+1)(2j_0+1)(2J_1+1)(2J_f+1)}$$

$$\times \sqrt{(2j'_1+1)(2j'_2+1)(2j'_0+1)(2J'_1+1)(2J'_f+1)}$$

$$\times \begin{pmatrix} 1/2 & 1/2 & k \\ m_{s_0} & -m'_{s_0} & -q \end{pmatrix} \begin{pmatrix} J & J & K' \\ M & -M' & -Q' \end{pmatrix}$$

$$\times \begin{pmatrix} \ell_1 & 1/2 & j_1 \\ m_1 & m_{s_1} & -m_{j_1} \end{pmatrix} \begin{pmatrix} \ell_2 & 1/2 & j_2 \\ m_2 & m_{s_2} & -m_{j_2} \end{pmatrix} \begin{pmatrix} \ell_0 & 1/2 & j_0 \\ 0 & m_{s_0} & -m_{j_0} \end{pmatrix}$$

$$\times \begin{pmatrix} \ell'_1 & 1/2 & j'_1 \\ m'_1 & m_{s_1} & -m'_{j_1} \end{pmatrix} \begin{pmatrix} \ell'_2 & 1/2 & j'_2 \\ m'_2 & m_{s_2} & -m'_{j_2} \end{pmatrix} \begin{pmatrix} \ell'_0 & 1/2 & j'_0 \\ 0 & m'_{s_0} & -m'_{j_0} \end{pmatrix}$$

$$\times \begin{pmatrix} J & j_1 & J_1 \\ M & m_{j_1} & -M_1 \end{pmatrix} \begin{pmatrix} J_1 & j_2 & J_f \\ M_1 & m_{j_2} & -M_f \end{pmatrix} \begin{pmatrix} J_0 & j_0 & J_f \\ M_0 & m_{j_0} & -M_f \end{pmatrix}$$

$$\times \begin{pmatrix} J & j'_1 & J'_1 \\ M' & m'_{j_1} & -M'_1 \end{pmatrix} \begin{pmatrix} J'_1 & j'_2 & J'_f \\ M'_1 & m'_{j'_2} & -M'_f \end{pmatrix} \begin{pmatrix} J_0 & j'_0 & J'_f \\ M_0 & m'_{j_0} & -M'_f \end{pmatrix} . \text{(E.4)}$$

The integration over the solid angles and the energy distribution of the electrons e_1^- and e_2^- can be carried out,

$$\int \mathrm{d}\mathbf{p}_1 \, a_{\ell_1 m_1}^{j_1 *} a_{\ell'_1 m'_1}^{j'_1} = \mathrm{e}^{\mathrm{i}(\sigma_{\ell_1}^{j_1} - \sigma_{\ell'_1}^{j'_1})} \delta_{\ell_1 \ell'_1} \delta_{m_1 m'_1} \Delta E_1 , \tag{E.5}$$

and

$$\int \mathrm{d}\mathbf{p}_2 \, b_{\ell_2 m_2}^{j_2 *} b_{\ell'_2 m'_2}^{j'_2} = \mathrm{e}^{\mathrm{i}(\sigma_{\ell_2}^{j_2} - \sigma_{\ell'_2}^{j'_2})} \delta_{\ell_2 \ell'_2} \delta_{m_2 m'_2} \Delta E_2 . \tag{E.6}$$

Note, that the phase difference can still be non-zero because without further assumptions we generally have $j_1 \neq j'_1$ and $j_2 \neq j'_2$.

Applying the above selection rules and defining $\Delta E_{12} = \Delta E_1 + \Delta E_2$ we obtain

$$B_e(K'Q', kq) = \frac{\sqrt{(2k+1)(2K'+1)}}{2J_0 + 1} \Delta E_{12} \sum_{\substack{MM'm_{s_1}m_{s_2} \\ m_{s_0}m'_{s_0}M_0}} \sum_{\substack{\ell_0 \ell_1 m_1 \ell_2 m_2 \\ j_0 m_{j_0} j_1 m_{j_1} j_2 m_{j_2} \\ J_1 M_1 J_f M_f}}$$

$$\times \sum_{\substack{\ell'_0 J'_1 M'_1 J'_f M'_f \\ j_0 m'_{j_0} j'_1 m'_{j_1} j'_2 m'_{j_2}}} (-1)^{J-M+1/2-m_{s_0}} \mathrm{e}^{\mathrm{i}(\sigma_{\ell_1}^{j_1} - \sigma_{\ell'_1}^{j'_1})} \mathrm{e}^{\mathrm{i}(\sigma_{\ell_2}^{j_2} - \sigma_{\ell'_2}^{j'_2})}$$

$$\times c_{\ell_0}^{j_0} \langle ([Jj_1]J_1 j_2) J_f \| V \| (J_0 j_0) J_f \rangle$$

$$\times c_{\ell'_0}^{j'_0 *} \langle ([Jj'_1]J'_1 j'_2) J'_f \| V \| (J_0 j'_0) J'_f \rangle$$

$$\times (-1)^{-\ell_1 + 1/2 - m_{j_1} - \ell_2 + 1/2 - m_{j_2} - \ell_0 + 1/2 - m_{j_0}}$$

$$\times (-1)^{-J + j_1 - M_1 - J_1 + j_2 - M_f - J_0 + j_0 - M_f}$$

$$\times (-1)^{-\ell_1 + 1/2 - m'_{j_1} - \ell_2 + 1/2 - m'_{j_2} - \ell'_0 + 1/2 - m'_{j_0}}$$

$$\times (-1)^{-J + j'_1 - M'_1 - J'_1 + j'_2 - M'_f - J_0 + j'_0 - M'_f}$$

$$\times \sqrt{(2j_1 + 1)(2j_2 + 1)(2j_0 + 1)(2J_1 + 1)(2J_f + 1)}$$

$$\times \sqrt{(2j_1' + 1)(2j_2' + 1)(2j_0' + 1)(2J_1' + 1)(2J_f' + 1)}$$

$$\times \begin{pmatrix} 1/2 & 1/2 & k \\ m_{s0} & -m_{s0}' & -q \end{pmatrix} \begin{pmatrix} J & J & K' \\ M & -M' & -Q' \end{pmatrix}$$

$$\times \begin{pmatrix} \ell_1 & 1/2 & j_1 \\ m_1 & m_{s_1} & -m_{j_1} \end{pmatrix} \begin{pmatrix} \ell_2 & 1/2 & j_2 \\ m_2 & m_{s_2} & -m_{j_2} \end{pmatrix} \begin{pmatrix} \ell_0 & 1/2 & j_0 \\ 0 & m_{s0} & -m_{j0} \end{pmatrix}$$

$$\times \begin{pmatrix} \ell_1 & 1/2 & j_1' \\ m_1 & m_{s_1} & -m_{j_1}' \end{pmatrix} \begin{pmatrix} \ell_2 & 1/2 & j_2' \\ m_2 & m_{s_2} & -m_{j_2}' \end{pmatrix} \begin{pmatrix} \ell_0' & 1/2 & j_0' \\ 0 & m_{s0}' & -m_{j0}' \end{pmatrix}$$

$$\times \begin{pmatrix} J & j_1 & J_1 \\ M & m_{j_1} & -M_1 \end{pmatrix} \begin{pmatrix} J_1 & j_2 & J_f \\ M_1 & m_{j_2} & -M_f \end{pmatrix} \begin{pmatrix} J_0 & j_0 & J_f \\ M_0 & m_{j0} & -M_f \end{pmatrix}$$

$$\times \begin{pmatrix} J & j_1' & J_1' \\ M' & m_{j_1}' & -M_1' \end{pmatrix} \begin{pmatrix} J_1' & j_2' & J_f' \\ M_1' & m_{j_2}' & -M_f' \end{pmatrix} \begin{pmatrix} J_0 & j_0' & J_f' \\ M_0 & m_{j0}' & -M_f' \end{pmatrix} . \tag{E.7}$$

Now, applying the orthogonality relations of the $3j$-symbols (A.8a), the summation over m_1, m_{s_1} and m_2, m_{s_2} can be carried out which gives the selection rules

$$j_1 = j_1' , \qquad m_{j_1} = m_{j_1}' , \tag{E.8}$$

and

$$j_2 = j_2' , \qquad m_{j_2} = m_{j_2}' . \tag{E.9}$$

Thus, the phase difference disappears which yields

$$B_e(K'Q', kq) = \frac{\sqrt{(2k+1)(2K'+1)}}{2J_0 + 1} \Delta E_{12} \sum_{\substack{MM'M_0 \\ m_{s0}m_{s0}'}} \sum_{\substack{\ell_0\ell_1\ell_2 j_0 m_{j0} \\ j_1 m_{j_1} j_2 m_{j_2} \\ J_1 M_1 J_f M_f}} \sum_{\substack{\ell_0' j_0' m_{j0}' \\ J_1' M_1' J_f' M_f'}}$$

$$\times c_{\ell_0}^{j_0} \langle ([Jj_1]J_1 j_2)J_f \| V \| (J_0 j_0)J_f \rangle$$

$$\times c_{\ell_0'}^{j_0'}{}^* \langle ([Jj_1]J_1' j_2)J_f' \| V \| (J_0 j_0')J_f' \rangle$$

$$\times (-1)^{J-M+1/2-m_{s0}+\ell_0'-\ell_0+j_0-j_0'+m_{j0}-m_{j0}'}$$

$$\times (-1)^{J_1'-J_1+M_1-M_1'+2M_f-2M_f'}$$

$$\times \sqrt{(2j_0+1)(2J_1+1)(2J_f+1)}$$

$$\times \sqrt{(2j_0'+1)(2J_1'+1)(2J_f'+1)}$$

$$\times \begin{pmatrix} 1/2 & 1/2 & k \\ m_{s0} & -m_{s0}' & -q \end{pmatrix} \begin{pmatrix} J & J & K' \\ M & -M' & -Q' \end{pmatrix}$$

$$\times \begin{pmatrix} \ell_0 & 1/2 & j_0 \\ 0 & m_{s0} & -m_{j0} \end{pmatrix} \begin{pmatrix} \ell_0' & 1/2 & j_0' \\ 0 & m_{s0}' & -m_{j0}' \end{pmatrix}$$

$$
\times \begin{pmatrix} J & j_1 & J_1 \\ M & m_{j_1} & -M_1 \end{pmatrix} \begin{pmatrix} J_1 & j_2 & J_f \\ M_1 & m_{j_2} & -M_f \end{pmatrix} \begin{pmatrix} J_0 & j_0 & J_f \\ M_0 & m_{j_0} & -M_f \end{pmatrix}
$$

$$
\times \begin{pmatrix} J & j_1 & J_1' \\ M' & m_{j_1} & -M_1' \end{pmatrix} \begin{pmatrix} J_1' & j_2 & J_f' \\ M_1' & m_{j_2} & -M_f' \end{pmatrix} \begin{pmatrix} J_0 & j_0' & J_f' \\ M_0 & m_{j_0}' & -M_f' \end{pmatrix}.
$$

$$(E.10)$$

Carrying out the sum over M, M', and m_{j_1} by using (A.25) the 2nd, 5th, and 8th $3j$-symbols can be contracted

$$
B_e(K'Q', kq) = \frac{\sqrt{(2k+1)(2K'+1)}}{2J_0+1} \Delta E_{12} \sum_{\substack{\ell_0\ell_1\ell_2 M_0 \\ m_{s_0}m_{s_0}'}} \sum_{\substack{j_0 m_{j_0} j_1 j_2 m_{j_2} \\ J_1 M_1 J_f M_f}} \sum_{\substack{\ell_0' j_0' m_{j_0}' \\ J_1' M_1' J_f' M_f'}}
$$

$$
\times c_{\ell_0}^{j_0} \langle([Jj_1]J_1 j_2)J_f \| V \| (J_0 j_0)J_f\rangle
$$

$$
\times c_{\ell_0'}^{j_0'\,*} \langle([Jj_1]J_1' j_2)J_f' \| V \| (J_0 j_0')J_f'\rangle
$$

$$
\times (-1)^{1/2-m_{s_0}+\ell_0'-\ell_0+j_0-j_0'+m_{j_0}-m_{j_0}'}
$$

$$
\times (-1)^{J-j_1+J_1'-J_1+M_1-2M_1'+2M_f-2M_f'}
$$

$$
\times \sqrt{(2j_0+1)(2J_1+1)(2J_f+1)}
$$

$$
\times \sqrt{(2j_0'+1)(2J_1'+1)(2J_f'+1)}
$$

$$
\times \begin{pmatrix} 1/2 & 1/2 & k \\ m_{s_0} & -m_{s_0}' & -q \end{pmatrix} \begin{pmatrix} J_1 & J_1' & K' \\ -M_1 & M_1' & -Q' \end{pmatrix} \begin{Bmatrix} J_1' & J_1 & K' \\ J & J & j_1 \end{Bmatrix}
$$

$$
\times \begin{pmatrix} \ell_0 & 1/2 & j_0 \\ 0 & m_{s_0} & -m_{j_0} \end{pmatrix} \begin{pmatrix} J_1 & j_2 & J_f \\ M_1 & m_{j_2} & -M_f \end{pmatrix} \begin{pmatrix} J_0 & j_0 & J_f \\ M_0 & m_{j_0} & -M_f \end{pmatrix}
$$

$$
\times \begin{pmatrix} \ell_0' & 1/2 & j_0' \\ 0 & m_{s_0}' & -m_{j_0}' \end{pmatrix} \begin{pmatrix} J_1' & j_2 & J_f' \\ M_1' & m_{j_2} & -M_f' \end{pmatrix} \begin{pmatrix} J_0 & j_0' & J_f' \\ M_0 & m_{j_0}' & -M_f' \end{pmatrix}.
$$

$$(E.11)$$

Further, summing over M_1, M_1', and m_{j_2} and again using (A.25) the 2nd, 4th, and 7th $3j$-symbols are contracted to

$$
B_e(K'Q', kq) = \frac{\sqrt{(2k+1)(2K'+1)}}{2J_0+1} \Delta E_{12} \sum_{\substack{\ell_1\ell_2 j_1 j_2 \\ M_0 m_{s_0} m_{s_0}'}} \sum_{\substack{\ell_0 j_0 m_{j_0} \\ J_1 J_f M_f}} \sum_{\substack{\ell_0' j_0' m_{j_0}' \\ J_1' J_f' M_f'}}
$$

$$
\times c_{\ell_0}^{j_0} \langle([Jj_1]J_1 j_2)J_f \| V \| (J_0 j_0)J_f\rangle
$$

$$
\times c_{\ell_0'}^{j_0'\,*} \langle([Jj_1]J_1' j_2)J_f' \| V \| (J_0 j_0')J_f'\rangle
$$

$$
\times (-1)^{1/2-m_{s_0}+\ell_0'-\ell_0+j_0-j_0'+m_{j_0}-m_{j_0}'}
$$

$$\times \ (-1)^{K'+J_1'-J_1+J-j_1-j_2+2M_f-M_f'}$$

$$\times \ \sqrt{(2j_0+1)(2J_1+1)(2J_f+1)}$$

$$\times \ \sqrt{(2j_0'+1)(2J_1'+1)(2J_f'+1)}$$

$$\times \ \begin{pmatrix} 1/2 & 1/2 & k \\ m_{s_0} & -m_{s_0}' & -q \end{pmatrix} \begin{pmatrix} J_f & J_f' & K' \\ -M_f & M_f' & Q' \end{pmatrix}$$

$$\times \ \begin{Bmatrix} J_1' & J_1 & K' \\ J & J & j_1 \end{Bmatrix} \begin{pmatrix} \ell_0 & 1/2 & j_0 \\ 0 & m_{s_0} & -m_{j_0} \end{pmatrix} \begin{pmatrix} J_0 & j_0 & J_f \\ M_0 & m_{j_0} & -M_f \end{pmatrix}$$

$$\times \ \begin{Bmatrix} J_f & J_f' & K' \\ J_1' & J_1 & j_2 \end{Bmatrix} \begin{pmatrix} \ell_0' & 1/2 & j_0' \\ 0 & m_{s_0}' & -m_{j_0}' \end{pmatrix} \begin{pmatrix} J_0 & j_0' & J_f' \\ M_0 & m_{j_0}' & -M_f' \end{pmatrix} .$$

$$(E.12)$$

Once more, applying (A.25) for carrying out the sum over M_f, M_f', and M_0 the 2nd, 4th, and 6th $3j$-symbols yield

$$B_e(K'Q',kq) = \frac{\sqrt{(2k+1)(2K'+1)}}{2J_0+1} \Delta E_{12} \sum_{\substack{\ell_1 \ell_2 j_1 j_2 \\ m_{s_0} m_{s_0}'}} \sum_{\substack{\ell_0 j_0 m_{j_0} \\ J_1 J_f}} \sum_{\substack{\ell_0' j_0' m_{j_0}' \\ J_1' J_f'}}$$

$$\times \ c_{\ell_0}^{j_0} \langle ([Jj_1]J_1j_2)J_f \| V \| (J_0j_0)J_f \rangle$$

$$\times \ c_{\ell_0'}^{j_0'*} \langle ([Jj_1]J_1'j_2)J_f' \| V \| (J_0j_0')J_f' \rangle$$

$$\times \ (-1)^{J_1'-J_1+J-j_1-j_2-J_f-J_f'-J_0}$$

$$\times \ (-1)^{K'+\ell_0'-\ell_0+j_0-j_0'-1/2+q}$$

$$\times \ \sqrt{(2j_0+1)(2J_1+1)(2J_f+1)}$$

$$\times \ \sqrt{(2j_0'+1)(2J_1'+1)(2J_f'+1)}$$

$$\times \ \begin{pmatrix} 1/2 & 1/2 & k \\ m_{s_0} & -m_{s_0}' & -q \end{pmatrix} \begin{pmatrix} j_0' & j_0 & K' \\ m_{j_0}' & -m_{j_0} & Q' \end{pmatrix}$$

$$\times \ \begin{pmatrix} \ell_0 & 1/2 & j_0 \\ 0 & m_{s_0} & -m_{j_0} \end{pmatrix} \begin{pmatrix} \ell_0' & 1/2 & j_0' \\ 0 & m_{s_0}' & -m_{j_0}' \end{pmatrix}$$

$$\times \ \begin{Bmatrix} j_0' & j_0 & K' \\ J_f & J_f' & J_0 \end{Bmatrix} \begin{Bmatrix} J_1' & J_1 & K' \\ J & J & j_1 \end{Bmatrix} \begin{Bmatrix} J_f & J_f' & K' \\ J_1' & J_1 & j_2 \end{Bmatrix} . \qquad (E.13)$$

Eventually, carrying out the sum over m_{s_0}, m_{s_0}', m_{j_0}, and m_{j_0}' the remaining four $3j$-symbols can be contracted to form a $9j$-symbol via (A.49) by introducing the artificial angular momentum b and its magnetic component β.

$$B_e(K'Q', kq) = \frac{\sqrt{(2k+1)(2K'+1)}}{2J_0+1} \Delta E_{12} \sum_{\substack{\ell_1\ell_2 j_1 j_2 \\ b\beta}} \sum_{\substack{\ell_0 j_0 \ell_0' j_0' \\ J_1 J_f J_1' J_f'}} (2b+1)$$

$$\times c_{\ell_0}^{j_0} \langle ([Jj_1]J_1 j_2) J_f \| V \| (J_0 j_0) J_f \rangle$$

$$\times c_{\ell_0'}^{j_0'}{}^* \langle ([Jj_1]J_1' j_2) J_f' \| V \| (J_0 j_0') J_f' \rangle$$

$$\times (-1)^{J_1 - J_1' - J + j_1 + j_2 + J_f + J_f' + J_0 + \ell_0 - j_0' + 1 - q}$$

$$\times \sqrt{(2j_0+1)(2J_1+1)(2J_f+1)}$$

$$\times \sqrt{(2j_0'+1)(2J_1'+1)(2J_f'+1)}$$

$$\times \begin{pmatrix} K' & b & k \\ Q' & \beta & -q \end{pmatrix} \begin{pmatrix} b & \ell_0 & \ell_0' \\ \beta & 0 & 0 \end{pmatrix} \begin{Bmatrix} K' & b & k \\ j_0 & \ell_0 & 1/2 \\ j_0' & \ell_0' & 1/2 \end{Bmatrix}$$

$$\times \begin{Bmatrix} J_1' & J_1 & K' \\ J & J & j_1 \end{Bmatrix} \begin{Bmatrix} J_f & J_f' & K' \\ J_1' & J_1 & j_2 \end{Bmatrix} \begin{Bmatrix} j_0' & j_0 & K' \\ J_f & J_f' & J_0 \end{Bmatrix} . \tag{E.14}$$

The second $3j$-symbol immediately gives $\beta = 0$. Thus, the first $3j$-symbol yields the important selection rule

$$q = Q' . \tag{E.15}$$

With this, the anisotropy parameter can be redefined as

$$B_e(K'kq) = B_e(K'Q', kq)\,\delta_{Q'q}$$

$$= \frac{\sqrt{(2k+1)(2K'+1)}}{2J_0+1} \Delta E_{12} \sum_{\substack{b\ell_1 j_1 \\ \ell_2 j_2}} \sum_{\substack{\ell_0 j_0 \ell_0' j_0' \\ J_1 J_f J_1' J_f'}} (2b+1)$$

$$\times c_{\ell_0}^{j_0} \langle ([Jj_1]J_1 j_2) J_f \| V \| (J_0 j_0) J_f \rangle$$

$$\times c_{\ell_0'}^{j_0'}{}^* \langle ([Jj_1]J_1' j_2) J_f' \| V \| (J_0 j_0') J_f' \rangle$$

$$\times (-1)^{J_1 - J_1' - J + j_1 + j_2 + J_f + J_f' + J_0 + \ell_0 - j_0' + 1 - q}$$

$$\times \sqrt{(2j_0+1)(2J_1+1)(2J_f+1)}$$

$$\times \sqrt{(2j_0'+1)(2J_1'+1)(2J_f'+1)}$$

$$\times \begin{pmatrix} K' & b & k \\ q & 0 & -q \end{pmatrix} \begin{pmatrix} b & \ell_0 & \ell_0' \\ 0 & 0 & 0 \end{pmatrix} \begin{Bmatrix} K' & b & k \\ j_0 & \ell_0 & 1/2 \\ j_0' & \ell_0' & 1/2 \end{Bmatrix}$$

$$\times \begin{Bmatrix} J_1' & J_1 & K' \\ J & J & j_1 \end{Bmatrix} \begin{Bmatrix} J_f & J_f' & K' \\ J_1' & J_1 & j_2 \end{Bmatrix} \begin{Bmatrix} j_0' & j_0 & K' \\ J_f & J_f' & J_0 \end{Bmatrix} . \tag{E.16}$$

Eventually, inserting the expansion coefficient

$$c_{\ell_0}^{j_0} = \langle \ell_0 0 | \mathbf{p}_0^{(+)} \rangle = \sqrt{\frac{2\ell_0 + 1}{4\pi |\mathbf{p}_0|^2}}\, i^{\ell_0}\, e^{i\sigma_{\ell_0}^{j_0}}\,, \tag{E.17}$$

which has been obtained in full analogy to the method used in Appendix D, e.g. see (D.7), the anisotropy parameter can be written as

$$B_e(K'kq) = \frac{\Delta E_{12}}{4\pi |\mathbf{p}_0|^2} \frac{\sqrt{(2k+1)(2K'+1)}}{2J_0 + 1} \sum_{\substack{b\ell_1 j_1 \\ \ell_2 j_2}} \sum_{\substack{\ell_0 j_0 \ell_0' j_0' \\ J_1 J_f J_1' J_f'}} (2b+1)$$

$$\times\, i^{\ell_0 - \ell_0'}\, e^{i(\sigma_{\ell_0}^{j_0} - \sigma_{\ell_0'}^{j_0'})}$$

$$\times\, \langle ([Jj_1]J_1 j_2) J_f \| V \| (J_0 j_0) J_f \rangle$$

$$\times\, \langle ([Jj_1]J_1' j_2) J_f' \| V \| (J_0 j_0') J_f' \rangle$$

$$\times\, (-1)^{J_1 - J_1' - J + j_1 + j_2 + J_f + J_f' + J_0 + \ell_0 - j_0' + 1 - q}$$

$$\times\, \sqrt{(2\ell_0 + 1)(2j_0 + 1)(2J_1 + 1)(2J_f + 1)}$$

$$\times\, \sqrt{(2\ell_0' + 1)(2j_0' + 1)(2J_1' + 1)(2J_f' + 1)}$$

$$\times \begin{pmatrix} K' & b & k \\ q & 0 & -q \end{pmatrix} \begin{pmatrix} b & \ell_0 & \ell_0' \\ 0 & 0 & 0 \end{pmatrix} \begin{Bmatrix} K' & b & k \\ j_0 & \ell_0 & 1/2 \\ j_0' & \ell_0' & 1/2 \end{Bmatrix}$$

$$\times \begin{Bmatrix} J_1' & J_1 & K' \\ J & J & j_1 \end{Bmatrix} \begin{Bmatrix} J_f & J_f' & K' \\ J_1' & J_1 & j_2 \end{Bmatrix} \begin{Bmatrix} j_0' & j_0 & K' \\ J_f & J_f' & J_0 \end{Bmatrix}. \tag{E.18}$$

References

Åberg, T., Howat, G. (1982) "Theory of the Auger Effect." In: S. Flügge, W. Mehlhorn (Eds.), *Handbuch der Physik* **31**, p. 469. Springer, Berlin.

Aksela, H., Mursu, J. (1996). *Phys. Rev. A* **54**, 2882.

Aksela, H., Aksela, S., Jen, J. S., Thomas, T. D. (1977). *Phys. Rev. A* **15**, 985.

Aksela, S., Aksela, H., Thomas, T. D. (1979). *Phys. Rev. A* **19**, 721.

Aksela, S., Kellokumpo, M., Aksela, H., Väyrynen, J. (1981). *Phys. Rev. A* **23**, 2374.

Aksela, H., Aksela, S., Hotokka, M., Jaentti, M. (1983a). *Phys. Rev. A* **28**, 287.

Aksela, S., Harkoma, M., Pohjola, M., Aksela, H. (1983b). *J. Phys. B: At. Mol. Phys.* **17**, 2227.

Aksela, H., Aksela, S., Patana, H. (1984a). *Phys. Rev. A* **30**, 858.

Aksela, H., Aksela, S., Pulkkinen, H. (1984b). *Phys. Rev. A* **30**, 865.

Aksela, H., Aksela, S., Pulkkinen, H. (1984c). Unpublished data, quoted in Tulkki *et al.* (1993).

Aksela, H., Aksela, S., Bancroft, G. M., Tan, K. H. (1986a). *Phys. Rev. A* **33**, 3867.

Aksela, H., Aksela, S., Pulkkinen, H., Bancroft, G. M., Tan, K. H. (1986b). *Phys. Rev. A* **33**, 3876.

Aksela, H., Aksela, S., Pulkkinen, H., Bancroft, G. M., Tan, K. H. (1988a). *Phys. Rev. A* **37**, 1798.

Aksela, S., Sairanen, O. P., Aksela, H., Bancroft, G. M., Tan, K. H. (1988b). *Phys. Rev. A* **37**, 2934.

Aksela, H., Bancroft, G. M., Olsson, B. (1992). *Phys. Rev. A* **46**, 1345.

Aksela, H., Sairanen, O. P., Aksela, S., Kivimäki, A., Naves de Brito, A., Nõmmiste, E. (1995). *Phys. Rev. A* **51**, 1291.

Aksela, H., Mursu J., Jauhiainen, J., Nõmmiste, E., Karvonen, J., Aksela, S. (1997). *Phys. Rev. A* **55**, 3532.

Amusia, M. Y. (1990). "Atomic Photoeffect." K. T. Taylor. (trans. Ed.), Plenum, New York, London.

Amusia, M. Y., Cherepkov, N. A. (1975). *Case Studies in Atomic Physics* **5**, 47.

Andersen, N., Bartschat, K., Broad, J. T., Hertel, I. V. (1997). *Phys. Rep.* **278**, 107.

Armen, G. B., Larkins, F. P. (1994). Unpublished, data published and cited by Saha (1994).

Asaad, W. N. (1963a). *Nucl. Phys.* **44**, 399.

Asaad, W. N. (1963b). *Nucl. Phys.* **44**, 415.

Asaad, W. N., Burhop, E. H. S. (1958). *Proc. Phys. Soc.* **71**, 369.

Asaad, W. N., Mehlhorn, W. (1968). *Z. Phys.* **217**, 304.

Atkins, P. W. (1970). "Molecular Quantum Mechanics." Clarendon Press, Oxford.

Auger, P. (1923). *Comm. Royal Acad. Sci. Paris* **177**, 169.

Auger, P. (1924). *Comm. Royal Acad. Sci. Paris* **178**, 929, and 1535.

Auger, P. (1926). *Comm. Royal Acad. Sci. Paris* **182**, 776.

Balashov, V. V., Bodrenko, I. V. (1999). *J. Phys. B: At. Mol. Opt. Phys.* **32**, L687.

Balashov, V. V., Bodrenko, I. V. (2000). *J. Phys. B: At. Mol. Opt. Phys.* **33**, 1473.

Barnett, A. R. (1982). *Comp. Phys. Comm.* **27**, 147.

Bartschat, K. (2003). In: G. F. Hanne, L. Malegat, H. Schmidt-Böcking (Eds.), *AIP Conf. Proc.* **697**, p. 213. Melville, New York.

Bartynski, R. A. (2003). In: G. F. Hanne, L. Malegat, H. Schmidt-Böcking (Eds.), *AIP Conf. Proc.* **697**, p. 111. Melville, New York.

Becker, U. (1990a). "The Physics of Electronic and Atomic Collisions." In: A. Dalgarno, R. S. Freund, P. M. Koch, M. S. Lubell, T. B. Lucatorto (Eds.), *AIP Conf. Proc.* **205**, p. 162. New York.

Becker, U. (1990b). "Synchrotron radiation experiments on atoms and molecules." In: A. Dalgarno, R. S. Freund, P. M. Koch, M. S. Lubell, T. B. Lucatorto (Eds.), *AIP Conf. Proc.* **205**, p. 160. New York.

Becker, U. (1994). Private communication.

Becker, U., Crowe, A. (2001). "Complete Scattering Experiments." Kluwer Academic/Plenum Publishers, New York.

Becker, U., Szostak, D., Kerkhoff, H. G., Kupsch, M., Langer, B., Wehlitz, R., Yagishita, A., Hayaishi, T. (1989). *Phys. Rev. A* **39**, 3902.

Bederson, B. (1969). *Comm. At. Mol. Phys.* **1**, 41, and 65.

Bederson, B., Miller, T. M. (1976). In: H. Kleinpoppen, M. R. C. McDowell (Eds.), "Electron and Photon Interactions with Atoms," p. 191. Plenum Press, New York.

Berezhko, E. G., Kabachnik, N. M. (1977). *J. Phys. B: At. Mol. Phys.* **10**, 2467.

Berezhko, E. G., Ivanov, V. K., Kabachnik, N. M. (1978a). *Phys. Lett. A* **66**, 474.

Berezhko, E. G., Kabachnik, N. M., Rostovsky, V. S. (1978b). *J. Phys. B: At. Mol. Opt. Phys.* **11**, 1749.

Bergmann, L., Schaefer, C. (1992). "Lehrbuch der Experimentalphysik." Vol. 4, "Teilchen." W. de Gruyter, Berlin.

Beyer, H. J., West, J. B., Ross, K. J., Ueda, K., Kabachnik, N. M., Hamdy, H., Kleinpoppen, H. (1995). *J. Phys. B: At. Mol. Opt. Phys.* **28**, L47.

Biedenharn, L. C., Gluckstern, R. L., Hull Jr., M. H., Breit, G. (1955). *Phys. Rev.* **97**, 542.

Blum, K. (1996). "Density Matrix Theory and Applications." 2nd Edn. Plenum Press, New York, London.

Blum, K., Lohmann, B., Taute, E. (1986). *J. Phys. B: At. Mol. Opt. Phys.* **19**, 3815.

Bolognesi, P., Coreno, M., De Fanis, A., Huetz, A., Rioual, S., Rouvellou, B., Avaldi, L. (2003). In: G. F. Hanne, L. Malegat, H. Schmidt-Böcking (Eds.), *AIP Conf. Proc.* **697**, p. 119. Melville, New York.

Bonhoff, K., Nahrup, S., Lohmann, B., Blum, K. (1996). *J. Chem. Phys.* **104**, 7921.

Bonhoff, S., Bonhoff, K., Schimmelpfennig, B., Nestmann, B. (1997). *J. Phys. B: At. Mol. Opt. Phys.* **30**, 2821.

Bonhoff, S. (1998a). *PhD Thesis*, University of Münster, Germany.

Bonhoff, K. (1998b). *PhD Thesis*, University of Münster, Germany.

Born, M., Wolf, E. (1970). "Principles of Optics." Pergamon Press, New York.

Brink, D. M., Satchler, G. R. (1962). "Angular Momentum." Oxford University Press, Oxford.

Buckmaster, H. A. (1964). *Can. J. Phys.* **42**, 386.

Buckmaster, H. A. (1966). *Can. J. Phys.* **44**, 2525.

Burhop, E. H. S., Asaad, W. N. (1972). "The Auger Effect." In: D. R. Bates, I. Estermann (Eds.), *Adv. At. Mol. Phys.* **8**, p. 164. Elsevier Academic Press, New York.

Burnett, G. C., Monroe, T. J., Dunnings, F. B. (1994). *Rev. Sci. Instrum.* **65**, 1893.

Bussert, W., Klar, H. (1983). *Z. Phys. A* **312**, 315.

Caldwell, C. D. (1990). In: T. A. Carlson, M. O. Krause, S. T. Manson (Eds.), "X-ray and Inner-Shell Processes." *AIP Conf. Proc.* **215**, p. 685. Melville, New York.

Caldwell, C. D., Krause, M. O. (1993). *Phys. Rev. A.* **47**, R759.

Carlson, J. A., Mullins, D. R., Beall, C. E., Yates, B. W., Taylor, J. W., Lindle, D. W., Pullen, B. P., Grimm, F. A. (1988). *Phys. Rev. Lett.* **60**, 1382.

Carlson, T. A., Mullin, D. R., Beall, C. E., Yates, B. W., Taylor, J. W., Lindle, D. W., Grimm, F. A. (1989). *Phys. Rev. A* **39**, 1170.

Carravetta, V., Ågren, H. (1987). *Phys. Rev. A* **35**, 1022.

Carre, B., d'Oliveira, P., Ferray, M., Fournier, P., Gounand, F., Cubayanes, D., Bizau, J. M., Wuilleumier, F. J. (1990). *Z. Phys. D* **15**, 117.

Chandra, N. (1989). *Phys. Rev. A* **40**, 752.

Chandra, N., Chakraborty, M. (1992). *J. Chem. Phys.* **97**, 236.

Chattarji, D. (1976). "The Theory of Auger Transitions." Academic Press, London, New York, San Francisco.

Chen, M. H. (1992). *Phys. Rev. A* **45**, 1684.

Chen, M. H. (1993). *Phys. Rev. A* **47**, 3733.

Chen, M. H. (1994). Data published and cited by Saha (1994).

Chen, M. H., Crasemann, B. (1973). *Phys. Rev. A* **8**, 7; and private communication cited by Hillig *et al.* (1974)

Chen, M. H., Larkins, F. P., Crasemann, B. (1990). *Atomic Data and Nuclear Data Tables* **45**, 1.

Cleff, B., Mehlhorn, W. (1971). *Phys. Lett. A* **37**, 3.

Cleff, B., Mehlhorn, W. (1974a). *J. Phys. B: At. Mol. Phys.* **7**, 593.

Cleff, B., Mehlhorn, W. (1974b). *J. Phys. B: At. Mol. Phys.* **7**, 605.

Clenshaw, C. W., Goodwin, E. T., Martin, D. W., Miller, G. F., Olver, F. W. J., Wilkinson, J. H. (1961). *National Physics Laboratory*. "Modern Computing Methods, Notes on Applied Science." 2nd Edn. Her Majesty's Stationary Office (HMSO), London. Vol. 16.

Combet Farnoux, F. (1992). *Phys. Scr. T* **41**, 28.

Condon, E. U., Shortley, G. H. (1935). "Theory of Atomic Spectra." Cambridge University Press, Cambridge.

Coplan, M. A., Cooper, J. W., Moore, J. H., Doering, J. P., van Boeyen, R. W. (2002). In: M. Schulz, D. H. Madison (Eds.), *AIP Conf. Proc.* **604**, p. 103. Melville, New York.

Cooper, J. W. (1989). *Phys. Rev. A* **39**, 3714.

Cowan, R. D. (1981). "The Theory of Atomic Structure and Spectra." University of California Press, Berkeley, Los Angeles, London.

de Gouw, J. A., van Eck, J., Peters, A. C., van der Weg, J., Heideman, H. G. M. (1995). *J. Phys. B: At. Mol. Opt. Phys.* **28**, 2127.

Dill, D., Starace, A. F., Manson, S. T. (1975). *Phys. Rev. A* **11**, 1596.

Dill, D., Swanson, J. R., Wallace, S., Dehmer, J. L. (1980). *Phys. Rev. Lett.* **45**, 1393.

Dorn, A., Nienhaus, J., Wetzstein, M., Winnewisser, C., Eichmann, U., Sandner, W., Mehlhorn, W. (1995). *J. Phys. B: At. Mol. Opt. Phys.* **28**, L225.

Douglas, B. E., Hollingsworth, C. A. (1985). "Symmetry in Bonding and Spectra." Academic Press, London.

Drescher, M., Khalil, T., Müller, N., Fritzsche, S., Kabachnik, N. M., Heinzmann, U. (2003). *J. Phys. B: At. Mol. Opt. Phys.* **36**, 3337.

Dyall, K. G., Grant, I. P., Johnson, C. T., Parpia, F. A., Plummer, E. P. (1989). *Comp. Phys. Comm.* **55**, 425.

Edmonds, A. R. (1974). "Angular Momentum in Quantum Mechanics." Princeton University Press, Princeton, NJ.

Eichler, J., Fritsch, W. (1976). *J. Phys. B: At. Mol. Opt. Phys.* **9**, 1477.

Einstein, A. (1905). *Ann. Physik* **17**, 132.

Erman, P., Sujkowski, Z. (1961). *Ark. Fys.* **20**, 209.

Fahlman, A., Nordberg, R., Nordling, C., Siegbahn, K. (1966). *Z. Phys.* **192**, 476.

Fano, U. (1957). *Rev. Mod. Phys.* **29**, 74.

Fano, U., Racah, G. (1959). "Irreducible Tensorial Sets." Academic Press, New York.

Farhat, A., Humphrey M., Langer, B., Berrah, N., Bozek, J. D., Cubaynes, D. (1997). *Phys. Rev. A* **56**, 501.

Ferrett, T. A., Piancastelli, M. N., Lindle, D. W., Heimann, P. A., Shirley, D. A. (1988). *Phys. Rev. A* **38**, 701.

Feuerstein, B., Grum-Grzhimailo, A. N., Bartschat, K., Mehlhorn, W. (1999). *J. Phys. B: At. Mol. Opt. Phys.* **32**, 3727.

Foldy, L., Wouthuysen, W. A. (1950). *Phys. Rev.* **78**, 29.

Frauenfelder, H., Henley, E. M. (1979). "Teilchen und Kerne." Oldenbourg, München, Wien.

Fritzsche, S. (1991). "RATR: An Input Description." University of Kassel, Germany.

Fritzsche, S. (1992). *PhD Thesis*, University of Kassel, Germany.

Fritzsche, S. (1993). *Phys. Rev. Lett. A* **180**, 262.

Fritzsche, S. (2001). *J. Electr. Spectr. & Relat. Phen.* **114–116**, 1155.

Fritzsche, S., Zscharnack, G., Musiol, G., Soff, G. (1991). *Phys. Rev. A* **44**, 388.

Fritzsche, S., Froese-Fischer, C., Dong, C. Z. (2000). *Comp. Phys. Comm.* **124**, 340.

Furness, J. B., Mc Carthy, I. E. (1973). *J. Phys. B: At. Mol. Opt. Phys.* **6**, 2280.

Geßner, O., Hikosaka, Y., Zimmermann, B., Hempelmann, A., Lucchese, R. R., Eland, J. H. D., Guyon, P. M., Becker, U. (2002). *Phys. Rev. Lett.* **88**, 193002.

Golovin, A. V. (1991). *Opt. Spectrosc. (USSR)* **71**, 537.

Grant, I. P. (1970). *Advan. Phys.* **19**, 747.

Grant, I. P., Mayers, D. F., Pyper, N. C. (1976). *J. Phys. B: At. Mol. Phys.* **9**, 2777.

Grant, I. P., McKenzie, B. J., Norrington, P. H., Mayers, D. F., Pyper, N. C. (1980). *Comp. Phys. Comm.* **21**, 207.

Grant, I. P. (1988). In: S. Wilson (Ed.), "Methods in Computational Chemistry", Vol. **2**, p. 1. Plenum Press, New York.

Grant, I. P., Parpia, F. A. (1992). Private communication.

Hahn, U., Semke, J., Merz, H., Kessler, J. (1985). *J. Phys. B: At. Mol. Phys.* **18**, L417.

Haken, H. Wolf, H. C. (1991). "Molekülphysik und Quantenchemie." Springer, Berlin.

Hanne, G. F., Malegat L., Schmidt-Böcking, H. (Eds.) (2003). *AIP Conf. Proc.* **697**. Melville, New York.

Hansen, J. E., Persson, W. (1987). *Phys. Scr.* **36**, 602.

Harnung, S. E., Schäffer, C. E. (1972). In: J. D. Dunitz, P. Hemmerich, J. A. Ibers, C. K. Jørgensen, J. B. Neilands, D. Reinen, R. J. P. Williams (Eds.), *Structure and Bonding*, Vol. **12**, p. 245. Springer, Berlin.

Heiser, F., Geßner, O., Viefhaus, J., Wieliczek, K., Hentges, R., Becker, U. (1997). *Phys. Rev. Lett.* **79**, 2435.

Hemmers, O., Heiser, F., Eiben, J., Wehlitz, R., Becker, U. (1993). *Phys. Rev. Lett.* **71**, 987.

Hergenhahn, U. (1996). "Winkelverteilung und Spin-Polarisation von Auger-Elektronen." *PhD Thesis*, Oberhofer, Berlin.

Hergenhahn, U., Becker, U. (1995a). *J. Electr. Spectr. & Relat. Phen.* **72**, 243.

Hergenhahn, U., Becker, U. (1995b). *J. Electr. Spectr. & Relat. Phen.* **76**, 225.

Hergenhahn, U., Kabachnik, N. M., Lohmann, B. (1991). *J. Phys. B: At. Mol. Opt. Phys.* **24**, 4759.

Hergenhahn, U., Lohmann, B., Kabachnik, N. M., Becker, U. (1993). *J. Phys. B: At. Mol. Opt. Phys.* **26**, L117.

Hergenhahn, U., Snell, G., Drescher, M., Schmidtke, B., Müller, N., Heinzmann, U., Wiedenhöft, M., Becker, U. (1999). *Phys. Rev. Lett.* **82**, 5020, and private communication.

Hertel, I. V., Stoll, W. (1977). *Adv. At. Mol. Phys.* **13**, 113. Elsevier Academic Press, New York.

Hillig, H., Cleff, B., Mehlhorn, W., Schmitz, W. (1974). *Z. Phys.* **268**, 225.

Hörnfeldt, O. (1962). *Ark. Fys.* **23**, 235.

Hörnfeldt, O., Fahlman, A., Nordling, C. (1962). *Ark. Fys.* **23**, 155.

Howat, G., Åberg, T., Goscinsky, O. (1978). *J. Phys. B: At. Mol. Phys.* **11**, 1575.

Huang, K. N. (1982). *Phys. Rev. A* **26**, 2274.

Huzinaga, S. (1965). *J. Chem. Phys.* **42**, 1293.

Kabachnik, N. M. (1981). *J. Phys. B: At. Mol. Phys.* **14**, L337.

Kabachnik, N. M. (2005). *J. Phys. B: At. Mol. Opt. Phys.* **38**, L19.

Kabachnik, N. M., Grum-Grzhimailo, A. N. (2001). *J. Phys. B: At. Mol. Opt. Phys.* **34**, L63.

Kabachnik, N. M., Lee, O. V. (1989). *J. Phys. B: At. Mol. Opt. Phys.* **22**, 2705.

Kabachnik, N. M., Lee, O. V. (1990a). *Z. Phys. D* **17**, 169.

Kabachnik, N. M., Lee, O. V. (1990b). *J. Phys. B: At. Mol. Opt. Phys.* **23**, 353.

Kabachnik, N. M., Sazhina, I. P. (1976). *J. Phys. B: At. Mol. Phys.* **9**, 1681.

Kabachnik, N. M., Sazhina, I. P. (1984). *J. Phys. B: At. Mol. Phys.* **17**, 1335.

Kabachnik, N. M., Sazhina, I. P. (1986). *Opt. Spectrosc.* **60**, 683.

Kabachnik, N. M., Sazhina, I. P. (1988). *J. Phys. B: At. Mol. Opt. Phys.* **21**, 267.

Kabachnik, N. M., Sazhina, I. P. (2002). *J. Phys. B: At. Mol. Opt. Phys.* **35**, 3591.

Kabachnik, N. M., Schmidt, V. (1995). *J. Phys. B: At. Mol. Opt. Phys.* **28**, 233.

Kabachnik, N. M., Ueda, K. (1995). *J. Phys. B: At. Mol. Opt. Phys.* **28**, 5013.

Kabachnik, N. M., Sazhina, I. P., Lee, I. S., Lee, O. V. (1988). *J. Phys. B: At. Mol. Opt. Phys.* **21**, 3695.

Kabachnik, N. M., Lohmann, B., Mehlhorn, W. (1991). *J. Phys. B: At. Mol. Opt. Phys.* **24**, 2249.

Kabachnik, N. M., Tulkki, J., Aksela, H., Ricz, S. (1994). *Phys. Rev. A* **49**, 4653.

Kämmerling, B., Schmidt, V., Mehlhorn, W., Peatman, W. B., Schaefers, F., Schroeter, T. (1989). *J. Phys. B: At. Mol. Opt. Phys.* **22**, L597.

Kämmerling, B., Krässig, B., Schmidt, V. (1990). *J. Phys. B: At. Mol. Opt. Phys.* **23**, 4487, and private communication.

Kämmerling, B., Krässig, B., Schwarzkopf, O., Ribeiro, J. P., Schmidt, V. (1992). *J. Phys. B: At. Mol. Opt. Phys.* **25**, L5.

Karim, K. R., Crasemann, B. (1985). *Phys. Rev. A* **31**, 709.

Karim, K. R., Chen, M. H., Crasemann, B. (1984). *Phys. Rev. A* **29**, 2605.

Kaur, S., Srivastava, R. (1999). *J. Phys. B: At. Mol. Opt. Phys.* **32**, 2323.

Kessler, J. (1985). "Polarized Electrons." 2nd Edn. Springer, Berlin.

Kettle, S. F. A. (1995). "Symmetry and Structure." Wiley, New York.

Klar, H. (1980). *J. Phys. B: At. Mol. Phys.* **13**, 4741.

Kleiman, U., Becker, U. (2005). *J. Electr. Spectr. & Relat. Phen.* **142**, 45.

Kleiman, U., Lohmann, B. (2003). *J. Electr. Spectr. & Relat. Phen.* **131–132**, 29.

Kleiman, U., Lohmann, B., Blum, K. (1999a). *J. Phys. B: At. Mol. Opt. Phys.* **32**, 309, and 4129.

Kleiman, U., Lohmann, B., Blum, K. (1999b). *J. Phys. B: At. Mol. Opt. Phys.* **32**, L219.

Kleinpoppen, H. (1997). Private communication.

H. Kleinpoppen, M. R. C. McDowell (Eds.) (1976). "Electron and Photon Interactions with Atoms." Plenum Press, New York.

Kleinpoppen, H., Lohmann, B., Grum-Grzhimailo, A., Becker, U. (2005). In: H. H. Stroke (Ed.), *Adv. At. Mol. & Opt. Phys.* **51**, p. 471. Elsevier Academic Press, New York.

Knyr, V. A., Nasyrov, V. V., Popov, Y. V. (2003). In: G. F. Hanne, L. Malegat, Schmidt-Böcking, H. (Eds.), *AIP Conf. Proc.* **697**, p. 76. Melville, New York.

Krause, M. O., Caldwell, C. D., Menzel, A., Benzaid, S., Jiménez-Mier, J. (1996). *J. Electr. Spectr. & Relat. Phen.* **79**, 241.

Kronast, W., Huster, R., Mehlhorn, W. (1986). *Z. Phys. D* **2**, 285.

Kuntze, R., Salzmann, M., Böwering, N., Heinzmann, U. (1993). *Phys. Rev. Lett.* **70**, 3716.

Kuntze, R., Salzmann, M., Böwering, N., Heinzmann, U., Ivanov, V. K., Kabachnik, N. M. (1994). *Phys. Rev. A* **50**, 489.

Lafosse, A., Lebech, M., Brenot, J. C., Guyon, P. M., Jagutzki, O., Spielberger, L., Vervloet, M., Houver, J. C., Dowek, D. (2000). *Phys. Rev. Lett.* **84**, 5987.

Lagutin, B. M., Petrov, I. D., Sukhorukov, V. L., Demekhin, P. V., Zimmermann, B., Mickat, S., Kammer, S., Schartner, K.-H., Ehresmann, A., Shutov, Y. A., Schmoranzer, H. (2003). *J. Phys. B: At. Mol. Opt. Phys.* **36**, 3251.

Landers, A., Weber, Th., Ali, I., Cassimi, A., Hattauss, M., Jagutzki, O., Nauert, A., Osipov, T., Staudte, A., Prior, M. H., Schmidt-Böcking, H., Cocke, C. L., Dörner, R. (2001). *Phys. Rev. Lett.* **87**, 013002.

Langer, B., Berrah, N., Farhat, A., Humphrey, M., Cubaynes, D., Menzel, A., Becker, U. (1997). *J. Phys. B: At. Mol. Opt. Phys.* **30**, 4255.

Larkins, F. P. (1977). *Atomic Data and Nuclear Data Tables* **20**, 311.

Larkins, F. P. (1990). *J. Electr. Spectr. & Relat. Phen.* **51**, 115.

Lee, O. V. (1990). Private communication.

Lehmann, J., Blum, K. (1997). *J. Phys. B: At. Mol. Opt. Phys.* **30**, 633.

Lehmann, J., Bonhoff, K., Bonhoff, S., Lohmann, B., Blum, K. (1997). In: J. L. Duggan, I. L. Morgan (Eds.), "Application of Accelerators in Research and Industry." *AIP Conf. Proc.* **392**, p. 63. New York.

Lindle, D. W., Ferrett, T. A., Heimann, P. A., Shirley, D. A. (1988). *Phys. Rev. A* **37**, 3808.

Löwdin, P. O. (1955). *Phys. Rev.* **97**, 1474.

Lohmann, B. (1984). *Diploma Thesis*, University of Münster, Germany.

Lohmann, B. (1988). *PhD Thesis*, University of Münster, Germany.

Lohmann, B. (1990). *J. Phys. B: At. Mol. Opt. Phys.* **23**, 3147.

Lohmann, B. (1991). *J. Phys. B: At. Mol. Opt. Phys.* **24**, 861.

Lohmann, B. (1992). *J. Phys. B: At. Mol. Opt. Phys.* **25**, 4163.

Lohmann, B. (1993). *J. Phys. B: At. Mol. Opt. Phys.* **26**, 1623.

Lohmann, B. (1996a). *Aust. J. Phys.* **49**, 365, and refs. therein.

Lohmann, B. (1996b). "Spin polarization of Auger electrons; recent developments." In: H. Kleinpoppen, M. C. Campell (Eds.), "Selected Topics in Electron Physics." *Peter-Farago-Symp. Conf. Proc.*, p. 119. Plenum Press, New York.

Lohmann, B. (1996c). "Recent developments in the theory of angular distribution and spin polarization of Auger electrons." In: "VIII. Int. Symp. on Pol. & Corr. in Elec. & Atomic Coll., Vancouver." *Can. J. Phys.* **74**, 962.

Lohmann, B. (1996d). *J. Phys. B: At. Mol. Opt. Phys.* **29**, L521.

Lohmann, B. (1997). "Spin polarization parameters for resonant Auger transitions." In: "IX Int. Symp. on Pol. & Corr. in Elec. & Atomic Coll. and Int. Symp. on (e,2e), Double Phot. & Rel. Topics," Ses. 8. Frascati, Italy.

Lohmann, B. (1998). *Habilitation Thesis*, University of Münster, Germany.

Lohmann, B. (1999a). *J. Phys. B: At. Mol. Opt. Phys.* **32**, L643.

Lohmann, B. (1999b). *Aust. J. Phys.* **52**, 397.

Lohmann, B., Fritzsche, S. (1994). *J. Phys. B: At. Mol. Opt. Phys.* **27**, 2919.

Lohmann, B., Fritzsche, S. (1996). *J. Phys. B: At. Mol. Opt. Phys.* **29**, 5711.

Lohmann, B., Larkins, F. P. (1994). *J. Phys. B: At. Mol. Opt. Phys.* **27**, L143.

Lohmann, B., Kleiman, U. (2001). In: J. Berakdar, J. Kirschner (Eds.), "Many Particle Spectroscopy of Atoms, Molecules, Clusters, and Surfaces." p. 173. Kluwer/Plenum, New York, London.

Lohmann, B., Kleiman, U. (2006). *J. Phys. B: At. Mol. Opt. Phys.* **39**, 271.

Lohmann, B., Hergenhahn, U., Kabachnik, N. M. (1993). *J. Phys. B: At. Mol. Opt. Phys.* **26**, 3327.

Lohmann, B., Fritzsche, S., Larkins, F. P. (1996). *J. Phys. B: At. Mol. Opt. Phys.* **29**, 1191.

Lohmann, B., Bonhoff, S., Bonhoff, K., Lehmann, J., Blum, K. (1997). In: J. L. Duggan, I. L. Morgan (Eds.), "Application of Accelerators in Research and Industry." *AIP Conf. Proc.* **392**, p. 59. New York.

Lohmann, B., Fritzsche, S., Andrä, H. J. (1998). "Absolut transition rates for Auger neutralization processes." Unpublished data. University of Münster, University of Kassel, Germany.

Lohmann, B., Srivastava, R., Kleiman, U., Blum, K. (2002). In: D. H. Madison, M. Schulz (Eds.), *AIP Conf. Proc.* **604**, p. 229. Melville, New York.

Lohmann, B., Zimmermann, B., Kleinpoppen, H., Becker, U. (2003a). In: B. Bederson, H. Walther (Eds.), *Adv. At. Mol. & Opt. Phys.* **49**, p. 217. Elsevier Academic Press, New York.

Lohmann, B., Langer, B., Snell, G., Kleiman, U., Canton, S., Martins, M., Becker, U., Berrah, N. (2003b). In: G. F. Hanne, L. Malegat, H. Schmidt-Böcking (Eds.), *AIP Conf. Proc.*, vol. **697**, p. 133. Melville, New York.

Lohmann, B., Langer, B., Snell, G., Kleiman, U., Canton, S., Martins, M., Becker, U., Berrah, N. (2005). *Phys. Rev. A* **71**, 020701(R).

Lombardi, M. (1975). *Phys. Rev. Lett.* **35**, 1172.

Lower, J., Panajotovic, R., Weigold, E. (2004). *Phys, Scr. T* **110**, 216.

Mayer-Kuckuk, T. (1977). "Atomphysik." Teubner, Stuttgart.

Mayer-Kuckuk, T. (1979). "Kernphysik." Teubner, Stuttgart.

McGuire, E. J. (1969). *Phys. Rev.* **185**, 1.

McGuire, E. J. (1970). *Phys. Rev. A* **2**, 273.

McGuire, E. J. (1975). *Sandia Research Lab.* Research report no. SAND-75-0443.

McKenzie, B. J., Grant, I. P., Norrington, P. H. (1980). *Comp. Phys. Comm.* **21**, 233.

Mehlhorn, W. (1968). *Phys. Lett. A* **26**, 166.

Mehlhorn, W. (1985). "Auger-Electron Spectrometry of Core Levels of Atoms." In: B. Crasemann (Ed.), "Atomic Inner Shell Physics." Plenum Press, New York.

Mehlhorn, W. (1990). In: T. A. Carlson, M. O. Krause, S. T. Manson (Eds.), "X-ray and Inner Shell Processes." *AIP Conf. Proc.* **215**, p. 465. New York.

Mehlhorn, W., Stalherm, D. (1968). *Z. Phys.* **217**, 294.

Mehlhorn, W., Taulbjerg, K. (1980). *J. Phys. B: At. Mol. Phys.* **13**, 445.

Meister, H. J., Weiss, H. F. (1968). *Z. Phys.* **216**, 165.

Menzel, A. (1994). *PhD Thesis*, Technical University of Berlin, Germany.

Menzel, A., Hemmers, O., Langer, B., Wehlitz, R., Becker, U. (1993). "BESSY annual report.", and private communication.

Merz, H. (1991). Private communication.

Merz, H., Semke, J. (1990). In: T. A. Carlson, M. O. Krause, S. T. Manson (Eds.), "X-ray and Inner Shell Processes." *AIP Conf. Proc.* **215**, p. 719. New York.

Messiah, A. (1979). "Quantenmechanik." de Gruyter, Berlin.

Meyer, M., von Raven, E., Sonntag, B., Hansen, J. E. (1991). *Phys. Rev. A* **43**, 177.

Meyer, M., Marquette, A., Grum-Grzhimailo, A. N., Kleiman, U., Lohmann, B. (2001). *Phys. Rev. A* **64**, 022703.

Michl, J., Thulstrup, E. W. (1986). "Spectroscopy with Polarized Light." V.C.H. Wiley, New York.

Moore, C. E. (1976). "Atomic energy levels." *Natl. Bur. Stand., U.S. Circ.* No. **467**, Vol. **1**. U.S. GPO, Washington, DC.

Motoki, S., Adachi, J., Hikosaka, Y., Ito, K., Sano, M., Soejima, K., Yagishita, A., Raseev, G., Cherepkov, N. A. (2000). *J. Phys. B: At. Mol. Opt. Phys.* **33**, 4193.

Mott, N. F., Massey, H. S. W. (1965). "The Theory of Atomic Collisions." 3rd Edn, Chap. **IX. 4**. Clarendon Press, Oxford.

Müller, N., David, R., Snell, G., Kuntze, R., Drescher, M., Böwering, N., Stoppmanns, P., Yu, S. W., Heinzmann, U., Viefhaus, J., Hergenhahn, U., Becker, U. (1995). *J. Electr. Spectr. & Relat. Phen.* **72**, 187.

Mursu, J., Aksela, H., Sairanen, O. P., Kivimäki, A., Nõmmiste, E., Ausmees, A., Svensson, S., Aksela, S. (1996). *J. Phys. B: At. Mol. Opt. Phys.* **29**, 4387.

Nahrup, S. (1995). *Diploma Thesis*, University of Münster, Germany.

Oenning, R. (1989). *Diploma Thesis*, University of Münster, Germany.

O'Keeffe, P., Aloïse, S., Meyer, M., Grum-Grzhimailo, A. N. (2003). *Phys. Rev. Lett.* **90**, 023002.

O'Keeffe, P., Aloïse, S., Fritzsche, S., Lohmann, B., Kleiman, U., Meyer, M., Grum-Grzhimailo, A. N. (2004). *Phys. Rev. A* **70**, 012705.

Ong, W., Russek, A. (1978). *Phys. Rev. A* **17**, 120.

Paripás, B., Víkor, G., Ricz, S. (1997). *J. Phys. B: At. Mol. Opt. Phys.* **30**, 403.

Parpia, F. A., Grant, I. P. (1992). Private communication.

Paulus, G. G., Lindner, F., Walther, H., Baltuska, A., Goulielmakis, E., Lezius, M., Krausz, F. (2003). *Phys. Rev. Lett.* **91**, 253004.

Persson, W., Wahlström, C.-G., Bertucelli, G., Di Rocco, H. O., Reyna Almandos, J. G., Gallardo, M. (1988). *Phys. Scr.* **38**, 347.

Petrini, D., Araújo, F. X. (1994). *Astron. Astrophys.* **282**, 315.

Racah, G. (1942). *Phys. Rev.* **62**, 438.

Ridder, D., Dieringer, J., Stolterfoht, N. (1976). *J. Phys. B: At. Mol. Phys.* **9**, L307.

Rose, M. E. (1957). "Elementary Theory of Angular Momentum." J. Wiley & Sons, New York.

Rotenberg, M., Bivins, R., Metropolis, N., Wooten Jr., J. K. (1959). "The $3\text{-}j$ and $6\text{-}j$ Symbols." Technology Press, Cambridge, MA.

Saha, H. P. (1994). *Phys. Rev. A* **49**, 894.

Sandner, W. (1985). "New Application of Electron-Electron Coincidences." In: H. Kleinpoppen, J. S. Briggs, H. O. Lutz (Eds.), "Fundamental Processes in Atomic Collision Physics." *NATO ASI Series B: Physics* **134**, p. 453. New York.

Sarkadi, L., Vajnai, T., Pálinkás, J., Kövér, A., Végh, J., Mukoyama, T. (1990). *J. Phys. B: At. Mol. Opt. Phys.* **23**, 3643.

Schimmelpfennig, B. (1994). *PhD Thesis*, University of Bonn, Germany.

Schimmelpfennig, B., Nestmann, B., Peyerimhoff, S. D. (1992). *J. Phys. B: At. Mol. Opt. Phys.* **25**, 1217.

Schimmelpfennig, B., Nestmann, B., Peyerimhoff, S. D. (1995). *J. Electr. Spectr. & Relat. Phen.* **74**, 173.

Schmidt, V. (1992). *Rep. Prog. Phys.* **55**, 1483.

Schmidt, V. (1997). "Electron Spectrometry of Atoms using Synchrotron Radiation." Cambridge University Press, Cambridge, MA.

Schmidtke, B., Drescher, M., Cherepkov, N. A., Heinzmann, U. (2000a). *J. Phys. B: At. Mol. Opt. Phys.* **33**, 2451.

Schmidtke, B., Khalil, T., Drescher, M., Müller, N., Kabachnik, N. M., Heinzmann, U. (2000b). *J. Phys. B: At. Mol. Opt. Phys.* **33**, 5225.

Schmidtke, B., Khalil, T., Drescher, M., Müller, N., Kabachnik, N. M., Heinzmann, U. (2001). *J. Phys. B: At. Mol. Opt. Phys.* **34**, 4293.

Schpolski, E. W. (1983). "Atomphysik." 12th Edn., VEB Deutscher Verlag der Wissenschaften, Berlin.

Schuch, R., Cederquist, M., Larsson, E., Lindroth, S., Schmidt, H. (Eds.) (2004). *Phys. Scr.* **T110**.

Shampine, L. F., Gordon, M. K. (1975). "Computer Solution of Ordinary Differential Equations: The Initial Value Problem." W. H. Freeman and Co., San Francisco.

Shaw Jr., R. W., Thomas, T. D. (1975). *Phys. Rev. A* **11**, 1491.

Shirley, D. A. (1973). *Phys. Rev. A* **7**, 1520.

Siegbahn, K., Nordling, C., Fahlman, A., Nordberg, R., Hamrin, K., Hedman, J., Johansson, G., Bergmark, T., Karlsson, S.-E., Lindgren, T., Lindberg, B. (1967). "ESCA – Atomic, Molecular and Solid State Structure Studied by Means of Electron Spectroscopy." *Nova Acta Regiae Soc. Sci. Upsaliensis* Ser. **IV**, Vol. **20**.

Siegbahn, K., Nordling, C., Johansson, G., Hedman, J., Heden, P. F., Hamrin, K., Gelius, U., Bergmark, T., Werme, L. O., Manne, R., Baer, Y. (1969). "ESCA Applied to Free Molecules." North-Holland, Amsterdam.

Slater, J. C. (1951). *Phys. Rev.* **81**, 385.

Snell, G., Hergenhahn, U., Drescher, M., Schmidtke, B,. Müller, N., Wiedenhöft, M., Becker, U., Heinzmann, U. (1996a). "BESSY annual report."

Snell, G., Drescher, M., Müller, N., Heinzmann, U., Hergenhahn, U., Viefhaus, J., Heiser, F., Becker, U., Brookes, N. B. (1996b). *Phys. Rev. Lett.* **76**, 3923.

Snell, G., Langer, B., Drescher, M., Müller, N., Zimmermann, B., Hergenhahn, U., Viefhaus, J., Heinzmann, U., Becker, U. (1999). *Phys. Rev. Lett.* **82**, 2480.

Snell, G., Viefhaus, J., Dunnings, F. B., Berrah, N. (2000). *Rev. Sci. Instrum.* **71**, 2608.

Snell, G., Hergenhahn, U., Müller, N., Drescher, M., Viefhaus, J., Becker, U., Heinzmann, U. (2001). *Phys. Rev. A* **63**, 032712.

Snell, G., Langer, B., Young, A. T., Berrah, N. (2002). *Phys. Rev. A* **66**, 022701.

Southworth, S., Becker, U., Truesdale, C. M., Kobrin, P. H., Lindle, D. W., Owaki, S., Shirley, D. A. (1983). *Phys. Rev. A* **28**, 261.

Srivastava, R., Blum, K., McEachran, R. P., Stauffer, A. D. (1996a). *J. Phys. B: At. Mol. Opt. Phys.* **29**, 3513.

Srivastava, R., Blum, K., McEachran, R. P., Stauffer, A. D. (1996b). *J. Phys. B: At. Mol. Opt. Phys.* **29**, 5947.

Starace, A. F. (1982). In: S. Flügge, W. Mehlhorn (Eds.), *Encyclopedia of Physics* **31**, p. 1. Springer, New York.

Stoppmanns, P., Schmiedeskamp, B., Vogt, B., Müller, N., Heinzmann, U. (1992). *Phys. Scr.* **T41**, 190.

Takahashi, M., Cave, J. P., Eland, J. H. D. (2000). *Rev. Sci. Instrum.* **71**, 1337.

Taouil, L., Duguet, A., Lahman-Bennani, A., Lohmann, B., Rasch, J., Whelan, C. T., Walters, H. R. J. (1999). *J. Phys. B: At. Mol. Opt. Phys.* **32**, L5.

Taylor, J. R. (1972). "Scattering Theory." J. Wiley & Sons, New York, London, Sydney, Toronto.

Theodosiou, C. E. (1987). *Phys. Rev. A* **36**, 3138.

Tinkham, M. (1964). "Group Theory and Quantum Mechanics." McGraw-Hill, New York.

Tulkki, J., Kabachnik, N. M., Aksela, H. (1993). *Phys. Rev. A* **48**, 1277, and private communication.

Tulkki, J., Aksela, H., Kabachnik, N. M. (1993). *Phys. Rev. A* **48**, 2957.

Turri, G., Lohmann, B., Langer, B., Snell, G., Becker, U., Berrah, N. (2007). *J. Phys. B: At. Mol. Opt. Phys.* **40**, 3453.

Ueda, K., West, J. B., Ross, K. J., Beyer, H. J., Kabachnik, N. M. (1998). *J. Phys. B: At. Mol. Opt. Phys.* **31**, 4801.

Varshalovich, D. A., Moskalev, A. N., Khersonskii, V. K. (1988). "Quantum Theory of Angular Momentum." World Scientific, Singapore.

Vanderpoorten, R. (1975). *J. Phys. B: At. Mol. Phys.* **8**, 926.

Väyrynen, J., Aksela, S. (1979). *J. Electr. Spectr. & Relat. Phen.* **16**, 423.

van der Laan, G., Thole, B. T. (1995). *Phys. Rev. B* **52**, 15355.

Viefhaus, J., Avaldi, L., Hentges, R., Wiedenhöft, M., Wieliczek, K., Becker, U. (1996). In: "17th Int. Conf. X-ray and Inner-Shell Processes." *Book of Abstracts*, p. 220. Hamburg.

Viefhaus, J., Zimmermann, B., Kleinpoppen, H., Becker, U. (1998). Private communication.

Vollmer, R., Etzkorn, M., Anilkumar, P. S., Ibach, H., Kirschner, J. (2003). *Phys. Rev. Lett.* **91**, 147201.

von Raven, E., Meyer, M., Pahler, M., Sonntag, B. (1990). *J. Electr. Spectr. & Relat. Phen.* **52**, 677.

Wagner, C. D. (1975). *Faraday Discuss. Chem. Soc.* **60**, 291.

Walters, D. L., Bhalla, C. P. (1971). *Atomic Data* **3**, 301.

Wehlitz, R., Pibida, L. S., Levin, J. C., Sellin, I. A. (1999). *Phys. Rev. A* **59**, 421.

Weigold, E., Lower, J., Berakdar, J., Mazevet, S. (2002). In: M. Schulz, D. H. Madison (Eds.), *AIP Conf. Proc.* **604**, p. 32. Melville, New York.

Wentzel, G. (1927). *Z. Phys.* **43**, 524.

Werme, L. O., Bergmark, T., Siegbahn, K. (1972). *Phys. Scr.* **6**, 141.

Werme, L. O., Bergmark, T., Siegbahn, K. (1973). *Phys. Scr.* **8**, 149.

Wigner, E. P. (1959). "Group Theory and its Application to the Quantum Mechanics of Atomic Spectra." Academic Press, New York.

Wöste, G., Fullerton, C., Blum, K., Thompson, D. (1994). *J. Phys. B: At. Mol. Opt. Phys.* **27**, 2625.

Ying-Nan, C. (1966). *J. Chem. Phys.* **45**, 2969.

Young, A. T., Feng, J., Arenholz, E., Padmore, H. A., Henderson, T., Marks, S., Hoyer, E., Schlueter, R., Kortright, J. B., Martynov, V., Steier, C., Portmann, G. (2001). *Nucl. Instrum. & Methods A* **467**, 549.

Zare, R. N. (1988). "Angular Momentum." J. Wiley & Sons, New York.

Zähringer, K., Meyer, H. D., Cederbaum, L. S. (1992). *Phys. Rev. A* **46**, 5643.

Zähringer, K., Meyer, H. D., Cederbaum, L. S., Tarantelli, F., Sgamellotti, A. (1993). *Chem. Phys. Lett.* **206**, 247.

Zheng, Q., Edwards, A. K., Wood, R. M., Mangan, M. A. (1995a). *Phys. Rev. A* **52**, 3940.

Zheng, Q., Edwards, A. K., Wood, R. M., Mangan, M. A. (1995b). *Phys. Rev. A* **52**, 3945.

Zimmermann, B. (2000). "Vollständige Experimente in der atomaren und molekularen Photoionisation." In: U. Becker (Ed.), *Studies of Vacuum Ultraviolet and X-Ray Processes* **13**. Wissenschaft und Technik Verlag, Berlin.

Zimmermann, B., Wilhelmi, O., Schartner, K.-H., Vollweiler, F., Liebel, H., Ehresmann, A., Lauer, S., Schmoranzer, B. M., Lagutin, B. M., Petrov, I. D., Sukhorukov, V. L. (2000). *J. Phys. B: At. Mol. Opt. Phys.* **33**, 2467.

Index

Springer Series on
ATOMIC, OPTICAL, AND PLASMA PHYSICS

Springer Series on
ATOMIC, OPTICAL, AND PLASMA PHYSICS

Printing: Krips bv, Meppel, The Netherlands
Binding: Stürtz, Würzburg, Germany